More Monthly Problem Gems

More Monthly Problem Gems is a sequel to *Monthly Problem Gems* (CRC Press, 2021). This book covers a broader range of math problems. In addition to analysis problems, problems from number theory, combinatorics, algebra, and geometry are included.

The book offers problems to promote creative techniques for problem-solving and undergraduate research. Each problem is selected for its natural charm, the connection with an authentic mathematical experience, originating from the ingenious work of professionals, and ready developments, all into well-shaped results of broader interest.

Each problem provides either a novel application of a familiar theorem or a lively discussion of multiple solutions. Special attention is paid to informal exploration of the essential assumptions, suggestive heuristic considerations, and roots of the motivations of the problem.

This text then presents a new type of problem-solving. It will challenge and stimulate math problem-solvers at varying degrees of proficiency. Since the selected problem gems contain sophisticated ideas and connect to important current research, this book is also geared toward graduate students in math and engineering.

Many of the problems in this book were originally offered in *The American Mathematical Monthly*.

Hongwei Chen earned his PhD from North Carolina State University in 1991. He is currently a professor of mathematics at Christopher Newport University. He has published more than 60 research papers in analysis and partial differential equations. Dr. Chen also authored *Monthly Problem Gems* and *Classical Analysis: An Approach through Problems* published by CRC Press and *Excursions in Classical Analysis* published by the Mathematical Association of America.

Roberto Tauraso holds a PhD in mathematics from Scuola Normale Superiore, Pisa, Italy. He is currently a professor of mathematical analysis at Tor Vergata University of Rome. His research interests include complex analysis, number theory, and combinatorics. He has published over 60 research papers in various mathematical journals.

More Monthly Problem Gems

Hongwei Chen and Roberto Tauraso

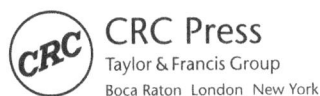

CRC Press
Taylor & Francis Group
Boca Raton London New York

CRC Press is an imprint of the
Taylor & Francis Group, an **informa** business

A CHAPMAN & HALL BOOK

Designed cover image: Roberto Tauraso

First edition published 2026
by CRC Press
2385 NW Executive Center Drive, Suite 320, Boca Raton, FL 33431

and by CRC Press
4 Park Square, Milton Park, Abingdon, Oxon, OX14 4RN

CRC Press is an imprint of Taylor & Francis Group, LLC

© 2026 Hongwei Chen and Roberto Tauraso

Library of Congress Cataloging-in-Publication Data
Names: Chen, Hongwei, author. | Tauraso, Roberto, author.
Title: More monthly problem gems / Hongwei Chen and Roberto Tauraso.
Description: First edition. | Boca Raton, FL ; Abingdon, Oxon : CRC Press, 2026. | Includes bibliographical references and index.
Identifiers: LCCN 2025020685 (print) | LCCN 2025020686 (ebook) | ISBN 9781041000150 (hardback) | ISBN 9781041003212 (paperback) | ISBN 9781003607809 (ebook)
Subjects: LCSH: Mathematical analysis--Problems, exercises, etc.
Classification: LCC QA301 .C437 2026 (print) | LCC QA301 (ebook) | DDC 510--dc23/eng/20250625
LC record available at https://lccn.loc.gov/2025020685
LC ebook record available at https://lccn.loc.gov/2025020686

ISBN: 978-1-041-00015-0 (hbk)
ISBN: 978-1-041-00321-2 (pbk)
ISBN: 978-1-003-60780-9 (ebk)

DOI: 10.1201/9781003607809

Typeset in CMR10 font
by KnowledgeWorks Global Ltd.

Publisher's note: This book has been prepared from camera-ready copy provided by the authors.

Contents

Preface

This book, a much-anticipated sequel to the first author's *Monthly Problem Gems* (MPG), features a new collection of 82 monthly problems and solutions in areas such as analysis, geometry, combinatorics, and number theory. Since 2017, both authors have started exchanging monthly problem solutions and shared enjoyment of problem-solving, and developed a mutual respect, which led to the realization of this book.

We will adopt the same format as MPG:

1. State the numbered monthly problem with its proposer(s).

2. Address the heuristic consideration of the solution. A careful discussion of the challenges and related mathematical ideas needed to solve the problem, focusing on mathematical insight and intuition to make sure the reader is on board.

3. Present one or more solutions based on the heuristics. Most are either our published solution in the *Monthly* or a solution we submitted to the *Monthly*, which differs from the featured one, aiming to train and develop typical inventive approaches.

4. Reveal some possible generalizations of the problem to promote further research. The materials vary in their degree of rigor and sophistication to accommodate readers with diverse interests. Each problem comes with a short, carefully selected list of additional problems for the most engaged readers. Some extensions may encourage deeper explorations and lead to valuable mathematics.

We hope, with some confidence, that this book conveys our deep appreciation for the *Monthly* problems.

This book is organized into six chapters as follows:

Chapter 1. Analysis. This chapter collects 15 monthly problems in analysis. These problems are concerned with roulette curves, special types of integrals, interesting series associated with famous numbers, power sums of real roots of a complex function, extrema of functional moments, and monotonic convex splines.

Chapter 2. Identities. We prove 12 identities which range from analysis and number theory to combinatorics and complex variables. Among the approaches used in the proofs, we will meet Wilf's "snake oil method", Euler's number-theoretic functions, q-series, integer partitions, elementary symmetric functions, Jacobi's triple product identity, formal power series, and differential operators.

Chapter 3. Geometry. Here the reader finds 10 problems exploring different results in Euclidean geometry, such as Anne's theorem, Napoleon's theorem, Heron's formula, Brianchon's theorem, and Newton's quadrilateral theorem. Also, several notable points of a triangle, including the Nagel point and Fermat point, show some unexpected properties.

Chapter 4. Combinatorics. We gathered 15 problems spanning topics from integer partitions and tilings to applications of the pigeonhole principle. Various methods have been used to unravel the interactions between combinatorics and other branches of mathematics, especially number theory. A particular focus is placed on recursions, generating functions, enumeration, arithmetics properties of combinatorial quantities, permutations, and graph theory.

Chapter 5. Number Theory. This chapter presents 16 monthly problems, which concentrate on the central area of number theory. Questions here run into divisibility, multiplication functions, congruence, quadratic forms, quadratic residues, and sums of squares. During the journey, we will revisit many important results such as Euler's criterion, Gauss quadratic reciprocity law, Legendre formula, Lucas theorem, and Wolstenholme's theorem.

Chapter 6. Potpourri. We selected 14 problems, which range from geometry and number theory to combinatorics and probability. The solutions touch on various topics, including asymptotic estimates of sums involving the floor function, characterization of palindrome numbers, test for existence of real zeros, enumerations of rationals, algorithmic puzzles, and Markov chains.

We try to present solutions in a coherent and captivating style, including some historical accounts, putting the problems in a broader mathematical context, and stimulating further inquiry. Some problems have concise solutions, while others may be quite complicated. Occasionally, *Mathematica* will help with solutions. Some problems are even related to the Riemann hypothesis and "$3n+1$ conjecture", affirming once again the vitality of monthly problems as a catalyst for research activity. In several cases, a solution requires a result that might not be well-known, so we provided all the details. For problems related to the authors' research, we have given either their extensions or references for further study. We hope this motivates undergraduates to find out "what comes after calculus and linear algebra" and also drives professional mathematicians in their everyday work.

Most of the ideas involved in the solutions should be comprehensible to the good undergraduate math student. The chapters and sections are laid out independently, so feel free to jump into any section and sample whatever tempts you. We are excited and delighted to work with monthly problems. We hope there is much here to please you. As problems in each new issue of the *Monthly* arrive, the reader may regard this book as a starter set, acquire a jumping-off point to new ideas, and extend one's profile of problems and solutions.

We are pleased to acknowledge those individuals who made this book possible. It couldn't have happened without the encouragement of Senior Editor Robert Ross of CRC Press and the valuable feedback from the anonymous reviewers. Thanks to the CRC Press staff for their help in making this book a reality and for all the assistance they offered. We express our deep appreciation to all individuals who contributed to the *Monthly* problems, from whom we have learned so much. The exceptional cover artwork by Adriano Tauraso has given this book its unique character, and for that, we are truly grateful. The first author also offers a special thanks to Professor Brian Bradie, who has carefully read the part of the earlier draft of the manuscript and corrected many errors. Lastly, we are indebted to our families for their love and constant support.

Naturally, all possible errors are our own responsibility. Comments, corrections, and suggestions from readers are always welcome. Please email at hchen@cnu.edu or tauraso@mat.uniroma2.it. Thank you in advance.

Chapter 1

Analysis

In this chapter, we select 15 *Monthly* problems for analysis. These problems are related to roulette curves, special types of integrals, interesting series associated with famous numbers, power sums of real roots of a complex function, extrema of functional moments, and monotonic convex splines. We try to present solutions in a cohesive and engaging manner, including some historical accounts of ideas and contexts. Some problems may touch upon current research topics. While some have concise solutions, others can be quite complicated. Occasionally, solutions are provided with the help of *Mathematica*. In several cases, a solution relies on a less well-known result, so we have included full details for clarity.

1.1 Rolling ellipse

Problem 12476 (Proposed by G. Fera, 131(6), 2024). Let C be one arch of the elliptic cycloid generated by the ellipse $x^2 + \frac{1}{4}(y-2)^2 = 1$. That is, let C be the curve traced by the vertex at the origin as the ellipse rolls without slipping along the x-axis for one revolution (see Figure 1.1). What is the area under C and above the x-axis?

Discussion.
The statement of the problem naturally suggests we revisit the *cycloid*, the path of one fixed point on the circumference of a circle as it rolls along a straight line with a constant angular velocity.

If we place the circle of radius a with the center initially at $(0, a)$ in the x-y plane with the origin O, then the center moves to $C = (at, a)$ after the circle has turned t radian. Let $P = (x(t), y(t))$. Then $\mathbf{r} = \overrightarrow{OP} = \overrightarrow{OC} - \overrightarrow{PC}$, and it follows that the parametric equation of the cycloid is given by

$$\mathbf{r}(t) = (x(t), y(t)) = (at, a) - (a \sin t, a \cos t) = (at - a \sin t, a - a \cos t).$$

DOI: 10.1201/9781003607809-1

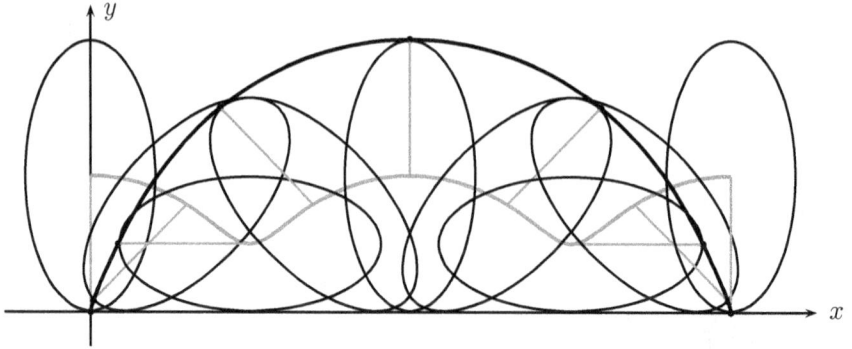

Figure 1.1: Elliptic cycloid with rolling ellipse

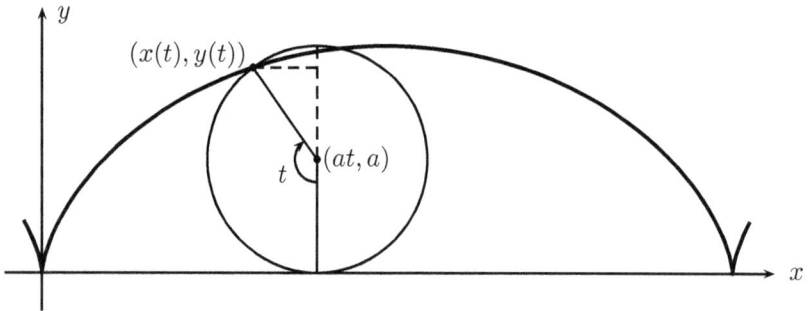

Figure 1.2: Cycloid with rolling circle

Here we see that the position point P relative to the center is given by $(d\sin t, d\cos t)$, where d is the distance between the point P and the center. Moreover, $x'(t) = y(t)$.

In general, when one curve C rolls on another curve without slipping, the path of one fixed point on C is called a *roulette*. In 2011, F. Kuczmarski proved that $x'(t) = y(t)$ is not a coincidence for the cycloid only. Indeed, in [62], he established the following general *Roulette Lemma*: Let $(x, y) = (f(t), g(t))$ be a parametric equation of a roulette rolling on the x-axis. Then $f'(t) = g(t)$.

Thus, based on the above observations, for the proposed problem, it suffices to find the y-coordinate for the center of the rolling ellipse.

Solution.
We show a more general result: If the ellipse is given by $x^2/a^2 + (y-b)^2/b^2 = 1$ with $a < b$, then the area under C and above the x-axis is $(a^2 + 2b^2)\pi$. Hence, the answer to the proposed problem is 9π.

To this end, we first derive a parametric equation for C. Let θ be the angle between the y-axis and the major axis of the rolling ellipse. To simplify,

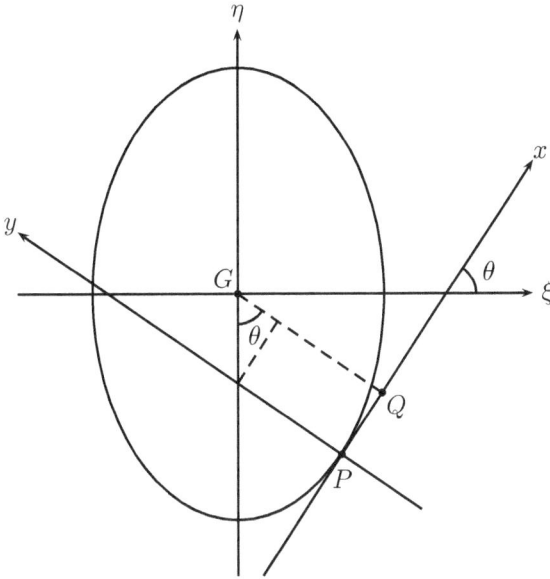

Figure 1.3: Parameterizing the center of rolling ellipse

we rotate the entire diagram so that the ellipse's major axis becomes vertical, and introduce a new Cartesian coordinate system (ξ, η) with the origin at the ellipse's center G, as shown in Figure 1.3.

The equation of the ellipse now is

$$\frac{\xi^2}{a^2} + \frac{\eta^2}{b^2} = 1. \tag{1.1}$$

Suppose the ellipse is tangent to the x-axis at $P = (\xi_P, \eta_P)$. Let the intersection point of the vertical line from G to the x-axis be Q. Then $Q = (y \sin \theta, -y \cos \theta)$, where $y = |\overline{GQ}|$. In the ξ-η plane, the slope of the x-axis is $\tan \theta$. By the slope-intercept form, the equation of the x-axis is

$$\eta = \xi \tan \theta - \frac{y}{\cos \theta}.$$

Plugging this into (1.1) yields a quadratic equation in ξ:

$$(a^2 \sin^2 \theta + b^2 \cos^2 \theta)\xi^2 - (2a^2 y \sin \theta)\xi + a^2(y^2 - b^2 \cos^2 \theta) = 0. \tag{1.2}$$

Since the tangency at P implies that (1.2) must have double roots, the discriminant of (1.2) yields

$$y = \sqrt{a^2 \sin^2 \theta + b^2 \cos^2 \theta},$$

which is the y-coordinate of the center of the ellipse in the x-y plane. Hence, the y-coordinate of the vertex of the ellipse in the x-y plane is $y - b \cos \theta$.

Applying the *Roulette Lemma*, we find the parametric equation of C as

$$(x, y) = \left(\int_0^\theta y(t)\, dt, y(\theta) \right).$$

where

$$y(\theta) = \sqrt{a^2 \sin^2 \theta + b^2 \cos^2 \theta} - b \cos \theta.$$

Hence, the area under C and above the x-axis is

$$\int_0^{2\pi} y\, dx = \int_0^{2\pi} \left(\sqrt{a^2 \sin^2 \theta + b^2 \cos^2 \theta} - b \cos \theta \right)^2 d\theta$$

$$= \int_0^{2\pi} \left(a^2 \sin^2 \theta + 2b^2 \cos^2 \theta - 2b \cos \theta \sqrt{a^2 \sin^2 \theta + b^2 \cos^2 \theta} \right) d\theta$$

$$= (a^2 + 2b^2)\pi - 2b \int_0^{2\pi} \cos \theta \sqrt{b^2 - (b^2 - a^2) \sin^2 \theta}\, d\theta = (a^2 + 2b^2)\pi.$$

$\square$

Remark. It would be interesting to find the x-coordinate of the parametric equation for the center of the rolling ellipse directly without using the *Roulette Lemma*. In terms of Figure 1.3, we have

$$x(\theta) = \text{the arclength of the ellipse from 0 to } \theta + |\overline{PQ}|.$$

This will force us to grind out the elliptical integral.

The idea of roulettes was first introduced by W. H. Besant in 1870 and later popularized by S. Wagon and his square-wheeled tricycle (see [55] and the more recent [58]). They have used a geometrical description: the road as a roulette curve and the wheel as a pedal curve. For example, the catenary, which is usually given as the solution to a hanging flexible cable, now is the locus of the focus of a parabola which rolls on the x-axis. Moreover, J. Steiner established two theorems relating the area and arclength of a roulette to those of a corresponding pedal. For more details, please refer again to F. Kuczmarski's paper [62].

Additional problems for practice.

1. Let the parabola $y = x^2/4$ roll without slipping along the x-axis. Show the parametric equation of the roulette traced by the focus is given by

$$(x, y) = (\ln(\sec \theta + \tan \theta), \sec \theta).$$

2. Find the locus of the vertex of the parabola $y = x^2$ as it rolls along the x-axis.

3. **Putnam 1974-A5.** Consider the two mutually tangent parabolas $y = x^2$ and $y = -x^2$. [These have foci at $(0, 1/4)$ and $(0, -1/4)$), and directrices $y = -1/4$ and $y = 1/4$, respectively.] The upper parabola rolls without slipping around the fixed lower parabola. Find the locus of the focus of the moving parabola.

4. Refer to Figure 1.3, show that

$$|\overline{PQ}| = \frac{|a^2 - b^2|\sin 2\theta}{2\sqrt{a^2\sin^2\theta + b^2\cos^2\theta}}.$$

5. **Problem 10254** (Proposed by E. Ehrhart, 99(8), 1992). The curve traced out by a fixed point of a closed convex curve as that curve rolls without slipping along a second curve will be called a "roulette". Let S be the area of one arch of a roulette traced out by an ellipse of area s rolling on a straight line. Prove or disprove that $S \geq 3s$, with equality only if the ellipse is a circle.

The featured solution is due to M. Klamkin, a prolific proposer and editor of professionally-challenging mathematical problems. His solution cited the Steiner theorems that appeared in [103, p. 201-203].

1.2   A limit involving a geometric mean of roots of factorials

Problem 12360 (Proposed by D. M. Bătinetu and N. Stanciu, 129(10), 2022). Evaluate

$$\lim_{n\to\infty} \frac{(n+1)^2}{x_{n+1}} - \frac{n^2}{x_n},$$

where $x_n = \sqrt[n]{\sqrt{2!}\sqrt[3]{3!}\cdots\sqrt[n]{n!}}$.

Discussion.
Pattern recognition sometimes plays a key role in problem solving. First ask yourself if you've seen a similar problem before, then revisit it to understand how it works, and try to figure out which "trick" in the old problem can be applied to your current problem. Our first solution illustrates how the arguments of *Monthly* 11935 [30, pp. 21-23] can be used to solve this problem effortlessly.

On the other hand, by the Stirling approximation

$$n! = \sqrt{2\pi n}(n/e)^n(1 + O(1/n)),$$

we have

$$\ln\left(\sqrt[n]{n!}\right) = \ln(n) - 1 + \frac{\ln(2\pi n)}{2n} + O\left(\frac{1}{n^2}\right).$$

The proposed problem naturally suggests establishing an asymptotic approximation for their average:

$$\ln x_n = \frac{1}{n}\left(\ln\left(\sqrt{2!}\right) + \ln\left(\sqrt[3]{3!}\right) + \cdots + \ln\left(\sqrt[n]{n!}\right)\right).$$

Indeed, we will prove that for some constant C,

$$\ln x_n = \ln(n) - 2 + \frac{\ln^2 n}{4n} + \frac{\ln(2\pi e)\ln(n)}{2n} + \frac{C}{n} + O\left(\frac{\ln(n)}{n^2}\right). \qquad (1.3)$$

This, together with the Stolz-Cesàro theorem, leads us to a second solution.

Solution I.
We show that the limit is e^2. To this end, we establish two facts first.
$\mathbf{F_1}$. $\lim_{n\to\infty} \frac{x_n}{n} = \frac{1}{e^2}$.
By the Stolz-Cesàro theorem, we have

$$\lim_{n\to\infty} \ln\left(\frac{x_n}{n}\right) = \lim_{n\to\infty} \frac{\sum_{k=2}^{n} \ln(\sqrt[k]{k!}) - n\ln(n)}{n}$$

$$\overset{SC}{=} \lim_{n\to\infty} \left(\ln(\sqrt[n+1]{(n+1)!}) - (n+1)\ln(n+1) + n\ln(n)\right)$$

$$= \lim_{n\to\infty} \left(\ln\left(\frac{\sqrt[n+1]{(n+1)!}}{n+1}\right) - n\ln\left(1 + \frac{1}{n}\right)\right)$$

$$= -1 - 1 = -2,$$

which is equivalent to $\mathbf{F_1}$. Here we have used the well-known fact that

$$\lim_{n\to\infty} \frac{\sqrt[n]{n!}}{n} = \frac{1}{e}.$$

$\mathbf{F_2}$. $\lim_{n\to\infty}(x_{n+1} - x_n) = \frac{1}{e^2}$.
Since $x_{n+1}^{n+1} = x_n^n \sqrt[n+1]{(n+1)!}$, as $n \to \infty$, using $\mathbf{F_1}$, we have

$$\frac{x_{n+1}}{x_n} = \left(\frac{\sqrt[n+1]{(n+1)!}}{x_n}\right)^{\frac{1}{n+1}} = \left(\frac{\sqrt[n+1]{(n+1)!}}{n+1} \cdot \frac{n+1}{n} \cdot \frac{n}{x_n}\right)^{\frac{1}{n+1}} \to 1.$$

Similarly, as $n \to \infty$, we have

$$\left(\frac{x_{n+1}}{x_n}\right)^n = \frac{x_{n+1}^{n+1}}{x_n^n} \frac{1}{x_{n+1}} = \frac{\sqrt[n+1]{(n+1)!}}{x_{n+1}} = \frac{\sqrt[n+1]{(n+1)!}}{n+1} \cdot \frac{n+1}{x_{n+1}} \to \frac{1}{e} \cdot e^2 = e.$$

Hence, using these facts and $\mathbf{F_1}$, as $n \to \infty$, we get

$$x_{n+1} - x_n = \frac{x_n}{n} \cdot \frac{\left(\frac{x_{n+1}}{x_n} - 1\right)}{\ln\left(1 + \left(\frac{x_{n+1}}{x_n} - 1\right)\right)} \cdot \ln\left(\frac{x_{n+1}}{x_n}\right)^n \to \frac{1}{e^2} \cdot 1 \cdot 1 = \frac{1}{e^2}.$$

Finally, using $\mathbf{F_1}$ and $\mathbf{F_2}$, we find that

$$\lim_{n\to\infty} \frac{(n+1)^2}{x_{n+1}} - \frac{n^2}{x_n} = \lim_{n\to\infty} \frac{(n+1)^2 x_n - n^2 x_{n+1}}{x_n x_{n+1}}$$

$$= \lim_{n\to\infty} \frac{n(n+1)}{x_n x_{n+1}} \left(\frac{(n+1)x_n}{n} - \frac{n x_{n+1}}{n+1} \right)$$

$$= \lim_{n\to\infty} \frac{n(n+1)}{x_n x_{n+1}} \left[\frac{x_n}{n} + \frac{x_{n+1}}{n+1} + (x_n - x_{n+1}) \right]$$

$$= e^4 \left(\frac{1}{e^2} + \frac{1}{e^2} - \frac{1}{e^2} \right) = e^2$$

as claimed. $\qquad\qquad\qquad\qquad\qquad\qquad\qquad\qquad\qquad\qquad\qquad\square$

Solution II.

To establish the asymptotic approximation (1.3), we first collect several facts that all follow from the Taylor expansion $-\ln(1-x) = x + x^2/2 + O(x^3)$ as $x \to 0$:

$$\ln(n) - \ln(n-1) = \frac{1}{n} + \frac{1}{2n^2} + O\left(\frac{1}{n^3}\right);$$

$$n\ln(n) - (n-1)\ln(n-1) = \ln(n) + 1 - \frac{1}{2n} + O\left(\frac{1}{n^2}\right);$$

$$\ln^2(n) - \ln^2(n-1) = \frac{2\ln(n)}{n} + \frac{\ln(n)}{n^2} + O\left(\frac{1}{n^2}\right);$$

$$\frac{\ln(n)}{n} - \frac{\ln(n-1)}{n-1} = -\frac{\ln(n)}{n^2} + O\left(\frac{1}{n^2}\right).$$

Next, let

$$a_n = n\ln(n) - 2n + \frac{\ln^2(n)}{4} + \frac{\ln(2\pi e)\ln(n)}{2}, \quad \text{and}$$

$$b_n = n\ln(x_n) - a_n = \sum_{k=1}^{n} \frac{\ln(k!)}{k} - a_n.$$

Applying the facts above, we have

$$a_n - a_{n-1} = \ln(n) - 1 + \frac{\ln(2\pi n)}{2n} + \frac{\ln(n)}{4n^2} + O\left(\frac{1}{n^2}\right).$$

Furthermore, by the Stirling approximation

$$\ln(n!) = n\ln(n) - n + \frac{\ln(2\pi n)}{2n} + O\left(\frac{1}{n}\right),$$

we obtain

$$b_n - b_{n-1} = \frac{\ln(n!)}{n} - (a_n - a_{n-1}) = -\frac{\ln(n)}{4n^2} + O\left(\frac{1}{n^2}\right).$$

This implies that $\sum_{n=1}^{\infty}(b_n - b_{n-1})$ converges, and so $\lim_{n\to\infty} b_n$ exists.

Let $C = \lim_{n\to\infty} b_n$ and $c_n = a_n + C$. Then $n \ln x_n - c_n = b_n - C \to 0$ as $n \to \infty$. Applying the Stolz-Cesàro theorem yields

$$\lim_{n\to\infty} \frac{\ln(x_n) - \frac{c_n}{n}}{\frac{\ln(n)}{n^2}} = \lim_{n\to\infty} \frac{b_n - C}{\frac{\ln(n)}{n}} = \lim_{n\to\infty} \frac{b_n - b_{n-1}}{\frac{\ln(n)}{n} - \frac{\ln(n-1)}{n-1}}$$

$$= \lim_{n\to\infty} \frac{-\frac{\ln(n)}{4n^2} + O\left(\frac{1}{n^2}\right)}{-\frac{\ln(n)}{n^2} + O\left(\frac{1}{n^2}\right)} = \frac{1}{4}.$$

This proves (1.3) as claimed. Now, from (1.3),

$$x_n = \frac{n}{e^2} \exp\left(\frac{\ln^2(n)}{4n} + \frac{\ln(2\pi e)\ln(n)}{2n} + \frac{C}{n} + O\left(\frac{\ln(n)}{n^2}\right)\right).$$

Using $e^{-x} = 1 - x + O(x^2)$ as $x \to 0$, we have

$$\frac{n^2}{x_n} = ne^2\left(1 - \frac{\ln^2(n)}{4n} - \frac{\ln(2\pi e)\ln(n)}{2n} - \frac{C}{n} + O\left(\frac{\ln^4(n)}{n^2}\right)\right)$$

$$= e^2\left(n - \frac{\ln^2(n)}{4} - \frac{\ln(2\pi e)\ln(n)}{2} - C + O\left(\frac{\ln^4 n}{n}\right)\right).$$

Thus,

$$\lim_{n\to\infty} \frac{(n+1)^2}{x_{n+1}} - \frac{n^2}{x_n} = \lim_{n\to\infty} e^2\left(1 - \frac{\ln^2(n+1) - \ln^2(n)}{4}\right.$$

$$\left. - \frac{\ln(2\pi e)(\ln(n+1) - \ln(n))}{2} + O\left(\frac{\ln^4(n)}{n}\right)\right) = e^2.$$

$$\square$$

Remark. If we assume the proposed limit exists, we can determine the value of the limit by using the Stolz-Cesàro theorem and $\mathbf{F_1}$ only. Indeed, we have

$$\lim_{n\to\infty} \frac{(n+1)^2}{x_{n+1}} - \frac{n^2}{x_n} = \lim_{n\to\infty} \frac{\frac{(n+1)^2}{x_{n+1}} - \frac{n^2}{x_n}}{n+1-n}$$

$$= \lim_{n\to\infty} \frac{\frac{n^2}{x_n}}{n} = \lim_{n\to\infty} \frac{n}{x_n} = e^2.$$

(1.3) provides an exquisite estimate, which is of interest in its own right. Numerically, we have $C = 1.54836\ldots$, but, we can't determine the value of C in a closed form.

The approaches used in the above solutions can be applied to a large class of sequences.

Additional problems for practice.

1. Does the sequence $x_n = \sqrt{1!\sqrt{2!\cdots\sqrt{n!}}}$ converge?

2. For $p \geq 1$, let

$$x_n(p) = n^{1-p}\left(\frac{\left(\sqrt[n+1]{(n+1)!}\right)^{2p}}{(n+1)^p} - \frac{\left(\sqrt[n]{n!}\right)^{2p}}{n^p}\right).$$

Find $\lim_{n\to\infty} x_n(p)$.

3. Let x_n be a positive sequence and $p \geq 0$. If $\lim_{n\to\infty} \frac{x_{n+1}}{n^{p+1}\cdot x_n} = q > 0$, show that

$$\lim_{n\to\infty} \frac{\sqrt[n+1]{x_{n+1}} - \sqrt[n]{x_n}}{n^p} = (p+1)qe^{-(p+1)}.$$

Let $x_n = n!$ and $p = 0$. This yields the classical limit

$$\lim_{n\to\infty} \left(\sqrt[n+1]{(n+1)!} - \sqrt[n]{n!}\right) = \frac{1}{e}.$$

4. Given a positive sequence x_n, define $(x_1)! = x_1, (x_{n+1})! = x_{n+1} \cdot (x_n)!$ for all $n \geq 1$. For $p \geq 0$, if $\lim_{n\to\infty} \frac{x_{n+1}-x_n}{n^p} = q > 0$, show that

$$\lim_{n\to\infty} \frac{(x_{n+1})!}{n^{p+1}(x_n)!} = \frac{q}{p+1}.$$

5. **Problem 12129** (Proposed by H. Ohtsuka, 126(7), 2019). Compute

$$\sqrt{2+\sqrt{2+\sqrt{2+\cdots+\sqrt{2-\sqrt{2+\cdots}}}}}$$

where the sequence of signs consists of $n-1$ plus signs followed by a minus sign and repeats with period n.

6. **Problem 12510** (Proposed by H. Ohtsuka, 132(2), 2025). For $c > 0$ and $n \geq 1$, let

$$R_n(c) = \sqrt{c+\sqrt{c+\sqrt{c+\cdots+\sqrt{c}}}}$$

with n radicals. Evaluate

$$\lim_{n\to\infty} \sum_{k=1}^{n}\prod_{j=k}^{n} \frac{1}{R_j(c)}.$$

1.3   A Lobachevsky-type integral

Problem 12351 (Proposed by S. Stewart, 129(9), 2022). Evaluate

$$\int_0^\infty \frac{\ln(\cos^2 x)\sin^3 x}{x^3(1+2\cos^2 x)}\,dx.$$

Discussion.
We begin our analysis by getting some feel for the proposed integral in two distinct directions. First, when routine methods such as substitution and integration by parts don't shed any insight, we naturally try to apply R. Feynman's "a different box of tools" — the parametric differentiation. Introduce

$$I(p) := \int_0^\infty \frac{\ln(1-p\sin^2 x)\sin^3 x}{x^3(3-2\sin^2 x)}\,dx.$$

The proposed integral is simply $I(1)$. For $p \in [0,1)$, using partial fractions and the geometric series, we have

$$\begin{aligned}
I'(p) &= \int_0^\infty -\frac{1}{(1-p\sin^2 x)(3-2\sin^2 x)}\frac{\sin^5 x}{x^3}\,dx \\
&= \int_0^\infty \left(\frac{\frac{2/3}{3p-2}}{1-\frac{2}{3}\sin^2 x} - \frac{\frac{p}{3p-2}}{(1-p^2\sin^2 x)}\right)\frac{\sin^5 x}{x^3}\,dx \\
&= \frac{1}{3p-2}\sum_{n=0}^\infty \left(\frac{2^n}{3^n}-p^n\right)\int_0^\infty \frac{\sin^{2n+3} x}{x^3}\,dx, \quad\quad (1.4)
\end{aligned}$$

where the interchange of the summation and integration is justified since the series is uniformly convergent. Our first solution will begin with (1.4).

Second, the sinc function $\mathrm{sinc}(x) := \frac{\sin x}{x}$ in the integrand reminds us of Lobachevsky's formula

$$\int_0^\infty \frac{\sin^2 x}{x^2}f(x)\,dx = \int_0^\infty \frac{\sin x}{x}f(x)\,dx = \int_0^{\pi/2} f(x)\,dx, \quad\quad (1.5)$$

where $f(x)$ is a continuous function with $f(x\pm\pi)=f(x)$ for all $x\geq 0$. This formula is quite unique because it uses the sinc function as a weight instead of the usual exponential function that appears in the Laplace transform. Letting $f(x)=1$ yields the classical Dirichlet formula:

$$\int_0^\infty \frac{\sin^2 x}{x^2}\,dx = \int_0^\infty \frac{\sin x}{x}\,dx = \frac{\pi}{2}.$$

Now, let us consider

$$f(x) = \frac{\ln(\cos^2 x)}{1+2\cos^2 x}.$$

Clearly, f is continuous and satisfies $f(x \pm \pi) = f(x)$ for all $x \geq 0$. So the proposed integral becomes the Lobachevsky-type:

$$I := \int_0^\infty \frac{\sin^3 x}{x^3} f(x)\, dx.$$

Since

$$\int_0^\infty \frac{\sin^3 x}{x^3}\, dx = \frac{3\pi}{8} \neq \frac{\pi}{2},$$

Lobachevsky's formula in the form of (1.5) no longer holds for the cubic power of sinc function. On the other hand, taking advantage of the π-periodic nature of f, we decompose the interval $[0, \infty)$ into intervals of length of $\pi/2$. Then

$$\int_0^\infty \frac{\sin^3 x}{x^3} f(x)\, dx = \sum_{n=0}^\infty \int_{n\pi/2}^{(n+1)\pi/2} \frac{\sin^3 x}{x^3} f(x)\, dx.$$

So we can also proceed with the proposed problem by first establishing

$$\int_0^\infty \frac{\sin^3 x}{x^3} f(x)\, dx = \int_0^{\pi/2} \left(1 - \frac{1}{2}\sin^2 x\right) f(x)\, dx. \qquad (1.6)$$

Our second solution will be based on (1.6).

Solution I.
Applying integration by parts twice to the integral in (1.4), we find

$$\int_0^\infty \frac{\sin^{2n+3} x}{x^3}\, dx = \frac{2n+3}{2} \int_0^\infty \frac{\sin^{2n+2} x \cos x}{x^2}\, dx$$

$$= \frac{(2n+3)(2n+2)}{2} \int_0^\infty \frac{\sin^{2n+1} x}{x}\, dx$$

$$- \frac{(2n+3)^2}{2} \int_0^\infty \frac{\sin^{2n+3} x}{x}\, dx.$$

Since

$$\sin^{2n+1} x = \frac{1}{4^n} \sum_{k=0}^n (-1)^k \binom{2n+1}{n-k} \sin(2k+1)x$$

and

$$\int_0^\infty \frac{\sin(\alpha x)}{x}\, dx = \frac{\pi}{2} \qquad \text{for every } \alpha > 0,$$

we find

$$\int_0^\infty \frac{\sin^{2n+1} x}{x}\, dx = \frac{1}{4^n} \sum_{k=0}^n (-1)^k \binom{2n+1}{n-k} \int_0^\infty \frac{\sin(2k+1)x}{x}\, dx$$

$$= \frac{1}{4^n} \sum_{k=0}^n (-1)^k \binom{2n+1}{n-k} \frac{\pi}{2} = \frac{\pi}{2^{2n+1}} \binom{2n}{n}.$$

This yields

$$\int_0^\infty \frac{\sin^{2n+3} x}{x^3}\, dx = \frac{\pi/4}{4^n} \left(1 + \frac{1/2}{n+1}\right) \binom{2n}{n}.$$

By (1.4), we obtain

$$I'(p) = \frac{\pi}{4} \frac{f(2/3) - f(p)}{3p - 2},$$

where

$$f(p) = \sum_{n=0}^\infty \binom{2n}{n} \left(\frac{p}{4}\right)^n + \frac{1}{2} \sum_{n=0}^\infty \frac{1}{n+1} \binom{2n}{n} \left(\frac{p}{4}\right)^n, \quad \text{for } |x| < 1.$$

Applying the binomial series of $(1 - p)^{-1/2}$ and its integration, we find

$$f(p) = \frac{1}{\sqrt{1-p}} + \frac{1}{1 + \sqrt{1-p}}.$$

Since $I(0) = 0$, we finally obtain

$$I(1) = I(0) + \int_0^1 I'(p)\, dp = \frac{\pi}{4} \int_0^1 \frac{f(2/3) - f(p)}{3p - 2}\, dp$$

$$= -\frac{\pi}{4} \int_0^1 \left(\frac{1}{t+1} + \frac{1}{\sqrt{3}t + 1}\right) dt \quad (\text{use } t = \sqrt{1-p})$$

$$= -\frac{\pi}{4} \left(\ln 2 + \frac{\sqrt{3}}{3} \ln(\sqrt{3} + 1)\right).$$

□

Solution II.
To prove (1.6), we split the right hand side sum into two parts by the parity of n to get

$$I = \sum_{k=0}^\infty \int_{k\pi}^{k\pi + \frac{\pi}{2}} \frac{\sin^3 x}{x^3} f(x)\, dx + \sum_{k=1}^\infty \int_{k\pi - \frac{\pi}{2}}^{k\pi} \frac{\sin^3 x}{x^3} f(x)\, dx. \qquad (1.7)$$

Let $x = k\pi + t$. Then

$$\int_{k\pi}^{k\pi + \frac{\pi}{2}} \frac{\sin^3 x}{x^3} f(x)\, dx = (-1)^k \int_0^{\pi/2} \frac{\sin^3 t}{(k\pi + t)^3} f(t)\, dt.$$

Similarly, let $x = k\pi - t$. Then

$$\int_{k\pi - \frac{\pi}{2}}^{k\pi} \frac{\sin^3 x}{x^3} f(x)\, dx = (-1)^{k-1} \int_0^{\pi/2} \frac{\sin^3 t}{(k\pi - t)^3} f(t)\, dt.$$

Using these two integrals and (1.7), we have

$$I = \int_0^{\pi/2} \frac{\sin^3 t}{t^3} f(t)\, dt + \sum_{k=1}^{\infty} \int_0^{\pi/2} (-1)^k \left(\frac{\sin^3 t}{(k\pi + t)^3} + \frac{\sin^3 t}{(t - k\pi)^3} \right) f(t)\, dt$$

$$= \int_0^{\pi/2} \sin^3 t \left[\frac{1}{t^3} + \sum_{k=1}^{\infty} (-1)^k \left(\frac{1}{(t + k\pi)^3} + \frac{1}{(t - k\pi)^3} \right) \right] f(t)\, dt.$$

Here the interchange of the order of the summation and integration is justified by the uniform convergence of the series. Next, recall that if $x \neq n\pi$ for $n \in \mathbb{Z}$, then

$$\frac{1}{\sin x} = \frac{1}{x} + \sum_{k=1}^{\infty} (-1)^k \left(\frac{1}{x - k\pi} + \frac{1}{x + k\pi} \right). \tag{1.8}$$

Differentiating (1.8) twice term-by-term yields

$$\frac{2}{\sin^3 x} - \frac{1}{\sin x} = \frac{2}{x^3} + \sum_{k=1}^{\infty} (-1)^k 2 \left(\frac{1}{(x - k\pi)^3} + \frac{1}{(x + k\pi)^3} \right). \tag{1.9}$$

Finally, applying (1.9) obtains

$$I = \int_0^{\pi/2} \sin^3 t \, \frac{1}{2} \left(\frac{2}{\sin^3 t} - \frac{1}{\sin t} \right) f(t)\, dt = \int_0^{\pi/2} \left(1 - \frac{1}{2} \sin^2 t \right) f(t)\, dt.$$

This proves (1.6) as claimed.

Using the formula [45, 4.385(3)]

$$\int_0^{\pi/2} \frac{\ln \cos x}{a^2 \cos^2 x + b^2 \sin^2 x}\, dx = \frac{\pi}{2ab} \ln \left(\frac{b}{a + b} \right) \qquad \text{for } a, b > 0,$$

after letting

$$f(x) = \frac{\ln(\cos^2 x)}{1 + 2 \cos^2 x}$$

we find

$$\int_0^{\pi/2} f(x)\, dx = 2 \int_0^{\pi/2} \frac{\ln \cos x}{3 \cos^2 x + \sin^2 x}\, dx = \frac{\pi}{\sqrt{3}} \ln \left(\frac{1}{\sqrt{3} + 1} \right).$$

Hence,

$$\int_0^\infty \frac{\ln(\cos^2 x)\sin^3 x}{x^3(1+2\cos^2 x)}\,dx = \int_0^{\pi/2} \frac{1}{2}(1+\cos^2 x)f(x)\,dx$$

$$= \frac{1}{4}\int_0^{\pi/2} f(x)\,dx + \frac{1}{2}\int_0^{\pi/2} \ln(\cos x)\,dx$$

$$= \frac{\pi}{4\sqrt{3}}\ln\left(\frac{1}{\sqrt{3}+1}\right) + \frac{1}{2}\cdot\left(-\frac{1}{2}\pi\ln 2\right)$$

$$= \frac{\pi}{4}\left[\frac{1}{\sqrt{3}}\ln\left(\frac{1}{\sqrt{3}+1}\right) - \ln 2\right]$$

$$= -\frac{\pi}{4}\left(\frac{\sqrt{3}}{3}\ln(\sqrt{3}+1) + \ln 2\right).$$

□

Remark. The proof of (1.6) has its own interest: it serves as a demonstration of how the proof extends Lobachevsky's formula (1.5) to higher powers. Indeed, similar to the proof of (1.6), using (1.8) and its derivatives, we have

$$\int_0^\infty \frac{\sin^4 x}{x^4}f(x)\,dx = \int_0^{\pi/2}\left(1 - \frac{2}{3}\sin^2 x\right)f(x)\,dx,$$

$$\int_0^\infty \frac{\sin^5 x}{x^5}f(x)\,dx = \int_0^{\pi/2}\left(1 - \frac{5}{6}\sin^2 x + \frac{1}{4}\sin^4 x\right)f(x)\,dx.$$

Recently, by using higher derivatives of the partial fraction expansion of $\csc x$ and their derivative polynomials, H. Chen has found Lobachevsky's formula in explicit form for all integer powers (see [33]). This provides a good example of how problem-solving can lead to a publication.

Additional problems for practice.

1. Show that

$$\int_0^{\pi/2} \frac{\ln\cos x}{a^2\cos^2 x + b^2\sin^2 x}\,dx = \frac{\pi}{2ab}\ln\left(\frac{b}{a+b}\right) \quad \text{for } a,b > 0,$$

2. **Problem 11423** (Proposed by G. Minton, 116(3), 2009). Show that if n and m are positive integers with $n \geq m$ and $n-m$ even, then $\int_0^\infty \frac{\sin^n x}{x^m}\,dx$ is a rational multiple of π.

3. **Problem 12375** (Proposed by H. Chen, 130(2), 2023). Let

$$I_n = \int_0^\infty \left(1 - x^2\sin^2\left(\frac{1}{x}\right)\right)^n\,dx.$$

Monthly 12288 [2021, 956] asked for a proof that $I_2 = \frac{\pi}{5}$. Prove that I_n is a rational multiple of π whenever n is a positive integer.

4. Let n and m be positive integers with $n > m \geq 2$ and $n - m$ odd. Evaluate

$$\int_0^\infty \frac{\sin^n x}{x^m} \, dx$$

in a closed form.

5. Show that

$$\int_0^\infty \frac{\arctan(\sin x)}{x} \, dx = \frac{\pi}{2} \ln(1 + \sqrt{2}).$$

6. In Lobachevsky's formula (1.5), if the condition $f(x + n\pi) = f(x)$ is replaced by $f(-x) = f(x)$ and $f(x + n\pi) = (-1)^n f(x)$, show that

$$\int_0^\infty \frac{\sin x}{x} f(x) \, dx = \int_0^{\pi/2} f(x) \cos x \, dx.$$

Using this result proves that

$$\int_0^\infty \frac{\sin(\tan x)}{x} \, dx = \frac{\pi}{2} \left(1 - \frac{1}{e} \right).$$

7. **Research Project**. Find a Lobachevsky's formula for all positive integer powers and f with arbitrary period T via Fourier analysis.
Recall some standard notations from Fourier analysis. Let $f, g \in L^1(\mathbb{R})$. We define

$$\mathcal{F}(\xi) = \int_{-\infty}^\infty f(x) e^{-2\pi i x \xi} \, dx$$

as the Fourier transform of f and

$$(f * g)(x) = \int_{-\infty}^\infty f(y) g(x - y) \, dy$$

as the convolution of f and g. Furthermore, let $T > 0$. Define

$$L^1[-T/2, T/2] = \{f(x)\,;\, f \text{ is } T\text{-periodic and absolutely integrable}\}.$$

For the set of such functions, the Fourier coefficient is defined by

$$\hat{f}(n) = \frac{1}{T} \int_{-T/2}^{T/2} e^{-2\pi i \frac{nx}{T}} f(x) \, dx, \quad n \in \mathbb{Z}.$$

(a) (Parseval Formula). Let $f \in L^1[-1/2, 1/2]$ and $g \in L^1(\mathbb{R})$. If $\text{supp}(g) \in [-M, M]$ for some positive constant M and is bounded variation in the neighborhoods of all $n \in \mathbb{Z}$ with $|n| \leq M$, show that

$$\int_{-\infty}^\infty f(x) \hat{g}(x) \, dx = \sum_{n \in \mathbb{Z}, \, |n|/T \leq M} \hat{f}(n) \frac{g(n^-) + g(n^+)}{2}.$$

(b) Define the normalized cardinal sine on $\mathbb{R}$ by

$$\mathrm{Sinc}(x) = \begin{cases} \frac{\sin(\pi x)}{\pi x}, & x \neq 0, \\ 1, & x = 0. \end{cases}$$

Find $\mathcal{F}^{-1}(\mathrm{Sinc}(x))$.

(c) Let $B(x) = \mathcal{F}^{-1}(\mathrm{Sinc}(x))$. If $f \in L^1[-1/2, 1/2]$ and k is a positive integer, show that

$$\int_{-\infty}^{\infty} \mathrm{Sinc}^k(x) f(x)\, dx = \sum_{|n|/T \leq k/2} \hat{f}(n)\, B^{*k}\left(\frac{n}{T}\right),$$

where $B^{*k} = \underbrace{B * B * \cdots * B}_{k \text{ times}}$. In particular, if $k = 1$ and $0 < T < 2$, then

$$\int_{-\infty}^{\infty} \mathrm{Sinc}(x) f(x)\, dx = B^*(0) \int_{-T/2}^{T/2} f(x)\, dx = \int_{-T/2}^{T/2} f(x)\, dx.$$

1.4 A quadratic series with the tail of $\zeta(2)$

Problem 12287 (Proposed by O. Furdui and A. Sîntămărian, 128(10), 2021). Prove

$$\sum_{n=1}^{\infty} \left(n \left(\sum_{k=n}^{\infty} \frac{1}{k^2} \right)^2 - \frac{1}{n} \right) = \frac{3}{2} - \frac{1}{2}\zeta(2) + \frac{3}{2}\zeta(3).$$

Discussion.
Let $T_n^2 := \sum_{k=n}^{\infty} 1/k^2$, the tail of $\zeta(2)$. The summand of the series $nT_n^2 - 1/n$ simply admits no special series that shed any insight. Naturally, we try to transform the summand into an equivalent but simpler form in a meaningful way. This can be accomplished by two different ways. The first method is based on the standard reformulate approach – Abel's summation formula, which is analogous to integration by parts. This technique has been used in Solution I of *Monthly* 11810 [30, pp. 71-72]. The second method is to decompose the summand into the form

$$nT_n^2 - \frac{1}{n} = t_n - t_{n-1} + s_n,$$

where s_n is a known convergent series. This idea originated from *Monthly* E3352 [84]: Show that

$$\sum_{n=0}^{\infty} \frac{1}{n!(n^4 + n^2 + 1)} = \frac{e}{2}.$$

Here the series itself is not telescoping, but

$$\frac{1}{n!(n^4 + n^2 + 1)} = \frac{1}{2}\left(\frac{n}{(n+1)!(n^2+n+1)} - \frac{n-1}{n!(n^2-n+1)} + \frac{1}{(n+1)!}\right).$$

Thus, the original series becomes the sum of a telescoping series and a well-known series.

Solution I.
Let the proposed series be S. For $n \geq 1$, let

$$a_n = n, \quad b_n = \left(\sum_{k=n}^{\infty}\frac{1}{k^2}\right)^2,$$

and $A_n = \sum_{k=1}^{n} a_k$. Then $A_n = n(n+1)/2$ and

$$b_n - b_{n+1} = \left(\frac{1}{n^2} + \sum_{k=n+1}^{\infty}\frac{1}{k^2}\right)^2 - \left(\sum_{k=n+1}^{\infty}\frac{1}{k^2}\right)^2$$

$$= \frac{1}{n^4} + \frac{2}{n^2}\sum_{k=n+1}^{\infty}\frac{1}{k^2}.$$

Applying Abel's summation formula:

$$\sum_{n=1}^{\infty} a_n b_n = \lim_{n\to\infty} A_n b_{n+1} + \sum_{n=1}^{\infty} A_n(b_n - b_{n+1}),$$

we have

$$S = \lim_{n\to\infty}\frac{1}{2}n(n+1)b_{n+1} + \sum_{n=1}^{\infty}\left(\frac{1}{2}n(n+1)\left(\frac{1}{n^4} + \frac{2}{n^2}\sum_{k=n+1}^{\infty}\frac{1}{k^2}\right) - \frac{1}{n}\right)$$

$$= \lim_{n\to\infty}\frac{1}{2}n(n+1)b_{n+1} + \frac{1}{2}(\zeta(2) + \zeta(3)) + \sum_{n=1}^{\infty}\left(\frac{n+1}{n}\sum_{k=n+1}^{\infty}\frac{1}{k^2} - \frac{1}{n}\right).$$

$$(1.10)$$

By the Stolz-Cesàro theorem,, we obtain

$$\lim_{n\to\infty}\frac{\sum_{k=n}^{\infty}1/k^2 - 1/n}{1/n^2} = \lim_{n\to\infty}\frac{-1/n^2 - 1/(n+1) + 1/n}{1/(n+1)^2 - 1/n^2} = \frac{1}{2},$$

and so

$$\sum_{k=n}^{\infty}\frac{1}{k^2} = \frac{1}{n} + \frac{1}{2n^2} + o(1/n^2) \qquad (1.11)$$

and

$$b_n = \left(\frac{1}{n} + \frac{1}{2n^2} + o(1/n^2)\right)^2 = \frac{1}{n^2} + \frac{1}{n^3} + o(1/n^3).$$

Therefore,
$$\lim_{n\to\infty} \frac{1}{2} n(n+1) b_{n+1} = \frac{1}{2}.$$

Reversing the order of summation yields

$$\sum_{n=1}^{\infty} \frac{1}{n} \sum_{k=n+1}^{\infty} \frac{1}{k^2} = \sum_{k=2}^{\infty} \frac{1}{k^2} \sum_{n=1}^{k-1} \frac{1}{n} = \sum_{k=2}^{\infty} \frac{1}{k^2} H_{k-1} = \sum_{k=1}^{\infty} \frac{1}{(k+1)^2} H_k = \zeta(3),$$

where the well-known Euler's identity is used in the last equality.

Applying Abel's summation formula again with $a_n = 1$ and

$$b_n = \sum_{k=n+1}^{\infty} \frac{1}{k^2} - \frac{1}{n},$$

in view of (1.11), we have

$$\sum_{k=n+1}^{\infty} \frac{1}{k^2} = \sum_{k=n}^{\infty} \frac{1}{k^2} - \frac{1}{n^2} = \frac{1}{n} - \frac{1}{2n^2} + o(1/n^2),$$

and

$$\sum_{n=1}^{\infty} \left(\sum_{k=n+1}^{\infty} \frac{1}{k^2} - \frac{1}{n} \right) = \lim_{n\to\infty} n b_{n+1} + \sum_{n=1}^{\infty} n \left(\frac{1}{(n+1)^2} - \frac{1}{n} + \frac{1}{n+1} \right)$$

$$= \lim_{n\to\infty} \left(-\frac{1}{2n} + o(1/n) \right) - \sum_{n=1}^{\infty} \frac{1}{(n+1)^2} = 1 - \zeta(2).$$

In summary, by (1.10), we find that

$$S = \frac{1}{2} + \frac{1}{2}(\zeta(2) + \zeta(3)) + \zeta(3) + 1 - \zeta(2) = \frac{3}{2} - \frac{1}{2}\zeta(2) + \frac{3}{2}\zeta(3)$$

as claimed.   □

Solution II.
We begin with

$$nT_n^2 - \frac{1}{n} = t_n - t_{n-1} + s_n, \quad t_0 = 0, \tag{1.12}$$

where t_n and s_n for $n \geq 1$ are to be determined. Let $t_n = \alpha_n T_{n+1}^2 + \beta_n T_{n+1}$, where α_n and β_n will be chosen later. Since $T_{n+1} = T_n - 1/n^2$, we have

$$t_n - t_{n-1} = \alpha_n \left(T_n - \frac{1}{n^2} \right)^2 + \beta_n \left(T_n - \frac{1}{n^2} \right) - \alpha_{n-1} T_n^2 - \beta_{n-1} T_n$$

$$= (\alpha_n - \alpha_{n-1}) T_n^2 - \frac{2\alpha_n}{n^2} T_n + \frac{\alpha_n}{n^4} + (\beta_n - \beta_{n-1}) T_n - \frac{\beta_n}{n^2}.$$

To match the coefficient of T_n^2 in (1.12), we set $\alpha_0 = 0$ and $\alpha_n - \alpha_{n-1} = n$ for $n \geq 1$. Then $\alpha_n = n(n+1)/2$ and

$$t_n - t_{n-1} = nT_n^2 - \frac{1}{n} + \left(\beta_n - \beta_{n-1} - \frac{n+1}{n}\right)T_n + \left(\frac{n+1}{2n^3} - \frac{\beta_n}{n^2} + \frac{1}{n}\right).$$

Comparing with (1.12) again, we set $\beta_0 = 0$ and $\beta_n = \beta_{n-1} + (n+1)/n$ for $n \geq 1$. We now have $\beta_n = n + H_n$, where $H_n = \sum_{k=1}^{n} 1/k$ is the nth harmonic number. Hence,

$$t_n - t_{n-1} = nT_n^2 - \frac{1}{n} + \left(\frac{1}{2n^2} + \frac{1}{2n^3} - \frac{H_n}{n^2}\right).$$

Summing this identity yields

$$\sum_{n=1}^{N}\left(nT_n^2 - \frac{1}{n}\right) = t_N + \sum_{n=1}^{N}\frac{H_n}{n^2} - \frac{1}{2}\sum_{n=1}^{N}\frac{1}{n^2} - \frac{1}{2}\sum_{n=1}^{N}\frac{1}{n^2}.$$

Letting $N \to \infty$ and using Euler's identity $\sum_{n=1}^{\infty} H_n/n^2 = 2\zeta(3)$, we find

$$\sum_{n=1}^{\infty}\left(nT_n^2 - \frac{1}{n}\right) = \lim_{N\to\infty} t_N + 2\zeta(3) - \frac{1}{2}\zeta(2) - \frac{1}{2}\zeta(3).$$

Moreover, by the definition of t_n and the selected α_n and β_n, we now have

$$t_N = \frac{N(N+1)}{2}T_{N+1}^2 + (N + H_N)T_{N+1}.$$

Applying (1.11) yields

$$\lim_{N\to\infty}\frac{N(N+1)}{2}T_{N+1}^2 = \frac{1}{2} \quad \text{and} \quad \lim_{N\to\infty} NT_{N+1} = 1.$$

Together with $\lim_{N\to\infty} H_N T_{N+1} = 0$ since $H_N = \ln(N) + \gamma + O(1/N)$, we finally obtain the claimed result. $\qquad\square$

Remark. When the summand of the series involves binomial coefficients, factorials, and products of rational functions, there is a very good chance the series can be solved by the powerful method of Wilf-Zeilberger via telescoping. For example, returning to Monthly E3352, we rewrite the series in the equivalent form

$$\sum_{n=0}^{\infty}\left(\frac{1}{n!(n^4 + n^2 + 1)} - \frac{1}{2n!}\right) = 0.$$

The Wilf-Zeilberger algorithm decomposes the summand into $b_{n+1} - b_n$ with

$$b_n = \frac{n^2}{2\,n!(n^2 - n + 1)}$$

which immediately implies that the above sum gives zero. See [80] for more examples of this kind. Similarly, we can rewrite this monthly problem as

$$\sum_{n=1}^{\infty} \left(2n \left(\sum_{k=n}^{\infty} \frac{1}{k^2} \right)^2 - \frac{2}{n} + \frac{1}{n^2} - \frac{3}{n^3} \right) = 3.$$

Can Wilf-Zeilberger algorithm handle this kind of summand?

Additional problems for practice.

1. Let $\zeta(z) = \sum_{n=1}^{\infty} 1/n^z$ be the Riemann zeta function. Show that

(a) $\displaystyle \sum_{n=1}^{\infty} \left(\sum_{k=n+1}^{\infty} \frac{1}{k^2} \right)^2 = 3\zeta(3) - \frac{5}{2}\zeta(4).$

(b) $\displaystyle \sum_{n=1}^{\infty} \frac{1}{n} \left(\sum_{k=n+1}^{\infty} \frac{1}{k^2} \right)^2 = 5\zeta(2)\zeta(3) - 9\zeta(5).$

(c) $\displaystyle \sum_{n=1}^{\infty} \frac{H_n}{n} \left(\sum_{k=n+1}^{\infty} \frac{1}{k^3} \right) = 2\zeta(2)\zeta(3) - \frac{7}{2}\zeta(5).$

2. **Problem 11805** (Proposed by G. Glebov and S. Fraser, 121(10), 2014).
   (a) Show that

$$\sum_{k=0}^{\infty} \frac{(-1)^k}{(3k+1)^3} + \sum_{k=0}^{\infty} \frac{(-1)^k}{(3k+2)^3} = \frac{5\pi^3\sqrt{3}}{243},$$

$$\sum_{k=0}^{\infty} \frac{(-1)^k}{(3k+1)^3} - \sum_{k=0}^{\infty} \frac{(-1)^k}{(3k+2)^3} = \frac{13}{18}\zeta(3).$$

   (b) Prove that

$$\zeta(3) = \frac{9}{13} \int_0^1 \frac{(\ln x)^2}{x^3+1}\, dx - \frac{18}{13} \sum_{k=0}^{\infty} \frac{(-1)^k}{(3k+2)^3}.$$

3. **Problem 12045** (Proposed by O. Furdui and A. Sîntămărian, 125(5), 2018). Prove that the series

$$\sum_{n=1}^{\infty} (-1)^{n-1} \left(n \left(\sum_{k=n+1}^{\infty} \frac{1}{k^2} \right) - 1 \right)$$

   converges to $\pi^2/16 - (1 + \ln 2)/2$.

4. **Problem 12091** (Proposed by C. I. Vălean, 125(2), 2019). Prove

$$2\sum_{i=1}^{\infty}\sum_{j=1}^{\infty}\sum_{k=1}^{\infty} \frac{i!j!k!}{ij(i+j+k)!}(H_{i+j+k} - H_k) = \zeta(3)$$

   where H_k is the kth harmonic number.

5. **Problem 12215** (Proposed by O. Furdui and A. Sîntămărian, 127(9), 2020). Calculate

$$\sum_{n=1}^{\infty} \left(\left(\frac{1}{n^2} + \frac{1}{(n+2)^2} + \frac{1}{(n+4)^2} + \cdots \right) - \frac{1}{2n} \right).$$

6. **Problem 12222** (Proposed by R. Tauraso, 127(10), 2020). Prove

$$\sum_{k=1}^{\infty} \frac{(-1)^k}{k^2} \sum_{n=k}^{\infty} \frac{1}{n2^n} = -\frac{13\zeta(3)}{24}.$$

7. **Problem 12254** (Proposed by C. Lupu (USA) and T. Lupu, 128(5), 2021). Prove

$$\sum_{n=0}^{\infty} \left(\frac{(-1)^n}{2n+1} \sum_{k=1}^{n} \frac{1}{n+k} \right) = \frac{3\pi}{8} \ln(2) - G$$

where G is Catalan's constant $\sum_{k=0}^{\infty} (-1)^k / (2k+1)^2$.

8. **Problem 12494** (Proposed by J. Santmyer, 131(9), 2024). Prove

$$\sum_{n=1}^{\infty} \sum_{m=1}^{\infty} \frac{1}{m^2 n + mn^2 + rmn} = \begin{cases} 2\zeta(3) & \text{if } r = 0, \\ \frac{1}{r}\left(H_r^2 + H_r^{(2)} \right) & \text{if } r \text{ is a positive integer.} \end{cases}$$

1.5 A trigonometric logarithmic integral

Problem 12274 (Proposed by R. Tauraso, 128(8), 2021). Evaluate

$$\int_0^1 \frac{\arctan x}{1+x^2} \left(\ln \left(\frac{2x}{1-x^2} \right) \right)^2 dx.$$

Discussion.
Let the proposed integral be I. Using the substitution $x = \tan(\theta/2)$, we have

$$I = \frac{1}{4} \int_0^{\pi/2} \theta \ln^2(\tan \theta) \, d\theta. \tag{1.13}$$

Beginning with this integral, we will present two solutions. The first solution transforms (1.13) into the well-known integral $\int_0^1 \ln^2 x / (1+x^2) \, dx$, and the second solution applies the Fourier series of $\ln(\tan \theta)$.

Solution I.
The substitution $u = \tan \theta$ in (1.13) readily leads to

$$I = \frac{1}{4} \int_0^{\infty} \frac{\arctan u \ln^2 u}{1+u^2} \, du.$$

We rewrite this integral as

$$I = \frac{1}{4}\left(\int_0^1 \frac{\arctan u \ln^2 u}{1+u^2}\,du + \int_1^\infty \frac{\arctan u \ln^2 u}{1+u^2}\,du \right).$$

Letting $u \to 1/u$ in the second integral yields

$$I = \frac{1}{4}\left(\int_0^1 \frac{\arctan u \ln^2 u}{1+u^2}\,du + \int_0^1 \frac{\arctan(1/u) \ln^2 u}{1+u^2}\,du \right).$$

Since

$$\arctan u + \arctan(1/u) = \frac{\pi}{2},$$

we finally find

$$I = \frac{\pi}{8}\int_0^1 \frac{\ln^2 u}{1+u^2}\,du = \frac{\pi}{8}\sum_{n=0}^\infty (-1)^n \int_0^1 u^{2n} \ln^2 u \, du$$

$$= \frac{\pi}{4}\sum_{n=0}^\infty \frac{(-1)^n}{(2n+1)^3} = \frac{\pi}{4}\cdot\frac{\pi^3}{32} = \frac{\pi^4}{128}.$$

□

Solution II.
Recall that

$$\ln(\sin\theta) = -\ln 2 - \sum_{n=1}^\infty \frac{\cos(2n\theta)}{n},$$

$$\ln(\cos\theta) = -\ln 2 - \sum_{n=1}^\infty \frac{(-1)^n \cos(2n\theta)}{n}.$$

From which we obtain

$$\ln(\tan\theta) = -2\sum_{n=1}^\infty \frac{\cos(2(2n-1)\theta)}{2n-1}, \quad \text{for } \theta \in (0, \pi/2).$$

Plugging this into (1.13), we have

$$I = \int_0^{\pi/2} \theta \left(\sum_{n=1}^\infty \frac{\cos(2(2n-1)\theta)}{2n-1} \right)^2 d\theta$$

$$= \int_0^{\pi/2} \theta \left(\sum_{n=1}^\infty \frac{\cos^2(2(2n-1)\theta)}{(2n-1)^2} \right) d\theta$$

$$+ 2\int_0^{\pi/2} \theta \left(\sum_{n=2}^\infty \sum_{m=1}^{n-1} \frac{\cos(2(2n-1)\theta)\cos(2(2m-1)\theta)}{(2n-1)(2m-1)} \right) d\theta.$$

Therefore

$$I = \sum_{n=1}^{\infty} \frac{1}{(2n-1)^2} \int_0^{\pi/2} \theta \cos^2(2(2n-1)\theta) \, d\theta$$

$$+ 2 \sum_{m=1}^{n-1} \frac{1}{(2n-1)(2m-1)} \int_0^{\pi/2} \theta \cos(2(2n-1)\theta) \cos(2(2m-1)\theta) \, d\theta.$$

Using trigonometric product to sum formulas and then integrating by parts, we find

$$\int_0^{\pi/2} \theta \cos(2n\theta) \cos(2m\theta) \, d\theta = \begin{cases} \frac{\pi^2}{16} & \text{if } n = m, \\ 0 & \text{if } n \neq m. \end{cases}$$

Hence,

$$I = \sum_{n=1}^{\infty} \frac{1}{(2n-1)^2} \cdot \frac{\pi^2}{16} = \frac{\pi^2}{16} \left(\zeta(2) - \frac{1}{4}\zeta(2) \right) = \frac{\pi^4}{128}.$$

□

Remark. Along the same lines in Solution I, we can derive a more general formula

$$\int_0^1 \frac{\arctan x}{1+x^2} \left(\ln \left(\frac{2x}{1-x^2} \right) \right)^{2n} dx = \frac{(2n)!}{8} \pi \beta(2n+1),$$

where $\beta(z) = \sum_{n=0}^{\infty} \frac{(-1)^n}{(2n+1)^z}$ is the Dirichlet beta function.

Additional problems for practice.

1. **CMJ Problem 1261** (Proposed by B. Bradie, 55(1), 2024). Evaluate the following integral for real $\alpha > -2$:

$$\int_0^{\infty} \frac{\arctan x}{1 + \alpha x + x^2} \, dx.$$

2. **Problem 12054** (Proposed by C. I. Vălean, 125(6), 2018). Prove that

$$\int_0^1 \frac{\arctan x}{x} \ln \left(\frac{1+x^2}{(1-x)^2} \right) dx = \frac{\pi^3}{16}.$$

For more challenging integrals of the same author see [100].

3. **Problem 12158** (Proposed by H. Grandmontagne, 127(1), 2020). Prove

$$\int_0^1 \frac{(\ln x)^2 \arctan x}{1+x} \, dx = \frac{21\pi\zeta(3)}{64} - \frac{\pi^2 G}{24} - \frac{\pi^3 \ln 2}{32},$$

where $G = \sum_{n=0}^{\infty} (-1)^n / (2n+1)^2$ is Catalan's constant.

4. **Problem 12221** (Proposed by N. Batir, 127(10), 2017). Prove

$$\int_0^1 \frac{\ln(x^6 + 1)}{x^2 + 1}\, dx = \frac{\pi \ln(6)}{2} - 3G,$$

where G is the Catalan's constant $\sum_{n=0}^{\infty}(-1)^n/(2n+1)^2$.

5. **Problem 12256** (Proposed by P. Bracken, 128(5), 2021). Prove

$$\int_0^1 \frac{\ln(1+x)\ln(1-x)}{x}\, dx = -\frac{5}{8}\zeta(3).$$

6. **Problem 12501** (Proposed by H. Grandmontagne, 131(10), 2024). Prove

$$\int_0^{\infty} \frac{(\ln(x/(1+x)))^4 \ln(x^3(1+x)^{17})}{1+x}\, dx = -240\,\zeta^2(3),$$

where $\zeta(3)$ is Apéry's constant $\sum_{n=1}^{\infty} 1/n^3$.

7. Let $\beta(z) = \sum_{n=0}^{\infty} \frac{(-1)^n}{(2n+1)^z}$ be the Dirichlet beta function (Note that $\beta(2) = G$). Prove

(a) $\displaystyle\int_0^1 \frac{\ln x \arctan x}{1+x^2}\, dx = \frac{7\zeta(3)}{16} - \frac{\beta(2)\pi}{4},$

(b) $\displaystyle\int_0^1 \frac{\ln x(\arctan x)^2}{1+x^2}\, dx = \frac{\beta(4)}{2} - \frac{\beta(2)\pi^2}{16},$

(c) $\displaystyle\int_0^1 \frac{x \arctan x(\ln x)^2}{1+x^2}\, dx = \frac{7\pi\zeta(3)}{64} + \beta(4) - \frac{\pi^3 \ln 2}{16},$

(d) $\displaystyle\int_0^1 \frac{\ln(x^2 + x + 1)}{1+x^2}\, dx = \frac{\pi}{6}\ln(2 + \sqrt{3}) - \frac{\beta(2)}{3}.$

1.6 An interesting series with a product of two central binomial coefficients

Problem 12381 (Proposed by H. Chen, 130(3), 2023). Prove

$$\sum_{n=1}^{\infty} \frac{\binom{2n}{n}\binom{4n}{2n}}{64^n(2n-1)} = 1 - \frac{1}{\pi}\left(\sqrt{2} + \ln(1 + \sqrt{2})\right).$$

Discussion.
The key idea which makes the problem feasible is that we can build up the desired summand step by step. We begin with the well-known generating

function of the sequence $\left\{\binom{2n}{n}\right\}_{n\geq 0}$

$$\sum_{n=0}^{\infty}\binom{2n}{n}x^n = \frac{1}{\sqrt{1-4x}}. \tag{1.14}$$

With some manipulation, then we apply the Wallis integral formula

$$\int_0^{\pi/2} \sin^{2n}x\,dx = \frac{\pi}{2^{2n+1}}\binom{2n}{n} \tag{1.15}$$

to associate with the factor $\binom{4n}{2n}$.

Solution.
Using the substitution $4x \to x^2$ in (1.14) yields

$$1 + \sum_{n=1}^{\infty}\frac{\binom{2n}{n}}{4^n}x^{2n} = \frac{1}{\sqrt{1-x^2}}.$$

Rearranging this as

$$\sum_{n=1}^{\infty}\frac{\binom{2n}{n}}{4^n}x^{2n-2} = \frac{1}{x^2\sqrt{1-x^2}} - \frac{1}{x^2},$$

then integrating with respect to x gives

$$\sum_{n=1}^{\infty}\frac{\binom{2n}{n}}{4^n(2n-1)}x^{2n-1} = \frac{1-\sqrt{1-x^2}}{x},$$

and so

$$\sum_{n=1}^{\infty}\frac{\binom{2n}{n}}{4^n(2n-1)}x^{2n} = 1 - \sqrt{1-x^2}.$$

Let $x = \sin^2 t$. Then

$$\sum_{n=1}^{\infty}\frac{\binom{2n}{n}}{4^n(2n-1)}\sin^{4n}t = 1 - \sqrt{1-\sin^4 t} = 1 - |\cos t|\sqrt{1+\sin^2 t}.$$

Integrating this series with respect to t from 0 to $\pi/2$ and then using the Wallis integral formula (1.15) yields

$$\sum_{n=1}^{\infty}\frac{\binom{2n}{n}}{4^n(2n-1)}\int_0^{\pi/2}\sin^{4n}t\,dt = \frac{\pi}{2}\sum_{n=1}^{\infty}\frac{\binom{2n}{n}\binom{4n}{2n}}{64^n(2n-1)}$$

$$= \int_0^{\pi/2}(1 - \cos t\sqrt{1+\sin^2 t})\,dt$$

$$= \frac{\pi}{2} - \int_0^1\sqrt{1+u^2}\,du \quad (\text{use } u = \sin t)$$

$$= \frac{\pi}{2} - \frac{1}{2}\left(\sqrt{2} + \ln(1+\sqrt{2})\right),$$

which is equivalent to the claimed identity. $\qquad\square$

Remark. Similarly, by applying the Wallis integral formula with odd exponent,

$$\int_0^1 \sin^{4k+1} t \, dt = \frac{16^k}{(4k+1)\binom{4k}{2k}}, \quad \text{for all } k \in \mathbb{N},$$

we can evaluate the series involving the ratio of the central binomial coefficients. We illustrate the idea by showing that

$$\sum_{n=1}^{\infty} \frac{4^n n \binom{2n}{n}}{(2n-1)^2(4n+1)\binom{4n}{2n}} = \frac{1}{9}(5\sqrt{2}-4).$$

Integrating the first series in the above Solution gives

$$\sum_{n=0}^{\infty} \frac{\binom{2n}{n}}{4^n(2n+1)} x^{2n+1} = \arcsin x.$$

Applying the substitution $x \to x^2$ and then shifting the index $n-1 \to n$ yields

$$\sum_{n=1}^{\infty} \frac{n\binom{2n}{n}}{4^n(2n-1)^2} x^{4n-2} = \frac{1}{2}\arcsin(x^2),$$

and so

$$\sum_{n=1}^{\infty} \frac{n\binom{2n}{n}}{4^n(2n-1)^2} x^{4n+1} = \frac{1}{2}x^3 \arcsin(x^2).$$

Let $x = \sin t$. Then

$$\sum_{n=1}^{\infty} \frac{n\binom{2n}{n}}{4^n(2n-1)^2} \sin^{4n+1} t = \frac{1}{2}\sin^3 t \arcsin(\sin^2 t).$$

Integrating with respect to t from 0 to $\pi/2$ and using the Wallis integral formula, we find that

$$\sum_{n=1}^{\infty} \frac{4^n n \binom{2n}{n}}{(2n-1)^2(4n+1)\binom{4n}{2n}} = \int_0^{\pi/2} \frac{1}{2}\sin^3 t \arcsin(\sin^2 t) \, dt$$

$$= \frac{1}{4}\int_0^1 \frac{u}{\sqrt{1-u}}\arcsin u \, du \quad (\text{use } u = \sin^2 t)$$

$$= \frac{1}{4}\int_0^1 \arcsin u \, d\left(-\frac{2}{3}(2+u)\sqrt{1-u}\right)$$

$$= \frac{1}{6}\int_0^1 (2+u)\sqrt{1-u}\frac{du}{\sqrt{1-u^2}}$$

$$= \frac{1}{6}\int_0^1 \frac{2+u}{\sqrt{1+u}} \, du = \frac{1}{9}(5\sqrt{2}-4),$$

as claimed.

It is worth taking a brief look at the background of the term "interesting series". In [65] marking the 60th anniversary of his first contribution to the *Monthly*, D. H. Lehmer studied two classes of series:

$$\sum_{n=1}^{\infty} a_n \binom{2n}{n} \qquad \text{and} \qquad \sum_{n=1}^{\infty} \frac{a_n}{\binom{2n}{n}}.$$

In his sense, the adjective "interesting" means that the sum of the series can be expressed in terms of known constants.

By manipulating the known power series through specialization, differentiation and integration, he found numerous intriguing results. Stimulated by his work, searching for interesting series associated with central binomial coefficients, Catalan numbers and harmonic numbers has become a very active research topic. Investigators have developed many different approaches including generating functions, Gauss hypergeometric functions and the Fourier-Legendre series. For example, by using Fourier-Legendre series expansions, The first author [32] derived many interesting series including these elegant Ramanujan-like formulas:

$$\sum_{n=0}^{\infty} \frac{(-1)^n}{16^n} \binom{2n}{n}^2 = \frac{\sqrt{2}\Gamma^2(1/4)}{4\pi^{3/2}},$$

$$\sum_{n=0}^{\infty} \frac{(-1)^n (4n+1)}{16^n (n+1)(1/2-n)} \binom{2n}{n}^3 = \frac{8}{\pi},$$

$$\sum_{n=0}^{\infty} \frac{(4n+1)}{256^n (n+1)^2 (1/2-n)^2} \binom{2n}{n}^4 = \frac{128}{3\pi^2}.$$

For more details on these kind of series, please refer also to [25, 31, 67] and the references listed therein. The interested reader is encouraged to pursue new results in this topic.

Additional problems for practice.

1. Let H_n be the n-th harmonic number, and let $h_n = \sum_{k=1}^{n} 1/(2k-1)$. For $|x| < 1/4$, show that

   (a) $\displaystyle\sum_{n=1}^{\infty} \binom{2n}{n} H_n x^n = \frac{2}{\sqrt{1-4x}} \ln\left(\frac{1+\sqrt{1-4x}}{2\sqrt{1-4x}}\right).$

   (b) $\displaystyle\sum_{n=1}^{\infty} \binom{2n}{n} h_n x^n = -\frac{1}{\sqrt{1-4x}} \ln(\sqrt{1-4x}).$

2. Let $H_n(2) = \sum_{k=1}^{n} 1/k^2$. For $|x| < 2$, show that

   $$\sum_{n=0}^{\infty} \frac{\binom{2n}{n} x^{2n+1}}{16^n (2n+1)} \sum_{k=0}^{n-1} \frac{1}{(2k+1)^2} = \frac{1}{3} \arcsin^3(x/2).$$

Use this result to prove

$$\sum_{n=1}^{\infty} \frac{\binom{2n}{n}}{16^n} \left(H_{2n}(2) - \frac{1}{4} H_n(2) \right) x^{2n} = \frac{\arcsin^2(x/2)}{\sqrt{4-x^2}}.$$

3. Let $H_n(2) = \sum_{k=1}^{n} 1/k^2$. Prove

(a) $$\sum_{n=1}^{\infty} \frac{\binom{2n}{n}^2 H_{2n}}{16^n(2n-1)} = \frac{6\ln 2 - 2}{\pi}.$$

(b) $$\sum_{n=1}^{\infty} \frac{\binom{2n}{n}^2 H_n}{16^n(2n-1)^2} = \frac{12 - 16\ln 2}{\pi}.$$

(c) $$\sum_{n=0}^{\infty} \frac{\binom{2n}{n}^2 H_n(2)}{16^n(2n-1)} = 4 - \frac{\pi}{3} - \frac{8}{\pi}.$$

(d) $$\sum_{n=0}^{\infty} \frac{\binom{2n}{n}^2 H_n(2)}{16^n(2n-1)^2} = -12 + \frac{2\pi}{3} + \frac{32}{\pi}.$$

4. For $|x| \le 4$, let $p^2 = (\sqrt{x^2+16} - 4)/x$. Show that

$$\sum_{n=1}^{\infty} \frac{(-1)^{n+1}}{n^2 \binom{4n}{2n}} x^{2n} = 8 \left(\operatorname{arctanh}^2(p) - \arctan^2(p) \right).$$

Use this formula to derive

$$\sum_{n=1}^{\infty} \frac{16^n}{n^2 \binom{4n}{2n}} = \pi^2 - 4\ln^2(\sqrt{2}-1).$$

Hint: Begin with the power series of $\arcsin^2 x$, then use the substitution $x \to xi$.

5. Prove

(a) $$\sum_{n=0}^{\infty} \frac{\binom{2n}{n}\binom{4n}{2n}}{64^n(2n+1)} = \frac{4}{\pi} \ln(1+\sqrt{2}).$$

(b) $$\sum_{n=0}^{\infty} \frac{\binom{2n}{n}\binom{4n}{2n}}{64^n(2n+3)} = \frac{4\sqrt{2} + 16\ln(1+\sqrt{2})}{15\pi}.$$

6. **Problem 12051** (Proposed by P. Ribeiro, 125(6), 2018). Prove

$$\sum_{n=0}^{\infty} \binom{2n}{n} \frac{1}{4^n(2n+1)^3} = \frac{\pi^3}{48} + \frac{\pi}{4} \ln^2 2.$$

7. **Problem 12094** (Proposed by P. F. Refolio, 126(2), 2019). Prove

$$\sum_{n=0}^{\infty} \frac{\binom{2n}{n}^2}{16^n (n+1)^3} = 16(\ln 2 - 1) + \frac{48 - 32G}{\pi},$$

where $G = \sum_{k=0}^{\infty} (-1)^k / (2k+1)^2$ is Catalan's constant.

8. **Problem 12180** (Proposed by P. F. Refolio, 127(4), 2020). Prove

$$\sum_{n=0}^{\infty} \frac{\binom{4n}{2n}^2}{2^{8n}(2n+1)} = \frac{2}{\pi} - \frac{\sqrt{2}\,C^2}{\pi^{3/2}} + \frac{\pi^{1/2}}{\sqrt{2}\,C^2},$$

where $C = \int_0^{\infty} t^{-1/4} e^{-t} = \Gamma(3/4)$.

9. **Problem 12337** (Proposed by H. Ohtsuka, 129(7), 2022). For $k \in \{0, 1, 2\}$, let

$$S_k = \sum \frac{(-4)^n}{2n+1} \binom{2n}{n}^{-1},$$

where the sum is taken over all nonnegative integers n that are congruent to k modulo 3.

Prove

(a) $S_0 = \dfrac{\ln(1+\sqrt{2})}{3\sqrt{2}} + \dfrac{\pi}{6}$;

(b) $S_1 = \dfrac{\ln(1+\sqrt{2})}{3\sqrt{2}} - \dfrac{\ln(2+\sqrt{3})}{2\sqrt{3}} - \dfrac{\pi}{12}$; and

(c) $S_2 = \dfrac{\ln(1+\sqrt{2})}{3\sqrt{2}} + \dfrac{\ln(2+\sqrt{3})}{2\sqrt{3}} - \dfrac{\pi}{12}$.

10. **Research Project – Lehmer's limit.** In [65], D. H. Lehmer studied the series

$$S_k(x) := \sum_{n=1}^{\infty} \frac{n^k}{\binom{2n}{n}} x^n.$$

In particular, recursively, he found that $S_1(2) = \pi+3$, $S_2(2) = 7\pi/2+11$, ..., and in general

$$S_k(2) = R_1(k)\pi + R_2(k),$$

where both $R_1(k)$ and $R_2(k)$ are rational numbers. Based on his calculations up to $k = 10$, he observed that the ratio $R_2(k)/R_1(k)$ provided an approximation to π. In [37], J. Dyson, N. E. Frankel, and M. L. Glasser applied the Gauss hypergeometric function $_2F_1$ combined with some other advanced tools to find $S_k(x)$ in non-recursive closed form and eventually proved that

$$\lim_{k\to\infty} \frac{R_2(k)}{R_1(k)} = \pi.$$

Can we have an elementary solution to this limit?

1.7   A log-tangent integral

Problem 12317 (Proposed by S. Stewart, 129(4), 2022). Prove

$$\int_0^{\pi/2} \frac{\sin(4x)}{\ln(\tan x)}\, dx = -14\frac{\zeta(3)}{\pi^2},$$

where $\zeta(3)$ is Apéry's constant.

Discussion.
The usual integration techniques don't work on this integral in the current form. So we transform it by substitution. Since

$$\sin(4x) = 2\sin(2x)\cos(2x) = 4\sin x \cos x(1 - 2\sin^2 x),$$

by using the substitution $u = \tan x$, then $t = u^2$, we obtain

$$I := \int_0^{\pi/2} \frac{\sin(4x)}{\ln(\tan x)}\, dx = \int_0^\infty \frac{4u(1-u^2)}{\ln u(1+u^2)^3}\, du = -4\int_0^\infty \frac{t-1}{\ln t(1+t)^3}\, dt \tag{1.16}$$

The challenge now comes from the $\ln t$ in the integrand. In the following, we present two solutions and demonstrate how to remove $\ln t$ by parametric integration and parametric differentiation, respectively.

Solution I – by parametric integration
To this end, observe that

$$\int_0^1 t^p\, dp = \frac{t-1}{\ln t}.$$

We have

$$I = -4\int_0^\infty \int_0^1 \frac{t^p}{(1+t)^3}\, dp\, dt = -4\int_0^1 \left(\int_0^\infty \frac{t^p}{(1+t)^3}\, dt\right) dp.$$

For $p \in (0,1)$, let $s = 1/(1+t)$. Then

$$\int_0^\infty \frac{t^p}{(1+t)^3} \, dt = \int_0^1 s^{1-p}(1-s)^p \, ds$$

$$= B(2-p, p+1) = \frac{\Gamma(2-p)\Gamma(p+1)}{\Gamma(3)}$$

$$= \frac{1}{2}p(1-p)\Gamma(1-p)\Gamma(p) \qquad (\text{use } \Gamma(1+x) = x\Gamma(x))$$

$$= \frac{1}{2}p(1-p)\frac{\pi}{\sin(p\pi)} \qquad \left(\text{use } \Gamma(p)\Gamma(1-p) = \frac{\pi}{\sin p\pi}\right),$$

and so

$$I = -2\pi \int_0^1 \frac{p(1-p)}{\sin(p\pi)} \, dp = -\frac{2}{\pi^2} \int_0^\pi \frac{x(\pi-x)}{\sin x} \, dx \qquad (\text{use } x = p\pi).$$

Applying the Fourier sine series of $x(\pi - x)$ on $[0, \pi]$ yields

$$x(\pi - x) = \frac{8}{\pi} \sum_{n=0}^\infty \frac{1}{(2n+1)^3} \sin((2n+1)x).$$

Hence

$$I = -\frac{16}{\pi^3} \sum_{n=0}^\infty \frac{1}{(2n+1)^3} \int_0^\pi \frac{\sin((2n+1)x)}{\sin x} \, dx$$

$$= -\frac{16}{\pi^3} \sum_{n=0}^\infty \frac{\pi}{(2n+1)^3} = -\frac{16}{\pi^2}\left(1 - \frac{1}{2^3}\right) \sum_{n=1}^\infty \frac{1}{n^3}$$

$$= -\frac{16}{\pi^2}\frac{7}{8}\zeta(3) = -14\frac{\zeta(3)}{\pi^2}.$$

Here, we have used the formula

$$\int_0^\pi \frac{\sin nx}{\sin x} \, dx = \begin{cases} 0, & \text{if } n \text{ is even}, \\ \pi, & \text{if } n \text{ is odd}. \end{cases}$$

for $n \in \mathbb{N}$, which follows from the fact that, for $n > 2$,

$$\int_0^\pi \frac{\sin nx}{\sin x} \, dx - \int_0^\pi \frac{\sin(n-2)x}{\sin x} \, dx = \int_0^\pi \frac{\sin nx - \sin(n-2)x}{\sin x} \, dx$$

$$= 2 \int_0^\pi \cos(n-1)x \, dx = 0.$$

$\square$

Solution II– by parametric differentiation

Here, the key observation is if we introduce

$$I(p) = \int_0^\infty \frac{t^p - 1}{\ln t (1+t)^3} \, dt,$$

since $d(t^p)/dp = t^p \ln t$, differentiating with respect to p will eliminate $\ln t$ from the integrand. Indeed, we have

$$I'(p) = \int_0^\infty \frac{t^p}{(1+t)^3}\, dt.$$

Using the substitution $s = 1/(1+t)$, the formula $\Gamma(1+x) = x\Gamma(x)$, and Euler's reflection formula, for $p \in (0,1)$, we obtain

$$I'(p) = \int_0^1 s^{1-p}(1-s)^p\, ds = B(2-p, p+1)$$

$$= \frac{\Gamma(2-p)\Gamma(p+1)}{\Gamma(3)} = \frac{p(1-p)}{2}\Gamma(p)\Gamma(1-p) = \frac{p(1-p)\pi}{2\sin(p\pi)},$$

and so

$$I(1) = I(0) + \int_0^1 I'(p)\, dp = \frac{\pi}{2}\int_0^1 \frac{p(1-p)}{\sin(p\pi)}\, dp$$

$$= \pi i \int_0^1 \frac{p(1-p)}{e^{ip\pi} - e^{-ip\pi}}\, dp = \pi i \sum_{k=0}^\infty \int_0^1 p(1-p)e^{-i(2k+1)p\pi}\, dp$$

$$= \frac{4}{\pi}\sum_{k=0}^\infty \frac{1}{(2k+1)^3} = \frac{7\zeta(3)}{2\pi^2}.$$

This leads to

$$\int_0^{\pi/2} \frac{\sin(4x)}{\ln(\tan x)}\, dx = -4I(1) = (-4)\cdot\frac{7\zeta(3)}{2\pi^2} = -14\frac{\zeta(3)}{\pi^2}.$$

$\square$

Remark. This proposed integral can also be solved either by converting into (1.16) and then using a contour integral or converting into $\int_0^\pi x(\pi-x)\csc x\, dx$ and then using the power series of $\csc x$. We leave the details to the readers.

   Parametric differentiation and integration provide a powerful approach to evaluate difficult integrals which often require complex contour integration. This method usually involves the following three operations:

1. Differentiation with respect to a parameter in the integral.

2. Integration with respect to a parameter in the integral.

3. Exchange the order of integrations.

For the justification of these three operations, please refer to Chapter 19 in [29, pp. 217-230].

Additional problems for practice.

1. Evaluate
$$\int_0^\infty \frac{\ln(1+x^4)}{1+x^2}\,dx.$$

2. Show that, for $p > 0$ and $q > 0$,
$$\int_0^\infty \frac{e^{-px}\cos\alpha x - e^{-qx}\cos\beta x}{x}\,dx = \frac{1}{2}\ln\left(\frac{q^2+\beta^2}{p^2+\alpha^2}\right).$$

3. **Problem 12228** (Proposed by H. Grandmontagne, 128(5), 2021). Prove
$$\int_0^1 \frac{(\ln(x))^2 \ln\left(2\sqrt{x}/(x^2+1)\right)}{x^2-1}\,dx = 2G^2,$$
   where G is Catalan's constant $\sum_{n=0}^\infty (-1)^n/(2n+1)^2$.

4. **Problem 12332** (Proposed by F. Holland, 129(6), 2022). Prove
$$\int_0^\infty \frac{\tanh^2 x}{x^2}\,dx = \frac{14\zeta(3)}{\pi^2},$$
   where $\zeta(3)$ is Apéry's constant $\sum_{k=1}^\infty 1/k^3$.

5. **Problem 12388** (Proposed by A. Garcia, 130(4), 2023). Let α be a real number. Evaluate
$$\int_0^\infty \frac{(\ln x)^2 \arctan(x)}{1 - 2(\cos\alpha)x + x^2}\,dx.$$
   Hint: First convert the integral to $\frac{\pi}{4}\int_0^\infty \frac{\ln^2 t}{1-2(\cos\alpha)t+t^2}\,dt$.

6. **Problem 12459** (Proposed by H. Grandmontagne, 131(4), 2024). Let α be a real number greater than 1. Evaluate
$$\int_0^\infty \frac{\text{Li}_2(-x^\alpha) + \text{Li}_2(-x^{-\alpha})}{1+x^\alpha}\,dx,$$
   where $\text{Li}_2(x)$ is the dilogarithm function, defined by $\text{Li}_2(x) = \sum_{k=1}^\infty x^k/k^2$ when $|x| < 1$ and extended by analytic continuation.

7. **Problem 12527** (Proposed by S. Attaoui, 132(4), 2025). Prove
$$\int_0^{\pi/2} \frac{\tanh(\tan^2\theta)}{\sin(2\theta)(1+\cosh(2\tan^2\theta))}\,d\theta = \frac{7\zeta(3)}{8\pi^2},$$
   where $\zeta(3)$ is Apéry's constant.

1.8   An arccosine integral

Problem 12344 (Proposed by B. Bradie, 129(8), 2022). Evaluate

$$\int_{-1}^{1} \frac{\arccos x}{x^2 + x + 1}\, dx.$$

Discussion.
First observe that $x^2 + x + 1$ becomes $x^2 - x + 1$ under the substitution $x \to -x$. Here we are unable to eliminate $\arccos x$ in the integrand by

$$\arccos(-x) = \pi - \arccos x.$$

Instead, we will transform the arccosine to arctangent because $\arctan x$ has a more friendly derivative.

Solution.
Let the desired integral be I. By the substitution $x = \frac{1-u}{1+u}$, we have

$$I = \int_0^\infty \frac{2 \arccos\left(\frac{1-u}{1+u}\right)}{u^2 + 3}\, du.$$

Using the identity

$$\arccos\left(\frac{1 - x^2}{1 + x^2}\right) = 2 \arctan x$$

for $x \geq 0$ and the substitution $t = \sqrt{u}$, we get

$$I = 4 \int_0^\infty \frac{\arctan(\sqrt{u})}{u^2 + 3}\, du = 8 \int_0^\infty \frac{t \arctan t}{t^4 + 3}\, dt.$$

To use parametric differentiation to evaluate this integral, we introduce

$$J(p) := \int_0^\infty \frac{t \arctan(pt)}{t^4 + 3}\, dt.$$

Then

$$J'(p) = \int_0^\infty \frac{t^2}{(t^4 + 3)(1 + p^2 t^2)}\, dt.$$

Let $a = \sqrt[4]{3}$. Note that

$$t^4 + 3 = (t^2 + \sqrt{2}ax + a^2)(t^2 - \sqrt{2}ax + a^2).$$

By partial fractions, we obtain

$$J'(p) = -\frac{p\pi}{2(1 + a^4 p^4)} + \frac{\pi}{2\sqrt{2}a} \frac{1 + a^2 p^2}{1 + a^4 p^4}.$$

Since $J(0) = 0$, we finally find that, with some simplifications,

$$I = 8J(1) = 8\int_0^1 J'(p)\,dp = -4\pi \int_0^1 \frac{p}{1+a^4 p^4}\,dp + \frac{4\pi}{\sqrt{2}a}\int_0^1 \frac{1+a^2 p^2}{1+a^4 p^4}\,dp$$

$$= -4\pi \left[\frac{\arctan(a^2 p^2)}{2a^2}\right]_0^1 + \frac{4\pi}{\sqrt{2}a}\left[\frac{\arctan(\sqrt{2}ap+1) + \arctan(\sqrt{2}ap-1)}{\sqrt{2}a}\right]_0^1$$

$$= \frac{2\pi\left(2\pi - 3\arctan(\sqrt{3+2\sqrt{3}})\right)}{3\sqrt{3}}.$$

□

Remark. The computation of $J'(p)$ via partial fractions is elementary but very tedious. This is why the *Monthly* used the phrase "we omit the details" when it featured the above solution. The interested reader may reveal the details with the aid of *Mathematica*.

Here we offer another derivation of $J'(p)$ via the reside theorem. To this end, let

$$f(z) = \frac{z^2}{(z^4+3)(1+p^2 z^2)}.$$

Consider the contour C_R: the interval $(-R, R)$ closed by an upper semicircle of radius R. For large R, the following three roots of $(z^4 + 3)(1 + p^2 z^2) = (t^2 + \sqrt{2}ax + a^2)(t^2 - \sqrt{2}ax + a^2)(1 + p^2 z^2)$

$$z_1 = \frac{a}{\sqrt{2}}(1+i), \quad z_2 = \frac{i}{p}, \quad z_3 = \frac{a}{\sqrt{2}}(1-i)$$

lie in the contour. Applying the residue theorem on the contour and then letting $R \to \infty$, we obtain

$$J'(p) = \frac{1}{2}\int_{-\infty}^{\infty} f(t)\,dt = \frac{1}{2}\lim_{R\to\infty}\oint_{C_R} f(z)\,dz = \frac{1}{2}\cdot 2\pi i \sum_{k=1}^{3}\mathrm{Res}(f, z_k)$$

$$= \pi i\left(\frac{t^2}{4t^3(1+p^2 t^2)}\bigg|_{t=z_1} + \frac{t^2}{2p^2 t(t^4+3)}\bigg|_{t=z_2} + \frac{t^2}{4t^3(1+p^2 t^2)}\bigg|_{t=z_3}\right)$$

$$= \pi i\left(\frac{1}{4z_1(1+p^2 a^2 i)} + \frac{ip}{2(1+a^4 p^4)} + \frac{1}{4z_3(1-p^2 a^2 i)}\right)$$

$$= -\frac{p\pi}{2(1+a^4 p^4)} + \frac{\pi}{2\sqrt{2}a}\frac{1+a^2 p^2}{1+a^4 p^4}$$

as desired.

Additional problems for practice.

1. Let $a > 1$. Evaluate

$$\int_0^{\infty} \frac{\arctan\sqrt{x^2+a^2}}{(1+x^2)\sqrt{x^2+a^2}}\,dx.$$

Hint. Use $\arctan x = \int_0^1 \frac{x}{1+x^2y^2}\,dy$.

2. Prove that

(a) $\displaystyle\int_0^\infty \frac{\arctan x}{1+x^2+x^4}\,dx = \frac{\pi^2}{8\sqrt{3}} - \frac{2}{3}G + \frac{\pi}{12}\ln(2+\sqrt{3})$,

(b) $\displaystyle\int_0^\infty \frac{\arctan(x^2)}{1+x^2+x^4}\,dx = \frac{\pi^2}{8\sqrt{3}} + \frac{\pi}{4}\ln(2+\sqrt{3}) - \frac{\pi}{2}\ln(1+\sqrt{2})$.

where $G = \sum_{n=0}^\infty (-1)^n/(2n+1)^2$ is the Catalan constant.

3. **Problem 11457** (Proposed by M. L. Glasser, 116(8), 2009). For real numbers a and b with $0 \le a \le b$, find

$$\int_a^b \arccos\left(\frac{x}{\sqrt{(a+b)x - ab}}\right)\,dx.$$

4. **Problem 12145** (Proposed by T. Amdeberhan and V. Moll, 126(9), 2019). Prove

$$\int_0^\infty \frac{\cos(x)\sin(\sqrt{1+x^2})}{\sqrt{1+x^2}}\,dx = \frac{\pi}{4}.$$

5. **Problem 12199** (Proposed by S. Sharma, 127(7), 2020). Prove

$$\int_0^\infty \frac{x\sinh(x)}{3+4\sinh^2(x)}\,dx = \frac{\pi^2}{24}.$$

6. **Problem 12417** (Proposed by M. Maesumi, 130(8), 2023). Consider the sphere S given by $x^2 + y^2 + (z-1)^2 = 1$, with north pole N at $(0,0,2)$. The stereographic projection of a point P at $(x,y,0)$ is the point, different from N, that is on the intersection of NP with S. Consider the region H in the xy-plane given by $0 \le xy \le c^2$, where $c > 0$.

What is the area of the stereographic projection of H to S?

Hint. Show that the area is equal to

$$2\pi - 16c^2 \int_0^\infty \frac{\arctan(t)(t^2 - 1)}{(c^2t^2 + 4t + c^2)^2}\,dt.$$

1.9 Two norms involving function moments

Problem 12318 (Proposed by M. Mehrabi, 129(4), 2022). Let a be a positive real number, and let S_a be the set of functions $f : [-a, a] \to \mathbb{R}$ such that $\int_{-a}^a (f(x))^2\,dx = 1$. Let

$$A(f) = \int_{-a}^a f(x)dx, \quad B(f) = \int_{-a}^a xf(x)dx \quad \text{and} \quad C(f) = \int_{-a}^a x^2 f(x)dx.$$

(a) What is $\sup\{A(f)^2 + B(f)^2 : f \in S_a\}$?

(b) What is $\sup\{A(f)^2 + B(f)^2 + C(f)^2 : f \in S_a\}$?

Discussion.
The integral $\int_a^b x^k f(x)\,dx$ is often called the *kth moment* of a function f. Moments play very important roles in many fields. For example, in physics, if f represents mass density, then $A(f), B(f)$ and $C(f)$ are the total mass, the center of mass, and the moment of inertia, respectively. In statistics, if f is a probability distribution, then $B(f)$ is the mean and $C(f) - B(f)^2$ is the variance.

Here we take another view with these moments $A(f), B(f)$ and $C(f)$. Let

$$L^2[-a, a] = \left\{ f : \int_{-a}^a (f(x))^2\,dx < \infty \right\}.$$

Then $\{1, x, x^2, \ldots\}$ constitutes a basis in $L^2[-a, a]$ with the inner product $(f, g) = \int_{-a}^a f(x)g(x)\,dx$. Let $f = \sum_{n=0}^\infty a_n x^n$. Then $A(f), B(f)$ and $C(f)$ can be determined in terms of a_0, a_1 and a_2. To simplify the process for finding a_n, in the following solution, we will replace the basis $\{1, x, x^2, \ldots\}$ by the Legendre polynomials $\{P_n(x)\}_{n \geq 0}$. It is well-known that $\{P_n(t)\}_{n \geq 0}$ are orthogonal in $L^2[-1, 1]$ with

$$\int_1^1 P_m(x) P_n(x)\,dx = \begin{cases} \frac{2}{2n+1}, & \text{if } m = n, \\ 0 & \text{otherwise.} \end{cases}$$

Moreover, if $g \in L^2[-1, 1]$, by Parseval's identity, we have

$$\int_{-1}^1 (g(x))^2\,dx = \sum_{n=0}^\infty \frac{2a_n^2}{2n + 1}, \tag{1.17}$$

where

$$a_n = \frac{2n + 1}{2} \int_{-1}^1 g(x) P_n(x)\,dx.$$

Solution.
(a) Note that $\sup\{A(f)^2 + B(f)^2 : f \in S_a\}$ is the smallest constant M such that

$$A(f)^2 + B(f)^2 \leq M\|f\|^2 \qquad \text{for all } f \in L^2[-1, 1].$$

From which with $g(x) = f(ax)$, we find

$$a^2 \left(\int_{-1}^1 g(x)\,dx \right)^2 + a^4 \left(\int_{-1}^1 xg(x)\,dx \right)^2 \leq aM \int_{-1}^1 (g(x))^2\,dx.$$

In view of $P_0(x) = 1$, $P_1(x) = x$ and (1.17), the above inequality is reduced to

$$2aa_0^2 + \frac{2a^3 a_1^2}{9} \leq M \sum_{n=0}^\infty \frac{a_n^2}{2n + 1}.$$

Now it suffices to find the smallest constant M such that for all $a_0, a_1 \in \mathbb{R}$

$$2aa_0^2 + \frac{2a^3a_1^2}{9} \leq M\left(a_0^2 + \frac{a_1^2}{3}\right),$$

which is equivalent to

$$(M - 2a)a_0^2 + \left(M - \frac{2a^3}{3}\right)\frac{a_1^2}{3} \geq 0.$$

This implies

$$M = \max\left\{2a, \frac{2a^3}{3}\right\}.$$

(b) Along the same lines in (a), we see that $\sup\{A(f)^2 + B(f)^2 + C(f)^2 : f \in S_a\}$ is the smallest constant M such that, for every $g \in L^2[-1, 1]$,

$$a^2\left(\int_{-1}^{1} g(x)\,dx\right)^2 + a^4\left(\int_{-1}^{1} xg(x)\,dx\right)^2 + a^6\left(\int_{-1}^{1} x^2g(x)\,dx\right)^2$$

$$\leq aM\int_{-1}^{1} (g(x))^2\,dx.$$

Since $P_0(x) = 1, P_1(x) = x, P_2(x) = (3x^2 - 1)/2$ and $x^2 = (P_0(x) + 2P_2(x)/3$, by (1.17), the above inequality becomes

$$4ac_0^2 + \frac{4a^3c_1}{9} + a^5\left(\frac{4c_2}{15} + \frac{2c_0}{3}\right)^2 \leq M\sum_{n=0}^{\infty} \frac{2c_n^2}{2n+1}.$$

Now it suffices to find the smallest constant M such that, for all $a_0, a_1, a_2 \in \mathbb{R}$,

$$\left(2a + \frac{2a^5}{9}\right)a_0^2 + \frac{2a^3}{9}a_1^2 + \frac{8a^5}{45}a_0a_2 + \frac{8a^5}{225}a_2^2 \leq M\left(a_0^2 + \frac{1}{3}a_1^2 + \frac{1}{5}a_2^2\right).$$

Let $X = (a_0, a_1, a_2)$ and

$$U = \begin{pmatrix} M - 2a - 2a^5/9 & 0 & -4a^5/45 \\ 0 & M/3 - 2a^3/9 & 0 \\ -4a^5/45 & 0 & M/5 - 8a^5/225 \end{pmatrix}.$$

The above inequality can be rewritten as $XUX^T \geq 0$. This holds for every X if and only if U is a positive semidefinite matrix. By the positive semidefinite matrix test, we must have

$$M \geq 2a + \frac{2a^5}{9}, \qquad M \geq \frac{2a^3}{3}, \qquad M \geq \frac{8a^5}{45}$$

and

$$\det\begin{pmatrix} M - 2a - 2a^5/9 & -4a^5/45 \\ -4a^5/45 & M/5 - 8a^5/225 \end{pmatrix} \geq 0.$$

Solving the above quadratic equation yields

$$M \geq M_+ := a + \frac{a^5}{5} + \frac{a}{15}\sqrt{9a^8 + 10a^4 + 225}.$$

Therefore,

$$M = \max\left\{2a + \frac{2a^5}{9}, \frac{2a^3}{3}, \frac{8a^5}{45}, M_+\right\} = M_+.$$

□

Remark. In the above solution, we have used the fact that

$$\text{span}\{1, x, x^2\} = \text{span}\{P_0(x), P_1(x), P_2(x)\}$$

to reduce the problem from L^2 to a finite dimensional subspace. The featured solution by K. Andersen [74] used a different orthonormal basis, which is obtained by applying the Gram-Schmidt process to the basis $\{1, x, x^2\}$. The same argument goes through in the general n-th order moments. The desired supremum will be determined based on the nonnegative eigenvalues of a symmetric matrix.

There is an elementary solution to Part (a) without invoking orthogonal polynomials. Indeed, for every $f(x) \in S_a$, using that $f(x) = f_e(x) + f_o(x)$ with

$$f_e(x) = \frac{1}{2}(f(x) + f(-x)), \quad f_o(x) = \frac{1}{2}(f(x) - f(-x)),$$

we have

$$\int_{-a}^{a}(f(x))^2 dx = \int_{-a}^{a}(f_e(x))^2 dx + \int_{-a}^{a}(f_o(x))^2 dx = 1,$$

and

$$A(f) = \int_{-a}^{a} f_e(x)dx, \quad B(f) = \int_{-a}^{a} xf_o(x)dx.$$

On the other hand, using the Cauchy-Schwarz inequality, we have

$$A^2(f) \leq \int_{-a}^{a} 1^2 dx \int_{-a}^{a}(f_e(x))^2 dx = 2a \int_{-a}^{a}(f_e(x))^2 dx,$$

$$B^2(f) \leq \int_{-a}^{a} x^2 dx \int_{-a}^{a}(f_o(x))^2 dx = \frac{2a^3}{3}\int_{-a}^{a}(f_o(x))^2 dx.$$

From which we have

$$\frac{1}{2a}A^2 + \frac{3}{2a^3}B^2 \leq \int_{-a}^{a}(f_e(x))^2 dx + \int_{-a}^{a}(f_o(x))^2 dx = 1.$$

Within this elliptical domain, the maximum of $A^2 + B^2$ is achieved at the end points of the major axis. Thus, we find

$$\sup\{A(f)^2 + B(f)^2 : f \in S_a\} = \max\{2a, 2a^3/3\}$$

as desired. Similarly, we have

$$C^2(f) \le \int_{-a}^{a} x^4 \, dx \int_{-a}^{a} (f_e(x))^2 \, dx = \frac{2a^5}{5} \int_{-a}^{a} (f_e(x))^2 \, dx.$$

Hence

$$A^2 + B^2 + C^2 \le \left(2a + \frac{2a^5}{5}\right) \int_{-a}^{a} (f_e(x))^2 \, dx + \frac{2a^3}{3} \int_{-a}^{a} (f_o(x))^2 \, dx.$$

This inequality seems too weak to recover the answer to Part (b).

Additional problems for practice.

1. Let $E = \{f \in L^2[-1,1] : f(-x) = f(x), \ x \in [-1,1]\}$. Find the distance from e^x to E.

2. Let $S = \{f(x) \in C^1[0,1] : f(0) = 0, \ f(1) = 1\}$. Show that

$$\inf \left\{ I(f) = \int_0^1 |f'(x) - f(x)| \, dx, \ f \in S \right\} = \frac{1}{e}.$$

3. **Putnam 2006-B5.** For each continuous function $f : [0,1] \to \mathbb{R}$, let $I(f) = \int_0^1 x^2 f(x) \, dx$ and $J(x) = \int_0^1 x(f(x))^2 \, dx$. Find the maximum value of $I(f) - J(f)$ over all such functions f.

4. **Problem 12142** (Proposed by C. Chiser, 126(9), 2019). Let $f : [a, b] \to \mathbb{R}$ be a twice continuously differentiable function satisfying $\int_a^b f(x) \, dx = 0$. Prove

$$\int_a^b (f''(x))^2 \, dx \ge \frac{980}{(8\sqrt{2}-1)^2} \cdot \frac{(f(a) + f(b))^2}{(b-a)^3}.$$

5. **Problem 12205** (Proposed by C. Chiser, 127(8), 2020). Find the minimum value of

$$\frac{\int_0^1 x^2 (f'(x))^2 \, dx}{\int_0^1 x^2 (f(x))^2 \, dx}$$

over all nonzero continuously differentiable functions $f : [0,1] \to \mathbb{R}$ with $f(1) = 0$.

6. **Problem 12308** (Proposed by C. Lupu, 129(3), 2022). What is the minimum value of $\int_0^1 (f'(x))^2 \, dx$ over all continuously differentiable functions $f : [0,1] \to \mathbb{R}$ such that $\int_0^1 f(x) \, dx = \int_0^1 x^2 f(x) \, dx = 1$?

1.10 Reciprocal power sums of real roots for a complex function

Problem 12479 (Proposed by M. Chamberland, 131(7), 2024). Evaluate $\sum_{n=1}^{\infty} r_n^{-6}$, where $r_1, r_2, \ldots$ are real roots of $2\cos\left(\sqrt{3}x\right) = -e^{-3x}$.

Discussion.
Let $p(z)$ be an n-th degree polynomial with nonzero roots r_k $(1 \le k \le n)$. Then $p(z) = a \prod_{k=1}^{n}(z - r_k)$. By logarithmic differentiation, we have

$$\frac{p'(z)}{p(z)} = \sum_{k=1}^{n} \frac{1}{z - r_k},$$

which enables us to find the reciprocal power sums of all roots. Indeed, for $m \in \mathbb{N}$, we have

$$\sum_{k=1}^{n} \frac{1}{r_k^m} = -\frac{1}{(m-1)!} \left(\frac{p'(z)}{p(z)}\right)^{(m-1)} \Bigg|_{z=0}.$$

This sheds light on our first solution.

The residue theorem is extremely useful in the evaluation of particularly troublesome real integrals. But this is not where this theorem use ends. To find a sum related to the roots of an entire function via residues, we cannot directly apply the residue theorem because an entire function has no poles (singularities) within the complex plane. However, you can utilize a clever trick by choosing a new function such that its residues are related to the form of roots you want to calculate. For example, let $g(z) = f'(z)/(z^6 f(z))$. Then if r is a simple root of $f(z)$, the residue of $g(z)$ at $z = r$ will be r^{-6}. By the residue theorem, the residue of $g(z)$ at $z = 0$ will be equal to the desired sum. Thus, to sum the series we need to compute the residue of $g(z)$ at $z = 0$. The details will be illustrated in the second solution.

Solution I.
Let $f(z) = 2\cos(\sqrt{3}z) + e^{-3z}$. Since $f(z)$ is an entire function, its roots are countable. Let $a_1, a_2, \ldots$ be the roots (real and complex) of $f(z)$. Note that the growth order of $f(z)$ is 1 and $f(0) = 3 \ne 0$, so by Hadamard's factorization theorem (see [92, p. 147]), we have

$$f(z) = e^{az+b} \prod_{n=1}^{\infty} \left(1 - \frac{z}{a_n}\right) e^{z/a_n},$$

where a and b are constants.

Letting $z = 0$ gives $e^b = 3$, so $b = \ln 3$. Applying logarithmic differentiation gives

$$\frac{f'(z)}{f(z)} = a + \sum_{n=1}^{\infty} \left(\frac{1}{z - a_n} + \frac{1}{a_n}\right). \tag{1.18}$$

Letting $z = 0$ in (1.18) gives $a = 2$. Next, differentiating (1.18) with respect to z five times yields

$$\left(\frac{f'(z)}{f(z)}\right)^{(5)} = -\sum_{n=1}^{\infty} \frac{5!}{(z - a_n)^6}. \tag{1.19}$$

Direct computation (for example, by *Mathematica*) gives

$$\left.\left(\frac{f'(z)}{f(z)}\right)^{(5)}\right|_{z=0} = -576.$$

This, together with setting $z = 0$ in (1.19), yields

$$\sum_{n=1}^{\infty} \frac{1}{a_n^6} = \frac{576}{5!} = \frac{24}{5}.$$

Next, to single out the real roots of $f(z)$, using $2\cos\theta = e^{i\theta} + e^{-i\theta}$, we rewrite $f(z) = 0$ as

$$f(z) = e^{i\sqrt{3}z} + e^{-i\sqrt{3}z} + e^{-3z} = 0.$$

Taking the conjugate of this equation shows that if z is a complex root of $f(z)$, so is its conjugate $\bar{z}$. Moreover, *Mathematica* experiments suggest that if r is a real root of $f(z)$, then ωr is a complex root of $f(z)$, where $\omega = e^{2\pi i/3} = \frac{1}{2}(-1 + i\sqrt{3})$, the primary unit root of $z^3 - 1$. This can be proved rigorously as follows. Direct computation shows

$$i\sqrt{3}\omega = 1 - \omega - 3, \quad -i\sqrt{3}\omega = 1 - \omega + i\sqrt{3}, \quad -3\omega = 1 - \omega - i\sqrt{3}.$$

Hence,

$$f(\omega z) = e^{i\sqrt{3}\omega z} + e^{-i\sqrt{3}\omega z} + e^{-3\omega z}$$
$$= e^{(1-\omega-3)z} + e^{(1-\omega+i\sqrt{3})z} + e^{(1-\omega-i\sqrt{3})z}$$
$$= e^{(1-\omega)z}\left(e^{-3z} + e^{i\sqrt{3}z} + e^{-i\sqrt{3}z}\right) = e^{(1-\omega)z}f(z).$$

This implies that if r_n ($n \in \mathbb{N}$) are all real roots of $f(z)$, then the total roots of $f(z)$ are given in the form of

$$\{r_n, \quad \omega r_n, \quad \omega^2 r_n \; : \; n \in \mathbb{N}\}.$$

Since $\omega^6 = (\omega^2)^6 = 1$, we find that

$$\sum_{n=1}^{\infty} \frac{1}{a_n^6} = \sum_{n=1}^{\infty} \frac{1}{r_n^6} + \sum_{n=1}^{\infty} \frac{1}{(\omega r_n)^6} + \sum_{n=1}^{\infty} \frac{1}{(\omega^2 r_n)^6} = 3\sum_{n=1}^{\infty} \frac{1}{r_n^6}.$$

Therefore,

$$\sum_{n=1}^{\infty} \frac{1}{r_n^6} = \frac{1}{3} \sum_{n=1}^{\infty} \frac{1}{a_n^6} = \frac{1}{3} \cdot \frac{24}{5} = \frac{8}{5}.$$

□

Solution II.

We consider the entire function

$$f(z) = 2\cos(\sqrt{3}z) + e^{-3z} = e^{i\sqrt{3}z} + e^{-i\sqrt{3}z} + e^{-3z}.$$

Letting $z = x + iy$, after separating real and imaginary parts, we have that $f(z) = 0$ if and only if

$$\begin{cases} 2\cosh(\sqrt{3}y)\cos(\sqrt{3}x) + e^{-3x}\cos(3y) = 0 \\ 2\sinh(\sqrt{3}y)\sin(\sqrt{3}x) + e^{-3x}\sin(3y) = 0. \end{cases}$$

If $y \neq 0$, the above system implies

$$4e^{6x} = \frac{\cos^2(3y)}{\cosh^2(\sqrt{3}y)} + \frac{\sin^2(3y)}{\sinh^2(\sqrt{3}y)}. \tag{1.20}$$

It is easy to see that for $a, b > 0$, and $y \neq 0$,

$$\cos^2(ay) \leq 1 < \cosh^2(by) \quad \text{and} \quad b^2\sin^2(ay) < a^2\sinh^2(by).$$

Therefore, if $x \geq 0$ and $y \neq 0$, then

$$4e^{6x} \geq 4 = 1 + \frac{3^2}{3} > \frac{\cos^2(3y)}{\cosh^2(\sqrt{3}y)} + \frac{\sin^2(3y)}{\sinh^2(\sqrt{3}y)}.$$

This inequality, together with (1.20), shows that, in the half-plane $\text{Re}(z) \geq 0$, if $f(z) = 0$ then $z = x > 0$. For $n \in \mathbb{N}$, let

$$\alpha_n = \frac{2n\pi}{\sqrt{3}}, \qquad \beta_n = \frac{(2n+1)\pi}{\sqrt{3}}, \qquad \gamma_n = \frac{2(n+1)\pi}{\sqrt{3}}.$$

We have

$$f(\alpha_n) = 2 + e^{-3\alpha_n} > 0, \quad f(\beta_n) = -2 + e^{-3\beta_n} < 0, \quad f(\gamma_n) = 2 + e^{-3\gamma_n} > 0.$$

By the intermediate value theorem, for every $n \in \mathbb{N}$, f has at least one root in (α_n, β_n) and (β_n, γ_n), respectively. Therefore f has infinitely many positive roots. Denote by $\{r_k\}_{k\geq 1}$ the sequence of such positive roots. Along the same lines in Solution I, we have

$$f(wz) = e^{(1-w)z}f(z) \tag{1.21}$$

where $\omega = e^{2\pi i/3}$, and obtain that the whole set of zeros of f is given by

$$Z = \{r_k,\ \omega r_k,\ \omega^2 r_k\}_{k \geq 1}.$$

Notice that the function $\dfrac{f'(z)}{z^m f(z)}$ has singularities at 0 and in Z and its residue

at $z_k \in Z$ is $\dfrac{1}{z_k^m}$ because f has only simple zeros. Hence, by the residue

theorem, if γ is a positively oriented simple closed curve which does not pass through any point in $Z \cup \{0\}$, and D is the domain surrounded by γ, then

$$\int_\gamma \frac{f'(z)}{z^m f(z)}\, dz = 2\pi i \left(\delta \operatorname{Res}\left(\frac{f'(z)}{z^m f(z)}, 0 \right) + \sum_{z_k \in D \cap Z} \frac{1}{z_k^m} \right) \qquad (1.22)$$

where $\delta = 1$ if $0 \in D$ and $\delta = 0$ otherwise.
Differentiating (1.21), we get

$$\omega f'(\omega z) = e^{(1-\omega)z}((1-\omega)f(z) + f'(z))$$

and therefore

$$\frac{\omega f'(\omega z)}{f(\omega z)} = \frac{e^{(1-\omega)z}((1-\omega)f(z) + f'(z))}{e^{(1-\omega)z} f(z)} = 1 - \omega + \frac{f'(z)}{f(z)}$$

which implies

$$\int_{\omega \gamma} \frac{f'(z)}{z^m f(z)}\, dz = \frac{1}{\omega^m} \int_\gamma \frac{\omega f'(\omega z)}{z^m f(\omega z)}\, dz = \frac{1-\omega}{\omega^m} \int_\gamma \frac{1}{z^m}\, dz + \frac{1}{\omega^m} \int_\gamma \frac{f'(z)}{z^m f(z)}\, dz.$$

Thus, by (1.22),

$$\int_{C_R} \frac{f'(z)}{z^m f(z)}\, dz - \int_{C_r} \frac{f'(z)}{z^m f(z)}\, dz = \left(1 + \frac{1}{\omega^m} + \frac{1}{\omega^{2m}} \right) \int_\gamma \frac{f'(z)}{z^m f(z)}\, dz$$

where C_R and C_r are circles counterclockwise oriented, of radius r and R respectively, with $0 < r < R$, and $\gamma = S_{r,R}$ is the bold curve pictured below. Evaluating the limits as $R \to +\infty$ and $r \to 0^+$, we obtain

$$-[z^{m-1}]\frac{f'(z)}{f(z)} = -\operatorname{Res}\left(\frac{f'(z)}{z^m f(z)}, 0 \right) = \frac{\omega^{2m} + \omega^m + 1}{\omega^{2m}} \sum_{k=1}^\infty \frac{1}{r_k^m}.$$

Since $\omega^{2m} + \omega^m + 1$ gives 3 when m is divisible by 3, and it is 0 otherwise, for $m = 3M$, we obtain

$$\sum_{k=1}^\infty \frac{1}{r_k^{3M}} = -\frac{1}{3}[z^{3M-1}]\frac{f'(z)}{f(z)}.$$

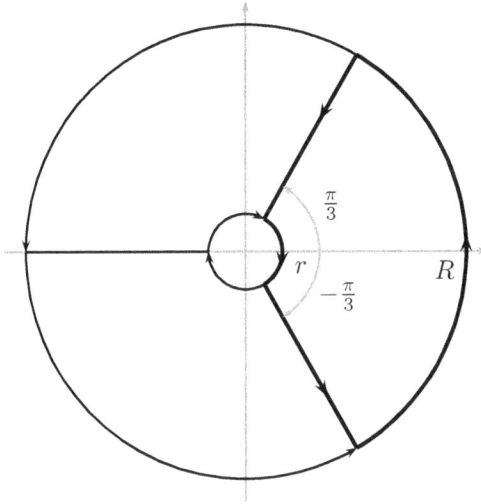

Figure 1.4: Path of integration

Since

$$\frac{f'(z)}{f(z)} = -1 - 4z^2 - \frac{24}{5}z^5 - \frac{212}{35}z^8 + O(z^{11}),$$

letting $M = 2$ immediately obtains

$$\sum_{k=1}^{\infty} \frac{1}{r_k^6} = \frac{8}{5}.$$

□

Remark. In view of the structure of the roots for f, we can find $\sum_{n=1}^{\infty} (r_n)^{-3k}$ in closed form for all $k \in \mathbb{N}$. For example, we have

$$\sum_{n=1}^{\infty} \frac{1}{r_n^3} = \frac{4}{3}, \quad \sum_{n=1}^{\infty} \frac{1}{r_n^9} = \frac{212}{105}, \quad \sum_{n=1}^{\infty} \frac{1}{r_n^{12}} = \frac{4912}{1925}, \quad \sum_{n=1}^{\infty} \frac{1}{r_n^{15}} = \frac{564944}{175175}.$$

Additional problems for practice.

1. Let $p(x)$ be an n-th degree polynomial with real roots $r_1 < r_2 < \ldots < r_n$. For $n > 3$, show that

$$p'\left(\frac{r_1 + r_2}{2}\right) \cdot p'\left(\frac{r_{n-1} + r_n}{2}\right) \neq 0.$$

2. Show that $f(z) = z\sin(z) - 1$ has only real roots.

3. **Problem 12077** (Proposed by M. A. Alekseyev, 125(10), 2018). Let $f(x)$ be a monic polynomial of degree n with distinct zeros $z_1, z_2, \ldots, z_n$. Prove

$$\sum_{k=1}^{n} \frac{z_k^{n-1}}{f'(z_k)} = 1.$$

4. (Series summation theorem) Let $f(z)$ be analytic in $\mathbb{C}$ except for some finite isolated poles, and let $|f(z)| < M/|z|^k$ for $k > 1$. Then

$$\sum_{n\in\mathbb{Z}} f(n) = -\sum \{\text{residues of } \pi\cot(\pi z)f(z) \text{ at } f\text{'s poles}\}.$$

Hint. Show that the residue of $\pi\cot(\pi z)f(z)$ at $z = n$ is

$$\lim_{z\to n} (z - n)\pi\cot(\pi z)f(z) = f(n).$$

5. Use the above series summation theorem to prove that

$$\sum_{n\in\mathbb{Z}} \frac{1}{(3n-1)^2} = \frac{4\pi^2}{27}.$$

6. Use the above series summation theorem to prove that, for $\alpha \neq 0$,

$$\sum_{n\in\mathbb{Z}} \frac{1}{n^2 + \alpha^2} = \frac{\pi}{\alpha}\coth(\pi\alpha).$$

7. Replacing $\pi\cot(\pi z)$ by $\pi\csc(\pi z)$ in Problem 4, show that

$$\sum_{n\in\mathbb{Z}} (-1)^n f(n) = -\sum \{\text{residues of } \pi\csc(\pi z)f(z) \text{ at } f\text{'s poles}\}.$$

For $\alpha \in \mathbb{R}\setminus\mathbb{Z}$, use this result to derive

$$\sum_{n\in\mathbb{Z}} \frac{(-1)^n}{(n+\alpha)^2} = \frac{\pi^2\cos(\alpha\pi)}{\sin^2(\alpha\pi)}.$$

1.11 A monotonic and convex quadratic spline

Problem 12467 (Proposed by L. László, 131(5), 2024). Given any real number c, it is not hard to see that there is a unique differentiable function $s : [1, \infty) \to \mathbb{R}$ such that (1) $s(n) = 1/n$ for all positive integers n, (2) s is quadratic or linear on $[n, n+1]$ for all positive integers n, and (3) the right derivative of

s at 1 is c. A function satisfying (1) and (2) is a *quadratic spline*. For what values of c is s decreasing and convex?

Discussion.
First, we present a simple and instructive way to establish the existence of a unique quadratic spline. For every positive integer n, we can explicitly find the quadratic polynomial $s_n(x) \in C^1[n, n+1]$ satisfying $s_n(n) = 1/n$, $s_n(n+1) = 1/(n+1)$, $s'_n(n) = d_n$ and $s'_n(n+1) = d_{n+1}$. Indeed, let

$$s_n(x) = \alpha_n + \beta_n(x - n) + \gamma_n(x - n)^2,$$

where α_n, β_n and γ_n are constants to be determined. It is easy to see that $\alpha_n = 1/n, \beta_n = d_n$. Moreover, $s'_n(n+1) = \beta_n + 2\gamma_n = d_{n+1}$ implies that $\gamma_n = (d_{n+1} - d_n)/2$. Hence, the unique quadratic spline $s(x)$ on $[n, n+1]$ is

$$s_n(x) = \frac{1}{n} + d_n(x - n) + \frac{1}{2}(d_{n+1} - d_n)(x - n)^2,$$

where

$$\frac{d_{n+1} + d_n}{2} = s(n+1) - s(n) = -\frac{1}{n(n+1)}.$$

This implies

$$d_{n+1} = -\frac{2}{n(n+1)} - d_n. \tag{1.23}$$

In particular, with $s'_1(1) = d_1 = c$, we have

$$s_1(x) = 1 + c(x - 1) - \frac{1}{2}(1 + 2c)(x - 1)^2.$$

If $s_1(x)$ is decreasing and convex, we find that $c \leq -1/2$. By *Mathematica*,

```
pw[x_]:=Piecewise[{{1+c(x-1)-(1+2c)(x-1)^2/2, 1<=x<=2},
{1/2-(1+c)(x-2)+(5/6+c)(x-2)^2, 2<x<=3},
{1/3+(2/3+c)(x-3)-(3/4+c)(x-3)^2, 3<x<=4},
{1/4-(5/6+c)(x-4)+(47/60+c)(x-4)^2, 4<x<=5}}];
Plot[{pw[x], 1/x}, {x,1,5}, PlotRange -> {0,1}]
```

The graph given in Figure 1.5 shows that even if s_1 is decreasing and convex this does not guarantee that whole spline is decreasing and convex. So the challenge lies in how to select c such that whole spline preserves the nature of decreasing and convex from s_1.

Solution.
For $s(x)$ being decreasing and convex, it suffices to ensure that for all $n \in \mathbb{N}, x \in (n, n+1)$,

$$s'_n(x) = d_n + (d_{n+1} - d_n)(x - n) \leq 0 \quad \text{and} \quad s''_n(x) = d_{n+1} - d_n \geq 0,$$

which is equivalent to

$$d_{n+1} \leq 0 \quad \text{and} \quad d_{n+1} - d_n \geq 0 \quad \text{for all } n \in \mathbb{N}. \tag{1.24}$$

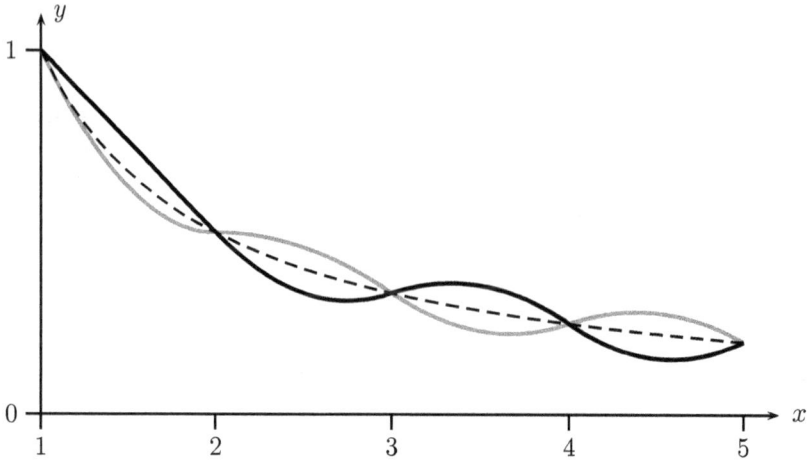

Figure 1.5: Two quadratic splines on $[1,5]$ with $c = -1/2$ (black) and $c = -1$ (gray)

Applying (1.23) repeatedly yields

$$c = d_1 = -\frac{2}{1 \cdot 2} - d_2 = -\frac{2}{1 \cdot 2} + \frac{2}{2 \cdot 3} + d_3$$

$$= -\frac{2}{1 \cdot 2} + \frac{2}{2 \cdot 3} - \frac{2}{3 \cdot 4} - d_4$$

$$= 2 \sum_{k=1}^{n-1} \frac{(-1)^k}{k(k+1)} + (-1)^{n+1} d_n. \qquad (1.25)$$

Let $c = 2 \sum_{k=1}^{\infty} \frac{(-1)^k}{k(k+1)}$. Then

$$c = 2 \left(1 - 2 \sum_{k=1}^{\infty} \frac{(-1)^{k+1}}{k} \right) = 2(1 - 2 \ln 2) = -0.77258872\ldots.$$

We now show that d_n satisfies (1.24). Indeed, for $n \geq 2$, by (1.25), we have

$$d_n = (-1)^{n+1} \left(c - 2 \sum_{k=1}^{n-1} \frac{(-1)^k}{k(k+1)} \right) = (-1)^{n+1} 2 \sum_{k=n}^{\infty} \frac{(-1)^k}{k(k+1)}$$

$$= (-1)^{n+1} 2 \left(\sum_{k=n}^{\infty} (-1)^k \int_0^1 (x^{k-1} - x^k) \, dx \right)$$

$$= (-1)^{n+1}2 \int_0^1 \left(\sum_{k=n}^{\infty} (-1)^k (x^{k-1} - x^k) \right) dx$$

$$= -2 \int_0^1 \frac{x^{n-1}(1-x)}{1+x} \, dx \le 0,$$

and

$$d_{n+1} - d_n = 2 \int_0^1 \frac{x^{n-1}(1-x)}{1+x} \, dx - 2 \int_0^1 \frac{x^n(1-x)}{1+x} \, dx$$

$$= 2 \int_0^1 \frac{x^{n-1}(1-x)^2}{1+x} \, dx \ge 0.$$

This proves (1.24) as desired. Moreover, in view of (1.25), if $c \ne 2(1 - 2\ln 2)$ then $d_n = s'(n)$ changes its sign infinitely many times as n goes to infinity, contradicting the fact that s is decreasing. Thus, $c = 2(1 - 2\ln 2)$ is the unique value such that $s(x)$ is decreasing and convex. □

Remark. By *Mathematica*, the decreasing and convex spline is given as follows:

```
c:=2(1-2Log[2]);
pw[x_]:=Piecewise[{{1+c(x-1)-(1+2c)(x-1)^2/2, 1<=x<=2},
{1/2-(1+c)(x-2)+(5/6+c)(x-2)^2, 2<x<=3},
{1/3+(2/3+c)(x-3)-(3/4+c)(x-3)^2, 3<x<=4},
{1/4-(5/6+c)(x-4)+(47/60+c)(x-4)^2, 4<x<=5}}];
Plot[pw[x], {x,1,5}, PlotRange -> {0,1}]
```

The resulting graph is shown in Figure 1.6.
By the alternating series estimation theorem, the tail of the series satisfies

$$\left| \sum_{k=n}^{\infty} (-1)^k a_k \right| < a_n.$$

It is interesting to see that in order to establish (1.24), we have derived a sharper estimate of the tail than the usual estimate above.

In contrast to Lagrange interpolation, spline interpolation is often preferred because it yields similar results but uses much lower degree polynomials. L. Schumaker in [83] studied the design of algorithms for interpolating general discrete data using C^1-quadratic splines to preserve the monotonicity and/or convexity of the data. His analysis culminates in an interactive algorithm that takes full advantage of the flexibility that quadratic splines permit.

Since a quadratic spline cannot ensure the same curvature at interior rods, the cubic spline is used more often in applications, especially in computer graphics. Here the second derivative of the cubic spline is a piecewise linear function, so the stability properties can be inferior to those of smoother

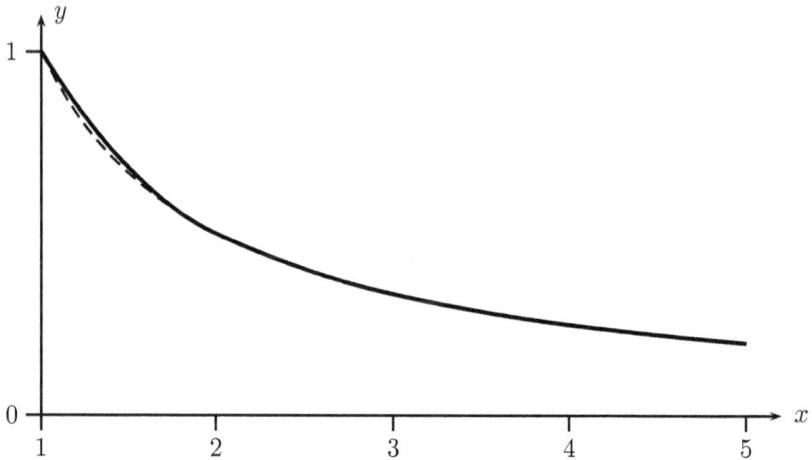

Figure 1.6: The spline on $[1, 5]$ with $c = 2(1 - 2\ln(2))$

kernels. In [79], E. Passow and J. Roulier revealed a necessary and sufficient condition for the existence of cubic differentiable interpolating splines that are monotone and convex. They also discuss the approximation properties of these splines when applied to the interpolation of functions with a preassigned degree of smoothness.

Additional problems for practice.

1. (Due to J. Tóth and J. Bukor) For $n \geq 1$, find the best possible constants α and β such that

$$\frac{1}{2n + \alpha} \leq \left| \sum_{k=n+1}^{\infty} \frac{(-1)^k}{k} \right| < \frac{1}{2n + \beta}.$$

2. **Problem 12359** (Proposed by P. Bracken, 129(10), 2022). Let n be a positive integer. Prove

$$\frac{-1 - \pi}{4n} - \frac{1}{8n^2} < \sum_{k=1}^{n} \frac{1}{(2k - 1)^2} - 2 \left(\sum_{k=1}^{n} \frac{(-1)^{k+1}}{2k - 1} \right)^2 < \frac{-1 + \pi}{4n} - \frac{1}{8n^2}.$$

3. Let $x_1 < x_2$. Show that there exists a quadratic polynomial $s(x)$ satisfying $s(x_i) = y_i$, $s'(x_i) = d_i$ for $i = 1, 2$ if and only if

$$\frac{d_1 + d_2}{2} = \frac{y_2 - y_1}{x_2 - x_1}.$$

If this condition fails, for every $x_1 < \xi < x_2$, show that, by choosing $s'(\xi)$, there exists a unique quadratic spline s that satisfies $s(x_i) = y_i$, $s'(x_i) = d_i$ for $i = 1, 2$.

4. Given the discrete data $\{(x_k, y_k)\}_{k=0}^n$. Let $s_i = (y_i - y_{i-1})/(x_i - x_{i-1})$ for $i = 1, 2, \ldots, n$. Show that:

   (i) if $y_i - y_{i-1} > 0$ and $s_i - s_{i-1} > 0$ for $i = 1, 2, \ldots, n$, then there exists an increasing C^2-interpolation.

   (ii) if $s_i - s_{i-1} > 0$ and $s_{i+1} - 2s_i + s_{i-1} > 0$ for $i = 1, 2, \ldots, n$, then there exists a convex C^2-interpolation.

5. Let $s(x)$ be a cubic spline function which interpolates $\{(x_k, y_k)\}_{k=0}^n$. If $f(x)$ is any twice continuously differentiable function through these points, then

$$\int_{x_0}^{x_n} (f''(x))^2 \, dx > \int_{x_0}^{x_n} (s''(x))^2 \, dx.$$

1.12 An estimate of a series involving the absolute value of sine

Problem 12389 (Proposed by G. Stoica, 130(4), 2023). Let

$$f(x) = \sum_{n=1}^{\infty} \frac{|\sin(nx)|}{n^2}.$$

Prove

$$\lim_{x \to 0^+} f(x)/(x \ln x) = -1.$$

Discussion.
It seems that there is no closed form for $f(x)$, although *Mathematica* gives

$$g(x) := \sum_{n=1}^{\infty} \frac{\sin(nx)}{n^2} = \frac{i}{2} \left(\mathrm{Li}_2(e^{-ix}) - \mathrm{Li}_2(e^{ix}) \right), \tag{1.26}$$

where $\mathrm{Li}_2(x) = \sum_{n=1}^{\infty} x^n/n^2$ is the dilogarithm function. Intuitively, when x is tiny, there exists a large number $N \in \mathbb{N}$ such that $nx \leq \pi$ and so $|\sin(nx)| = \sin(nx)$ for all $n \leq N$. For example, if $x = 0.01$ then $N = 314$. So from the limit point of view, we have $f(x) \sim g(x)$ as $x \to 0$. Our first solution is based on this observation. In particular, we will find a closed form of $g(x)$ other than (1.26). On the other hand, we try to establish very precise (meticulous) upper and lower bounds for $f(x)$ via the sequence. This process will be demonstrated in the second solution.

Solution I.
For $x > 0$, we rewrite

$$f(x) = \sum_{nx \leq \pi} \frac{|\sin(nx)|}{n^2} + \sum_{nx > \pi} \frac{|\sin(nx)|}{n^2}$$

$$= \sum_{n=1}^{\infty} \frac{\sin(nx)}{n^2} + \sum_{nx > \pi} \frac{|\sin(nx)| - \sin(nx)}{n^2}.$$

Let $N(x)$ be the first n such that $nx > \pi$. Then

$$\sum_{nx > \pi} \frac{|\sin(nx)| - \sin(nx)}{n^2} \leq \sum_{N}^{\infty} \frac{2}{n^2} \leq 2 \int_{N-1}^{\infty} \frac{dx}{x^2} = \frac{2}{N-1},$$

so we obtain

$$\left| \frac{\sum_{nx > \pi} \frac{|\sin(nx)| - \sin(nx)}{n^2}}{x \ln x} \right| \leq \frac{2}{(N-1)x |\ln x|} \leq \frac{2}{(\pi - x) |\ln x|},$$

and

$$\lim_{x \to 0^+} \frac{\sum_{nx > \pi} \frac{|\sin(nx)| - \sin(nx)}{n^2}}{x \ln x} = 0.$$

Now, it suffices to show that

$$\lim_{x \to 0^+} \frac{\sum_{n=1}^{\infty} \frac{\sin(nx)}{n^2}}{x \ln x} = \lim_{x \to 0^+} \frac{g(x)}{x \ln x} = -1.$$

To calculate this limit by L'Hôpital's rule with $g(x)$ given by (1.26), we are unable to calculate $g''(x)$ via term by term differentiation. So we need another closed form for $g(x)$ which is easy to find its derivatives. To this end, we first establish

$$\sum_{n=1}^{\infty} \frac{\cos(nx)}{n} = -\frac{1}{2} \ln[2(1 - \cos x)] \qquad (0 < x < 2\pi). \qquad (1.27)$$

Notice that, for $0 < r < 1$,

$$\sum_{n=0}^{\infty} r^n \cos(nx) = \mathrm{Re} \left(\sum_{n=0}^{\infty} (re^{ix})^n \right) = \frac{1 - r \cos x}{1 - 2r \cos x + r^2}.$$

This implies that

$$\sum_{n=1}^{\infty} r^n \cos(nx) = \frac{1 - r \cos x}{1 - 2r \cos x + r^2} - 1 = \frac{r \cos x - r^2}{1 - 2r \cos x + r^2}.$$

Dividing by r, then integrating with respect to r yields

$$\sum_{n=1}^{\infty} \frac{r^n \cos(nx)}{n} = -\frac{1}{2} \ln(1 - 2r \cos x + r^2).$$

By Dirichlet's test, this series converges at $r = 1$. Thus, (1.27) follows from letting $r = 1$ in the above equation. Integrating (1.27) with respect to x, we obtain another closed form of

$$g(x) = \sum_{n=1}^{\infty} \frac{\sin(nx)}{n^2} = -\frac{1}{2} \int_0^x \ln[2(1 - \cos t)]\, dt.$$

Now, applying L'Hôpital's rule twice, we find

$$\lim_{x \to 0^+} \frac{\sum_{n=1}^{\infty} \frac{\sin(nx)}{n^2}}{x \ln x} = \lim_{x \to 0^+} \frac{-\frac{1}{2} \int_0^x \ln[2(1 - \cos t)]\, dt}{x \ln x}$$

$$= \lim_{x \to 0^+} \frac{-\frac{1}{2} \ln[2(1 - \cos x)]\, dt}{\ln x + 1}$$

$$= -\frac{1}{2} \lim_{x \to 0^+} \frac{x \sin x}{1 - \cos x} = -1$$

as claimed. $\qquad\qquad\qquad\qquad\qquad\qquad\qquad\qquad\qquad\qquad\quad$ □

Solution II.
Since f is a nonnegative function, by using upper and lower limits, we show that

$$\lim_{k \to \infty} \frac{f(x_k)}{x_k |\ln(x_k)|} = 1$$

for every positive sequence $x_k \to 0$ as $k \to \infty$.

To estimate the upper limit, let $M = \lceil 1/x_k \rceil \geq 1$. Since $|\sin(t)| \leq |t|$ and $\sum_{n=1}^{M} 1/n \leq \ln(M) + 1$, we obtain

$$f(x_k) = \sum_{n=1}^{M} \frac{|\sin(nx_k)|}{n^2} + \sum_{n=M+1}^{\infty} \frac{|\sin(nx_k)|}{n^2} \leq \sum_{n=1}^{M} \frac{x_k}{n} + \sum_{n=M+1}^{\infty} \frac{1}{n^2}$$

$$\leq x_k(\ln(M) + 1) + \frac{1}{M} \leq x_k \left(\ln\left(\frac{1}{x_k} + 1\right) + 1 \right) + x_k,$$

where we have used $\sum_{n=M+1}^{\infty} \frac{1}{n^2} \leq \int_M^{\infty} 1/x^2\, dx = 1/M$. Therefore,

$$\limsup_{k \to \infty} \frac{f(x_k)}{x_k |\ln(x_k)|} \leq \lim_{k \to \infty} \frac{\ln(1 + 1/x_k) + 2}{|\ln(x_k)|} = 1. \qquad (1.28)$$

On the other hand, to estimate the lower limit, we choose $\delta \in (0, 1)$. By the mean value theorem, we have $\sin(t) = \cos(\xi)t$ for $\xi \in (0, t)$, from which we obtain $\sin(t) \geq \cos(\delta)t$ for $0 \leq t \leq \delta$ since $\cos x$ is decreasing in $(0, \pi)$. Next, let $M = \lceil \delta/(2x_k) \rceil \geq 1$. Then for every $1 \leq n \leq M$,

$$0 < nx_k \leq Mx_k \leq \left(\frac{\delta}{2x_k} + 1 \right) x_k = \frac{\delta}{2} + x_k \leq \delta$$

for sufficiently large values of k. Applying $\sum_{n=1}^{M} 1/n \geq \ln(M)$ yields

$$f(x_k) \geq \sum_{n=1}^{M} \frac{|\sin(nx_k)|}{n^2} \geq \cos(\delta) \sum_{n=1}^{M} \frac{x_k}{n}$$

$$\geq \cos(\delta) x_k \ln(M) \geq \cos(\delta) x_k \ln \left(\frac{\delta}{2x_k} \right).$$

Therefore, for any $\delta \in (0, 1)$,

$$\liminf_{k \to \infty} \frac{f(x_k)}{x_k |\ln(x_k)|} \geq \cos(\delta) \lim_{k \to \infty} \frac{\ln(\delta/(2x_k))}{|\ln(x_k)|} = \cos(\delta).$$

Since δ is arbitrary and $\cos(\delta) \to 1$ as $\delta \to 0$, we get

$$\liminf_{k \to \infty} \frac{f(x_k)}{x_k |\ln(x_k)|} \geq 1.$$

This, together with (1.28), leads to

$$1 \leq \liminf_{k \to \infty} \frac{f(x_k)}{x_k |\ln(x_k)|} \leq \limsup_{k \to \infty} \frac{f(x_k)}{x_k |\ln(x_k)|} \leq 1,$$

which implies

$$\lim_{k \to \infty} \frac{f(x_k)}{x_k |\ln(x_k)|} = 1,$$

as claimed. $\square$

Remark. Here we give another way to estimate the lower limit. Let $\epsilon \in (0, 1/2)$. Since $\sin(x)/x$ is decreasing on $[0, \pi/2]$, for $0 < t \leq \epsilon$,

$$\frac{\sin(t)}{t} \geq \frac{\sin(\epsilon)}{\epsilon}.$$

For any $0 < x < \epsilon$, let $N = \lfloor \epsilon/x \rfloor$. Then for all $n \leq N$,

$$nx \leq Nx = \lfloor \epsilon/x \rfloor x \leq \frac{\epsilon}{x} x = \epsilon,$$

and so

$$\frac{\sin(nx)}{nx} \geq \frac{\sin(\epsilon)}{\epsilon}.$$

From here, we obtain

$$f(x) \geq \frac{\sin(\epsilon)}{\epsilon} \sum_{n=1}^{N} \frac{nx}{n^2} = \frac{x\sin(\epsilon)}{\epsilon} \sum_{n=1}^{N} \frac{1}{n} \geq \frac{x\sin(\epsilon)}{\epsilon} \ln(N).$$

Therefore,

$$\frac{f(x)}{x|\ln(x)|} \geq \frac{\sin(\epsilon)}{\epsilon} \frac{\ln(\lfloor \epsilon/x \rfloor)}{|\ln(x)|} \geq \frac{\sin(\epsilon)}{\epsilon} \frac{\ln(\lfloor \epsilon/x \rfloor - 1)}{|\ln(x)|} = \frac{\sin(\epsilon)}{\epsilon} \frac{\ln(\epsilon - x) - \ln(x)}{|\ln(x)|},$$

which implies

$$\liminf_{k \to \infty} \frac{f(x)}{x|\ln(x)|} \geq \frac{\sin(\epsilon)}{\epsilon}.$$

Now, letting $\epsilon \to 0^+$ yields $\liminf_{k \to \infty} \frac{f(x)}{x|\ln(x)|} \geq 1$ again.

Additional problems for practice.

1. For $0 < \alpha \leq \pi$, show that

$$\sum_{n=-\infty}^{\infty} \frac{\sin^2(n\alpha + \theta)}{(n\alpha + \theta)^2} = \frac{\pi}{\alpha}.$$

2. **Problem 11185** (Proposed by R. Bruck and R. Mortini, 112(9), 2005). Find all natural numbers n and positive real numbers α such that the integral

$$I(\alpha, n) = \int_0^{\infty} \ln\left(1 + \frac{\sin^n x}{x^\alpha}\right) dx$$

converges.

3. Let $f : [0, 1] \to \mathbb{R}$ be differentiable, strictly increasing and convex. If $f'(1) < 2f(1)$, show that

$$\lim_{n \to \infty} \frac{n+1}{f^{2n+1}(1)} \int_0^1 f^{2n+1}(x)\, dx = \frac{f(1)}{2f'(1)}.$$

4. Let $\{x_n\}$ be a sequence with $\sum_{n=1}^{\infty} |x_n| \leq 1$. Find

$$\sup \left(\sum_{n=1}^{\infty} \sin(n)(x_n - x_{n+1}) \right).$$

5. Let $n \geq 3$ be a positive integer and

$$S = \{(a_1, a_2, \ldots, a_n) : a_k > 0 \text{ for all } 1 \leq k \leq n.\}.$$

Find the supremum and the infimum of the set

$$\left\{ \sum_{k=1}^{n} \frac{a_k}{a_k + a_{k+1} + a_{k+2}}, \quad (a_1, a_2, \ldots, a_n) \in S \right\}$$

with $a_{n+1} = a_1$ and $a_{n+2} = a_2$.

6. (Due to C. Popescu and D. Schwartz). Let $f : [0, \infty) \to [0, \infty)$ be continuous and satisfy that $\lim_{x \to \infty} f(x) \int_0^x f(t)\, dt = l > 0$. Prove

$$\lim_{x \to \infty} \sqrt{x} f(x)$$

exists and find its value in terms of l.

7. **Problem 12362** (Proposed by A. Garcia, 129(10), 2022). Evaluate

$$\lim_{n \to \infty} \int_0^{\pi/2} \frac{n}{(\sqrt{2} \cos x)^n + (\sqrt{2} \sin x)^n}\, dx.$$

1.13   An integral with many solutions

Problem 12407 (Proposed by A. K. Parcha, 130(7), 2023). Let r be a positive real number. Evaluate

$$\int_0^\infty \frac{x^{r-1}}{(1 + x^2)(1 + x^{2r})}\, dx.$$

Discussion.
This is a pretty straightforward *Monthly* problem. It can be solved in many different ways. In the following we will present three solutions. Each solution converts the desired integral into an elementary integral after an appropriate substitution. In particular, the idea used in Solution III can be applied to compute more general integrals such as

$$\int_0^\infty \frac{x^{pq-1}}{(1 + x^2)(1 + x^{2p})^q}\, dx \quad \text{for } q > 0.$$

Solution I.
We show the integral value is $\frac{\pi}{4r}$. To this end, rewrite the desired integral as

$$\int_0^\infty \frac{x^{r-1}}{(1 + x^2)(1 + x^{2r})}\, dx = \int_0^1 \frac{x^{r-1}}{(1 + x^2)(1 + x^{2r})}\, dx + \int_1^\infty \frac{x^{r-1}}{(1 + x^2)(1 + x^{2r})}\, dx.$$

Applying the substitution $x \to 1/x$ in the second integral yields

$$\int_1^\infty \frac{x^{r-1}}{(1 + x^2)(1 + x^{2r})}\, dx = \int_0^1 \frac{x^{r+1}}{(1 + x^2)(1 + x^{2r})}\, dx.$$

Therefore,

$$\int_0^\infty \frac{x^{r-1}}{(1+x^2)(1+x^{2r})}\,dx = \int_0^1 \frac{x^{r-1}+x^{r+1}}{(1+x^2)(1+x^{2r})}\,dx$$

$$= \int_0^1 \frac{x^{r-1}(1+x^2)}{(1+x^2)(1+x^{2r})}\,dx = \int_0^1 \frac{x^{r-1}}{1+x^{2r}}\,dx$$

$$= \frac{1}{r}[\arctan(x^r)]_0^1 = \frac{\pi}{4r}.$$

This proves the claimed result.   □

Solution II.

Let the desired integral be I. By the substitution $x = (\tan\theta)^{1/r}$ with $p = 2/r$, we have

$$I = \frac{1}{r}\int_0^{\pi/2} \frac{1}{1+\tan^p\theta}\,d\theta.$$

This integral is well-known and often appears in various math contests (for example, Putnam Problem 1980-A3 with $p = \sqrt{2}$) because the integral value is surprisingly independent of p. To see this, the substitution $\theta \to \pi/2 - \theta$ yields

$$\int_0^{\pi/2} \frac{1}{1+\tan^p\theta}\,d\theta = \int_0^{\pi/2} \frac{1}{1+\cot^p\theta}\,d\theta = \int_0^{\pi/2} \frac{\tan^p\theta}{1+\tan^p\theta}\,d\theta.$$

Hence,

$$I = \frac{1}{2r}\left(\int_0^{\pi/2} \frac{1}{1+\tan^p\theta}\,d\theta + \int_0^{\pi/2} \frac{\tan^p\theta}{1+\tan^p\theta}\,d\theta\right)$$

$$= \frac{1}{2r}\int_0^{\pi/2} \frac{1+\tan^p\theta}{1+\tan^p\theta}\,d\theta = \frac{1}{2r}\int_0^{\pi/2} 1\,d\theta = \frac{\pi}{4r}.$$

□

Solution III.

Let the desired integral be I. We associate I with the related integral

$$J = \int_0^\infty \frac{x^{r+1}}{(1+x^2)(1+x^{2r})}\,dx.$$

Then

$$I + J = \int_0^\infty \frac{x^{r-1}+x^{r+1}}{(1+x^2)(1+x^{2r})}\,dx$$

$$= \int_0^\infty \frac{x^{r-1}(1+x^2)}{(1+x^2)(1+x^{2r})}\,dx = \int_0^\infty \frac{x^{r-1}}{1+x^{2r}}\,dx = \frac{\pi}{2r},$$

Moreover, applying the substitution $x = 1/u$ in I, we find

$$I = \int_0^\infty \frac{u^{r+1}}{(1+u^2)(1+u^{2r})}\, du = J.$$

From which we easily obtain $I = \frac{\pi}{4r}$ again.   □

Remark. The change of variable $x \to \tan x$ converts I into

$$I = \int_0^{\pi/2} \frac{1}{(\tan^r x + \cot^r x)\tan x}\, dx.$$

In [45], Entry 3.688.14 states that

$$\int_0^{\pi/2} \frac{1}{(\tan^p x + \cot^p x)^q \tan x}\, dx = \frac{\sqrt{\pi}}{2^{2q+1}p}\frac{\Gamma(q)}{\Gamma(q+1/2)} \qquad \text{(for } q > 0\text{).} \quad (1.29)$$

Applying this formula to I with $p = r$ and $q = 1$ gives an incorrect answer

$$I = \frac{\sqrt{\pi}}{8r}\frac{\Gamma(1)}{\Gamma(3/2)} = \frac{1}{4r}.$$

Moreover, recall that, if $q \in \mathbb{N}$,

$$\Gamma\left(q + \frac{1}{2}\right) = \frac{(2q-1)!!}{2^q}\sqrt{\pi}.$$

This implies the answer in (1.29) is independent of π. By *Mathematica*, we find

$$\int_0^{\pi/2} \frac{1}{(\tan^2 x + \cot^2 x)^3 \tan x}\, dx = \frac{\pi}{64}, \qquad \int_0^{\pi/2} \frac{1}{(\tan^3 x + \cot^3 x)^3 \tan x}\, dx = \frac{\pi}{96}.$$

Now, using the idea in Solution III, we revise (1.29), for $q > 0$, as

$$\int_0^{\pi/2} \frac{1}{(\tan^p x + \cot^p x)^q \tan x}\, dx = \frac{1}{4p}\frac{\Gamma^2(q/2)}{\Gamma(q)} = \frac{\sqrt{\pi}}{2^{q+1}p}\frac{\Gamma(q/2)}{\Gamma((q+1)/2)}.$$

To this end, let

$$I(p,q) := \int_0^{\pi/2} \frac{1}{(\tan^p + \cot^p x)^q \tan x}\, dx.$$

We first rewrite

$$I(p,q) = \int_0^{\pi/2} \frac{\tan^{pq-1} x}{(1+\tan^{2p} x)^q}\, dx.$$

Next, applying the substitution $u = \tan x$, we obtain

$$I(p,q) = \int_0^\infty \frac{u^{pq-1}}{(1+u^{2p})^q(1+u^2)}\, du.$$

Now, we associate $I(p, q)$ with

$$J(p, q) := \int_0^\infty \frac{u^{pq+1}}{(1 + u^{2p})^q (1 + u^2)} \, du.$$

Therefore,

$$
\begin{aligned}
I + J &= \int_0^\infty \frac{u^{pq-1}(1 + u^2)}{(1 + u^{2p})^q (1 + u^2)} \, du \\
&= \int_0^\infty \frac{u^{pq-1}}{(1 + u^{2p})^q} \, du = \frac{1}{2p} \int_0^\infty \frac{t^{q/2-1}}{(1 + t)^q} \, dt,
\end{aligned}
$$

where $t = u^{2p}$ has been used in the last integral. Recall the Euler beta function

$$B(\alpha, \beta) = \int_0^\infty \frac{t^{\alpha-1}}{(1 + t)^{\alpha+\beta}} \, dt.$$

Let $\alpha = \beta = q/2$. Then

$$I + J = \frac{1}{2p} B\left(\frac{q}{2}, \frac{q}{2}\right) = \frac{1}{2p} \frac{\Gamma^2(q/2)}{\Gamma(q)}$$

Applying the substitution $u \to 1/u$ in I yields

$$I(p, q) = \int_0^\infty \frac{u^{pq+1}}{(1 + u^{2p})^q (1 + u^2)} \, du = J(p, q).$$

Hence,

$$I(p, q) = \frac{1}{2}(I + J) = \frac{1}{4p} \frac{\Gamma^2(q/2)}{\Gamma(q)}.$$

By Legendre's duplication formula $\Gamma(z)\Gamma(z+1/2) = 2^{1-2z}\sqrt{\pi}\,\Gamma(2z)$, we obtain

$$I(p, q) = \frac{\sqrt{\pi}}{2^{q+1}p} \frac{\Gamma(q/2)}{\Gamma((q + 1)/2)},$$

as claimed.

Additional problems for practice.

For $r > 0$, let us define the integral function

$$I(r) := \int_0^\infty \frac{x^{r-1}}{(1 + x^2)(1 + x^{2r})} \, dx.$$

1. Evaluate

$$\int_0^{10} \frac{\sqrt{\ln(2025 - x)}}{\sqrt{\ln(2025 - x)} + \sqrt{\ln(2015 + x)}} \, dx.$$

2. Evaluate

$$\int_{-1}^1 \frac{dx}{(e^x + 1)(1 + x^2)}.$$

3. (4-th Solution). Let

$$J = \int_0^\infty \frac{x^{r-1}}{1+x^{2r}}\, dx.$$

Show that $I(r) = \frac{J}{2} = \frac{\pi}{4r}$.

4. (5-th Solution). Let

$$T = \int_0^\infty \frac{x\arctan(x^r)}{(1+x^2)^2}\, dx.$$

Show that $I(r) = \frac{2T}{r}$ and $T = \frac{\pi}{8}$.

5. (6-th Solution). Let

$$F(p) = \int_0^\infty \frac{dx}{(1+x^2)(1+x^{2p})}.$$

Show that $F'(p) = 0$ and $I(r) = \frac{F(1/r)}{r}$.

6. (7-th Solution). Let $p = 2/r$. Show that

$$I(r) = \frac{1}{r}\int_0^{\pi/2} \frac{\cos^p\theta\, d\theta}{\cos^p\theta + \sin^p\theta} = \frac{1}{r}\int_0^{\pi/2} \frac{\sin^p\theta\, d\theta}{\cos^p\theta + \sin^p\theta}.$$

Using this result proves $I(r) = \frac{\pi}{4r}$.

7. Show that

$$I(p,q) = \int_0^{\pi/2} \frac{1}{(\tan^p x + \cot^p x)^q}\frac{dx}{\cot x} = \int_0^{\pi/2} \frac{1}{(\tan^p x + \cot^p x)^q}\frac{dx}{\sin 2x}.$$

8. (Due to A. Sofo). For $s \geq 0, r \neq 0$, show that

$$\int_0^\infty \frac{x^{r-1}\ln^{2s}(x)}{(1+x^2)(1+x^{2r})}\, dx = \frac{(-1)^s E_{2s}}{4^{s+1}|r|^{2s+1}}\pi^{2s+1}$$

where E_{2s} are the Euler numbers.

9. For $q > 0$, show that

$$\int_0^\infty \frac{\arctan x}{x}\left(\frac{x^p}{1+x^{2p}}\right)^{2q} dx = \frac{\pi^{3/2}}{2^{2q+2}p}\frac{\Gamma(q)}{\Gamma(q+1/2)}.$$

1.14 An inequality on the Wallis integrals

Problem 12413 (Proposed by S. Lee, 130(8), 2023). For a positive real number r, let $I_r = \int_0^{\pi/2} \sin^r \theta \, d\theta$. Prove

$$\frac{1}{(r+1)^2} + I_{r+1}^2 < \left(\frac{r+3}{r+2}\right)^2 I_r^2$$

for all $r \geq 1$.

Discussion.
In analysis, I_r are often called the Wallis integrals for $r \in \mathbb{N}$. It is well-known, see for example [23, p. 113], that

$$I_{2n} = \frac{\pi}{2^{2n+1}} \binom{2n}{n} \quad \text{and} \quad I_{2n+1} = \frac{2^{2n}}{2n+1} \binom{2n}{n}^{-1}.$$

In general, for $r \in \mathbb{R}$, I_r can be calculated in terms of the Euler beta-gamma functions. In fact, let $t = \sin^2 \theta$. Then

$$I_r = \frac{1}{2} \int_0^1 t^{\frac{r-1}{2}} (1-t)^{-\frac{1}{2}} \, dt = \frac{1}{2} B\left(\frac{r+1}{2}, \frac{1}{2}\right) = \frac{\sqrt{\pi}}{2} \frac{\Gamma((r+1)/2)}{\Gamma(1+r/2)}. \quad (1.30)$$

We will present three solutions. The first two solutions are based on the formula (1.30). The third solution will link I_r to the centroid of mass with a particular density distribution. The idea was originated from S. Lee, the proposer of this problem.

Solution I.
By (1.30), using $\Gamma(1 + (r+1)/2) = \frac{r+1}{2}\Gamma((r+1)/2)$, we find

$$I_r I_{r+1} = \frac{\pi}{2(r+1)}.$$

Since I_r is decreasing in r, we have

$$\frac{1}{(r+1)^2} + I_{r+1}^2 < \frac{1}{(r+1)^2} + I_{r+1} I_r = \frac{1}{(r+1)^2} + \frac{\pi}{2(r+1)}$$

$$= \frac{\pi}{2(r+1)}\left(1 + \frac{2}{\pi(r+1)}\right) < I_r I_{r+1}\left(1 + \frac{1}{\pi(1+r)}\right)^2$$

$$< \left(1 + \frac{1}{3(1+r)}\right)^2 I_r^2.$$

Here the coefficient of I_r^2 is actually tighter than the required inequality because if $r > -1/2$ we have

$$1 + \frac{1}{3(1+r)} = \frac{r+4/3}{r+1} < \frac{r+3}{r+2}.$$

□

Solution II.

We show that the inequality indeed holds for all $r > 0$. Recall Wendel's inequality (see [101]): for $0 < s < 1$ and $x > 0$, we have that

$$x^{1-s} \leq \frac{\Gamma(x+1)}{\Gamma(x+s)} \leq (x+s)^{1-s}.$$

For $r > 0$, applying this inequality with $x = r/2$ and $s = 1/2$, and $\Gamma(1+x) = x\Gamma(x)$ for $x > 0$, we find

$$I_r^2 = \frac{\pi}{4} \left(\frac{\Gamma(r/2+1)}{\Gamma(r/2+1/2)} \right)^{-2} \geq \frac{\pi}{4} \left[\left(\frac{r}{2} + \frac{1}{2} \right)^{1/2} \right]^{-2} = \frac{\pi}{2(r+1)}$$

and

$$I_{r+1}^2 = \frac{\pi}{4} \left(\frac{\Gamma((r+2)/2)}{\Gamma((r+3)/2)} \right)^2 = \frac{\pi}{(r+1)^2} \left(\frac{\Gamma(r/2+1)}{\Gamma(r/2+1/2)} \right)^2$$

$$\leq \frac{\pi}{(r+1)^2} \frac{r+1}{2} = \frac{\pi}{2(r+1)}.$$

Now it suffices to prove that

$$\frac{1}{(r+1)^2} + \frac{\pi}{2(r+1)} < \left(\frac{r+3}{r+2} \right)^2 \frac{\pi}{2(r+1)}.$$

This is equivalent to

$$2(\pi - 1)r^2 + (7\pi - 8)r + (5\pi - 8) > 0,$$

which is evident for $r > 0$.   □

Solution III.

Let $D = \{(x, y) : x \geq 0, y \geq 0, \text{and } x^2 + y^2 \leq 1\}$. We define a density function on D by $\rho(x, y) = y^r$. In the polar coordinates: $(x, y) = (s\cos\theta, s\sin\theta)$, the mass of D is

$$m = \iint_D y^r \, dA = \int_0^{\pi/2} \int_0^1 s^{r+1} \sin^r \theta \, ds d\theta = \frac{1}{r+2} I_r.$$

Similarly, we have

$$\iint_D x\rho(x, y) \, dA = \int_0^{\pi/2} \int_0^1 s^{r+2} \cos\theta \sin^r \theta \, ds d\theta = \frac{1}{(r+1)(r+3)}$$

and

$$\iint_D y\rho(x, y) \, dA = \int_0^{\pi/2} \int_0^1 s^{r+2} \sin^{r+1} \theta \, ds d\theta = \frac{1}{r+3} I_{r+1},$$

from which the center of mass of D is given by

$$x_c = \frac{1}{m} \iint_D x\rho(x,y)\,dA = \frac{r+2}{(r+1)(r+3)I_r},$$

$$y_c = \frac{1}{m} \iint_D y\rho(x,y)\,dA = \frac{r+2}{r+3} \cdot \frac{I_{r+1}}{I_r}.$$

Since (x_c, y_c) is strictly inside D, we must have

$$x_c^2 + y_c^2 = \frac{(r+2)^2}{(r+1)^2(r+3)^2 I_r^2} + \frac{(r+2)^2}{(r+3)^2} \cdot \frac{I_{r+1}^2}{I_r^2} < 1.$$

Multiplying $(r+3)^2 I_r^2/(r+2)^2$ on both sides yields the required inequality

$$\frac{1}{(r+1)^2} + I_{r+1}^2 < \left(\frac{r+3}{r+2}\right)^2 I_r^2$$

$\square$

Remark. Using $I_{r+1}^2 < I_r^2$, we can prove a stronger version: for $r \geq 0$,

$$\frac{1}{(r+1)^2} + I_r^2 < \left(\frac{r+3}{r+2}\right)^2 I_r^2.$$

Since

$$\left(\frac{r+3}{r+2}\right)^2 - 1 = \frac{2}{r+2} + \left(\frac{1}{r+2}\right)^2 > \frac{2}{r+2} \geq \frac{1}{r+1},$$

it suffices to prove

$$\frac{1}{(r+1)^2} \leq \frac{1}{r+1} I_r^2,$$

which is equivalent to $I_r^2 \geq 1/(r+1)$. This easily follows from

$$I_r^2 > I_r I_{r+1} = \frac{\pi}{2(r+1)} > \frac{1}{r+1}.$$

Based on the approach used in Solution II, the inequality can also be improved as

$$\frac{\alpha}{(r+1)^2} + I_{r+1}^2 < \left(\frac{r+3}{r+2}\right)^2 I_r^2,$$

where $\alpha \leq 5\pi/8 = 1.963495\ldots$.

Additional problems for practice.

1. Prove $\lim_{r \to \infty} r I_r^2 = \pi/2$.

2. Show

$$\frac{1}{(r+1)^2} + I_{r+1}^2 < \left(\frac{r+3}{r+2}\right)^2 I_r^2$$

holds for all $r > -1$.

3. For $x > 0$, prove
$$\frac{\Gamma(x+1)}{\Gamma(x+1/2)} < \frac{2x+1}{\sqrt{4x+3}}.$$

4. For $x \geq 1$, prove
$$\sqrt{x+1/4} < \frac{\Gamma(x+1)}{\Gamma(x+1/2)} \leq \alpha\sqrt{x+1/4},$$
where $\alpha = 4/\sqrt{5\pi} = 1.009253\ldots$.

5. Let f be positive, increasing and integrable on $[a, b]$. Define
$$x_c = \frac{\int_a^b xf(x)\,dx}{\int_a^b f(x)\,dx}.$$
Prove that
$$\int_a^{x_c} f(x)\,dx \leq \int_{x_c}^b f(x)\,dx.$$

6. **Problem 12312** (Proposed by M. Tchernookov, 129(3), 2022). Find all continuous functions $f : [0, \infty) \to \mathbb{R}$ such that, for all positive x,
$$f(x)\left(f(x) - \frac{1}{x}\int_0^x f(t)\,dt\right) \geq (f(x) - 1)^2.$$

7. **Problem 12340** (Proposed by A. Garcia, 129(7), 2022). Let $g : [0, 1] \to \mathbb{R}$ be continuous. Prove that
$$\lim_{n\to\infty} \frac{n}{2^n}\int_0^1 \frac{g(x)}{x^n + (1-x)^n}\,dx = Cg(1/2)$$
for some constant C (independent of g) and determine the value of C.

1.15 A sequence defined by inequalities

Problem 12504 (Proposed by O. Kouba, 132(1), 2025). Let P be the set of real sequences $(a_1, a_2, \ldots)$ such that $a_n > 0$ and $a_{n+1} + n \leq 2\sqrt{(n+1)a_n}$ for all n. Given $(a_1, a_2, \ldots)$, let $b_n = a_n - n - 1$.

(a) Prove that if $(a_1, a_2, \ldots) \in P$, then the sequence $(b_1, b_2, \ldots)$ is nonincreasing and converges to 0.

(b) For which real numbers x does there exists a sequence $(a_1, a_2, \ldots) \in P$ with $a_1 = x$?

(c) Prove that if $(a_1, a_2, \ldots) \in P$, then $b_n = O(1/\ln(n))$.

Discussion.
Except for the log-concave sequence (see [30, pp. 222-225]), we rarely encounter a sequence which is defined by inequalities. So let's take a closer look at the sequences in P. Clearly, $\{a_n = n+1\}_{n \geq 1} \in P$. The AM-GM inequality, together with the assumption, yields

$$2\sqrt{na_{n+1}} \leq a_{n+1} + n \leq 2\sqrt{(n+1)a_n}.$$

Squaring both sides yields $na_{n+1} \leq (n+1)a_n$, which implies that the sequence $\{a_n/n\}_{n \geq 1}$ is nonincreasing. Moreover, we will show that if $\{a_n\}_{n \geq 1} \in P$ then $a_n \geq n+1$. This fact will play a key role in our following solution.

Solution.
(a) Note that $a_n = b_n + n + 1$. We rewrite $a_{n+1} + n \leq 2\sqrt{(n+1)a_n}$ as

$$2(n+1) + b_{n+1} \leq 2\sqrt{(n+1)(n+1+b_n)},$$

then square both sides to get

$$b_{n+1}^2 \leq 4(n+1)(b_n - b_{n+1}). \tag{1.31}$$

This shows that b_n is nonincreasing. Next, based on the fact that a_n/n is nonincreasing which we proved in Discussion above, let $\lim_{n \to \infty} \frac{a_n}{n} = l$. It is evident that $l \geq 0$. Since

$$\frac{a_{n+1}}{n} + 1 \leq 2\sqrt{\left(1 + \frac{1}{n}\right)\frac{a_n}{n}},$$

letting $n \to \infty$ yields $l + 1 \leq 2\sqrt{l}$ or $(1 - \sqrt{l})^2 \leq 0$. Hence $l = 1$, $a_n \geq n$ for all $n \geq 1$, and consequently $b_n = a_n - n - 1 \geq -1$ for all $n \geq 1$. So the monotone convergence theorem implies that $\lim_{n \to \infty} b_n$ exists. By (1.31), we have

$$b_n - b_{n+1} \geq \frac{b_{n+1}^2}{4(n+1)}.$$

$\lim_{n \to \infty} b_n = 0$ now follows from the convergence of $\sum_{n=1}^{\infty}(b_n - b_{n+1})$.
(b) We show that $(x, a_2, \ldots) \in P$ if and only if $x \geq 2$. Based on the results of (a), if $(a_1, a_2, \ldots) \in P$, then $a_n - n - 1$ is nonincreasing and converges to 0. In particular, we have $a_1 - 2 = x - 2 \geq 0$. Thus, $x \geq 2$. On the other hand, for any $a_1 = x \geq 2$, let

$$a_n = n + 1 + (x - 2)\alpha^{n-1}, \tag{1.32}$$

where $0 < \alpha \leq 1$ is a constant to be determined. Note that $b_n = (x - 2)\alpha^{n-1}$.

By (1.31), $(x, a_2, \ldots) \in P$ if and only if

$$(x - 2)\alpha^{n+1} + 4(n + 1)\alpha \leq 4(n + 1). \tag{1.33}$$

When $n = 0$, solving $(x - 2)\alpha + 4\alpha - 4 = 0$ yields $\alpha = \frac{4}{x+2}$. For this α and the corresponding a_n defined by (1.32), rewrite (1.33) as

$$(x - 2)\alpha^{n+1} \leq 4(n + 1)(1 - \alpha) = 4(n + 1)\frac{x - 2}{x + 2},$$

which is equivalent to

$$\alpha^{n+1} \leq \frac{4(n + 1)}{x + 2}.$$

Since for every $n \geq 1$,

$$\alpha^{n+1} \leq \alpha^2 = \frac{4^2}{(x + 2)^2} \leq \frac{4}{x + 2} \leq \frac{4(n + 1)}{x + 2},$$

This shows that (1.33) holds for every $n \geq 1$, and so $(x, a_2, \ldots) \in P$.
(c) If $a_1 = 2$, we have $a_n = n + 1$ and so $b_n = 0$. For $x > 2$, define a sequence $\{c_n\}_{n \geq 1}$ by

$$c_1 = x, \quad c_{n+1} = 2\sqrt{(n + 1)c_n} - n \text{ for all } n \geq 1.$$

By (b), the sequence $(c_1, c_2, \ldots) \in P(x)$. Moreover, by (a), $r_n := c_n - n - 1$ is nonincreasing, converges to 0, and satisfies

$$r_{n+1}^2 = 4(n + 1)(r_n - r_{n+1}). \tag{1.34}$$

Hence

$$\frac{r_n}{r_{n+1}} = \left(1 + \frac{r_{n+1}}{4(n + 1)}\right) \leq \left(1 + \frac{r_2}{8}\right) = \frac{1}{2}\left(1 + \sqrt{\frac{x}{2}}\right).$$

This, together with (1.34), gives

$$\frac{1}{n + 1} = \frac{4r_n}{r_{n+1}}\left(\frac{1}{r_{n+1}} - \frac{1}{r_n}\right) \leq (2 + \sqrt{2x})\left(\frac{1}{r_{n+1}} - \frac{1}{r_n}\right).$$

Telescoping this yields

$$H_n - 1 \leq (2 + \sqrt{2x})\left(\frac{1}{r_n} - \frac{1}{r_1}\right) \leq (2 + \sqrt{2x})\frac{1}{r_n},$$

where H_n is the nth harmonic number. This proves that

$$r_n = O(1/H_n) = O(1/\ln(n)).$$

If $(x, a_2, \ldots) \in P$, we have $a_1 = c_1 = x$. If $a_n \leq c_n$, then

$$a_{n+1} \leq 2\sqrt{(n + 1)a_n} - n \leq 2\sqrt{(n + 1)c_n} - n = c_{n+1}.$$

By induction, we conclude that $a_n \leq c_n$ for all $n \geq 1$. By (a) again, we have $n + 1 \leq a_n \leq n + 1 + r_n$ for all $n \geq 1$. Hence

$$b_n = a_n - n - 1 = O(1/\ln(n)).$$

□

Remark. In the proof of (b), if we set the equality in (1.33) with $n = 1$, then $\alpha = 2/(1 + \sqrt{x/2})$, which also satisfies (1.33) for all $n \in \mathbb{N}$. Moreover, let $a_n = n + 1 + 4/H_n$. It is easy to check that $\{a_n\}_{n \geq 1} \in P$ by (1.31). Here we have $b_n = 4/H_n$. So the upper bound for $b_n = O(1/\ln(n))$ is optimal.

Additional problems for practice.

1. **Open Problem.** Replacing the inequality in Problem 12504 by

$$a_{n+1} + n \leq \frac{4(n+1)a_n}{a_n + n + 1},$$

do the properties (a), (b) and (c) in the problem for the sequences $(a_n)_{n \geq 1} \in P$ still hold?

2. **Putnam 1946-B4.** For each positive integer n, put

$$p_n = (1 + 1/n)^n, \quad P_n = (1 + 1/n)^{n+1}, \quad h_n = \frac{2p_n P_n}{p_n + P_n}.$$

Prove that $h_1 < h_2 < \cdots < h_n < \cdots$.

3. **Putnam 1948-A3.** Let $\{a_n\}_{n \geq 1}$ be a decreasing sequence of positive numbers with limit 0 such that

$$b_n = a_n - 2a_{n+1} + a_{n+2} \geq 0$$

for all n. Prove that $\sum_{n=1}^{\infty} nb_n = a_1$.

4. **Putnam 1994-A1.** Suppose that a sequence $\{a_n\}_{n \geq 1}$ satisfies $0 < a_n \leq a_{2n} + a_{2n+1}$ for all $n \geq 1$. Prove that the series $\sum_{n=1}^{\infty} a_n$ diverges.

5. **Putnam 2000-A6.** Let $f(x)$ be a polynomial with integer coefficients. Define a sequence $\{a_n\}_{n \geq 0}$ of integers such that $a_0 = 0$ and $a_{n+1} = f(a_n)$ for all $n \geq 0$. Prove that if there exists a positive integer m for which $a_m = 0$ then either $a_1 = 0$ or $a_2 = 0$.

6. **Putnam 2001-B6.** Assume that $\{a_n\}_{n \geq 1}$ is an increasing sequence of positive real numbers such that $\lim \frac{a_n}{n} = 0$. Must there exist infinitely many positive integers n such that

$$a_{n-i} + a_{n+i} < 2a_n \quad \text{for } i = 1, 2, \ldots, n - 1?$$

7. **Putnam 2016-A2.** Given a positive integer n, let $M(n)$ be the largest integer m such that

$$\binom{m}{n-1} > \binom{m-1}{n}.$$

Evaluate

$$\lim_{n \to \infty} \frac{M(n)}{n}.$$

Chapter 2

Identities

In this chapter, we compiled 12 identities which range from analysis and number theory to combinatorics and complex variables. We try to present a broad variety of strategies in proofs even to solve the same type of problems. Among them, we will meet Wilf's "snake oil method", Euler's number-theoretic functions, q-series, integer partitions, elementary symmetric functions, Jacobi's triple product identity, formal power series and differential operators. For problems related to the authors' research, we have given either their extensions or references, and hope this serves as a motivation for the reader to pursue further independent study.

2.1 Two identities concerning Fibonacci and Lucas numbers

Problem 12160 (Proposed by H. Ohtsuka and R. Tauraso, 127(2), 2020). Let F_n be the nth Fibonacci number, and let L_n be the nth Lucas number. Prove

$$\sum_{k=0}^{n} \binom{2n+1}{n-k} F_{2k+1} = 5^n \quad \text{and} \quad \sum_{k=0}^{n} \binom{2n+1}{n-k} L_{2k+1} = \sum_{k=0}^{n} \binom{2k}{k} 5^{n-k}$$

for all $n \in \mathbb{N}$.

Discussion.
Let m be an integer and $s(n)$ be a combinatorics sum for each integer $n \geq m$. To condense $s(n)$ into a closed form, if induction is out of the question, we usually turn to the method of generating functions. This method enables us to find a simple formula for $\sum_{n \geq m} s(n)x^n$, whose coefficients are the sequence that we are investigating. In particular, if $s(n)$ involves sums of binomial coefficients, H. Wilf offered the so-called "*snake oil method*" (see [102, Section 4.3]) as follows:

DOI: 10.1201/9781003607809-2

(a) Let $s(n) = \sum_k g(n,k)$. Introduce the generating function

$$F(x) = \sum_{n \geq m} s(n)x^n = \sum_{n \geq m} \sum_k g(n,k)x^n.$$

(b) Interchange the order of the two summations to obtain

$$F(x) = \sum_k \left(\sum_{n \geq m} g(n,k)x^n \right).$$

(c) Find the inner sum in simple closed form.

(d) Identify the coefficients of the generating function of the answer.

The success of this method depends on favorable outcomes of steps (c) and (d). For this purpose it will be luck for us to encounter some well-known series for the interchanged sum.

Before proceeding to the desired identities by the snake oil method, we recall the following known identity: for any nonnegative integer m,

$$g_m(x) := \sum_{n=0}^{\infty} \binom{2n+m}{n} x^n = \frac{(C(x))^m}{\sqrt{1-4x}} \tag{2.1}$$

where $C(x) = 2/(1 + \sqrt{1-4x})$ is the generating function of the Catalan numbers (see for example (5.72) in [47]). We will use (2.1) in step (c) of the solution. For the sake of completeness we give a sketch of the proof of (2.1). We have that

$$g_0(x) = \sum_{n=0}^{\infty} \binom{2n}{n} x^n = \sum_{n=0}^{\infty} \binom{-1/2}{n}(-4x)^n = \frac{1}{\sqrt{1-4x}}$$

and

$$g_1(x) = \sum_{n=0}^{\infty} \binom{2n+1}{n} x^n = \frac{1}{2} \sum_{n=0}^{\infty} \binom{2n+2}{n+1} x^n = \frac{g_0(x)-1}{2x}$$

which implies that $g_1(x) = C(x)/\sqrt{1-4x}$. Moreover, for $m \geq 1$,

$$g_{m+1}(x) - 1 = \sum_{n=1}^{\infty} \binom{2n+m+1}{n} x^n$$

$$= \sum_{n=1}^{\infty} \left(\binom{2n+m}{n} + \binom{2(n-1)+m+2}{n-1} \right) x^n$$

$$= g_m(x) - 1 + x g_{m+2}(x),$$

and by solving the above linear recursion of second order with the initial values $g_0(x)$ and $g_1(x)$, we arrive at (2.1).

One more challenge of the desired problem is the extra F_{2k+1} and L_{2k+1} in the summands, beside the binomial coefficients. Because F_{2k+1} and L_{2k+1} both share Binet type formulas, this suggests the use of the two variable version of the snake oil method.

Solution.
Let $f(x, y)$ be the generating function of the sequence $\left\{ \sum_{k=0}^{n} \binom{2n+1}{n-k} y^{2k+1} \right\}_{n \geq 0}$,

$$f(x, y) = \sum_{n=0}^{\infty} x^n \sum_{k=0}^{n} \binom{2n+1}{n-k} y^{2k+1}.$$

Then, after switching the order of summation and replacing $n - k$ with n, by (2.1) we obtain,

$$\begin{aligned}
f(x, y) &= \sum_{k=0}^{\infty} y^{2k+1} \sum_{n=k}^{\infty} \binom{2n+1}{n-k} x^n \\
&= \sum_{k=0}^{\infty} x^k y^{2k+1} \sum_{n=0}^{\infty} \binom{2n + (2k+1)}{n} x^n \\
&= \frac{1}{\sqrt{1-4x}} \sum_{k=0}^{\infty} x^k y^{2k+1} (C(x))^{2k+1} \\
&= \frac{yC(x)}{\sqrt{1-4x}\,(1 - xy^2 C^2(x))}.
\end{aligned}$$

By specializing y to the *golden ratio* $\phi_+ := (1 + \sqrt{5})/2$ and $\phi_- := 1 - \phi_+ = (1 - \sqrt{5})/2$, respectively, the function f reduces to

$$f(x, \phi_+) = \frac{1}{2(1 - 5x)} \left(\frac{1}{\sqrt{1-4x}} + \sqrt{5} \right)$$

and

$$f(x, \phi_-) = \frac{1}{2(1 - 5x)} \left(\frac{1}{\sqrt{1-4x}} - \sqrt{5} \right).$$

Finally, in view of Binet's formula

$$F_{2k+1} = \frac{\phi_+^{2k+1} - \phi_-^{2k+1}}{\sqrt{5}} \quad \text{and} \quad L_{2k+1} = \phi_+^{2k+1} + \phi_-^{2k+1},$$

we find

$$\sum_{n=0}^{\infty} x^n \sum_{k=0}^{n} \binom{2n+1}{n-k} F_{2k+1} = \frac{f(x, \phi_+) - f(x, \phi_-)}{\sqrt{5}} = \frac{1}{1 - 5x}$$

$$= \sum_{n=0}^{\infty} 5^n x^n,$$

and

$$\sum_{n=0}^{\infty} x^n \sum_{k=0}^{n} \binom{2n+1}{n-k} L_{2k+1} = f(x, \phi_+) + f(x, \phi_-) = \frac{1}{\sqrt{1-4x}} \cdot \frac{1}{(1-5x)}$$

$$= \sum_{n=0}^{\infty} x^n \sum_{k=0}^{n} \binom{2k}{k} 5^{n-k}$$

as claimed.   □

Remark. Another approach would be to consider the sequences

$$x_{\pm}(n) = 2 \sum_{k=0}^{n} \binom{2n+1}{n-k} \phi_{\pm}^{2k+1}$$

and show that they both satisfy the recursion

$$x_{\pm}(n+1) = 5x_{\pm}(n) + \binom{2n}{n}.$$

From which, it is not difficult to find that

$$x_{\pm}(n) = \pm 5^n \sqrt{5} + \sum_{k=0}^{n} \binom{2k}{k} 5^{n-k}$$

and show the identities by using $\phi_+^n = (L_n + \sqrt{5} F_n)/2$. We leave the details to the interested reader.

Additional problems for practice.

1. (Due to S. Marivan) For all $n, a, b \in \mathbb{N}$, show that

$$\sum_{k=0}^{n} \binom{n}{k} F_{4ak+b} = F_{2an+b} \cdot L_{2a}^n.$$

2. For all $n \in \mathbb{N}$, prove

$$\sum_{k=0}^{\lfloor n/2 \rfloor} \binom{n}{2k+1} 5^k = 2^{n-1} F_n \quad \text{and} \quad \sum_{k=0}^{\lfloor n/2 \rfloor} \binom{n}{2k} 5^k = 2^{n-1} L_n.$$

3. For all $n \in \mathbb{N}$, prove

$$\sum_{k=0}^{2n+1} \binom{2n+1}{k} F_k^2 = 5^n F_{2n+1} \quad \text{and} \quad \sum_{k=0}^{2n+2} \binom{2n+2}{k} L_k^2 = 5^{n+1} L_{2n+2}.$$

4. **Putnam 2021-A2.** Let k be a nonnegative integer. Evaluate

$$\sum_{j=0}^{k} \binom{k+j}{j} 2^{k-j}.$$

5. **Problem 12032** (Proposed by D. Galante and A. Plaza, 125(3), 2018). For a positive integer n, compute

$$\sum_{j=0}^{n}\sum_{k=j}^{n}(-1)^{k-j}\binom{k}{2j}\binom{n}{k}2^{n-k}.$$

Hint. Show the generating function is $(1-x)/(1-2x+2x^2)$. The answer leads to the sequence $A146559$ in OEIS ($\texttt{https://oeis.org/A146559}$).

6. **Problem 11274** (Proposed by D. E. Knuth, 114(2), 2007). Prove that for nonnegative integers m and n

$$\sum_{k=0}^{m}2^{k}\binom{2m-k}{m+n} = 4^{m} - \sum_{j=1}^{n}\binom{2m+1}{m+j}.$$

7. **Problem 11343** (Proposed by D. Beckwith, 115(2), 2008). Show that when n is a positive integer,

$$\sum_{k\geq 0}\binom{n}{k}\binom{2k}{k} = \sum_{k\geq 0}\binom{n}{2k}\binom{2k}{k}3^{n-2k}.$$

8. **Problem 11545** (Proposed by M. Kauers and S. Ko, 118(1), 2011). Find a closed-form expression for

$$\sum_{k=0}^{n}(-1)^{k}\binom{2n}{n+k}s(n+k,k),$$

where s refers to the (signed) Stirling numbers of the first kind.

9. **Problem 11791** (Proposed by M. Štofka, 121(7), 2014). Show that for $n \geq 1$,

$$\sum_{k=1}^{n}\binom{6n+1}{6k-2}B_{6k-2} = -\frac{6n+1}{6},$$

where B_n denotes the nth Bernoulli number.

10. **Problem 12262** (Proposed by L. Zhou, 128(6), 2021). For a nonnegative integer m, let

$$A_m = \sum_{k=0}^{\infty}\left(\frac{1}{(6k+1)^{2m+1}} - \frac{1}{(6k+5)^{2m+1}}\right).$$

Prove $A_0 = \pi\sqrt{3}/6$ and, for $m \geq 1$,

$$2A_m + \sum_{n=1}^{m}\frac{(-1)^{n}\pi^{2n}}{(2n)!}A_{m-n} = \frac{(-1)^{m}(4^{m}+1)\sqrt{3}}{2(2m)!}\left(\frac{\pi}{3}\right)^{2m+1}.$$

2.2 Summing Catalan numbers

Problem 12440 (Proposed by H. Ohtsuka, 131(2), 2024). Let $C_n = \frac{1}{n+1}\binom{2n}{n}$, the nth Catalan number. Prove

$$\sum_{k=0}^{n-1} C_k = 2 \sum_{k=1}^{n} \binom{2n}{n-k} \sin\left(\frac{(4k+1)\pi}{6}\right).$$

Discussion.
We first notice that the right-hand side can be written as

$$\sum_{k=1}^{n} r_k \binom{2n}{n-k}$$

where $r_{3k} = 1, r_{3k+1} = 1$, and $r_{3k+2} = -2$.
The statement of the problem naturally suggests us to proceed by induction. In view of the binomial sums on each side of the desired identity, we will begin with the right side. Let $P(n)$ be the proposition which states the desired identity holds for n. The basic identity

$$\binom{n+1}{k} = \binom{n}{k-1} + \binom{n}{k}, \tag{2.2}$$

enables us to prove the truth of $P(n+1)$ based on the truth of $P(n)$. We can also proceed with this problem by using formal power series. Notice that the right side is in the form of Cauchy product:

$$[x^n]\left(\sum_{k=1}^{\infty} r_k x^k\right) \cdot \left(\sum_{i=0}^{2n} \binom{2n}{i} x^i\right) = \sum_{k=1}^{n} r_k \binom{2n}{n-k}$$

It then suffices to prove that

$$[x^n]\left(\sum_{k=1}^{\infty} r_k x^k\right) \cdot \left(\sum_{i=0}^{2n} \binom{2n}{i} x^i\right) = \sum_{k=0}^{n-1} C_k. \tag{2.3}$$

Based on the above discussion, we will present two solutions below.

Solution I.
We prove the desired identity holds for every positive integer n by induction. When $n = 1$, we have

$$\sum_{k=0}^{1-1} C_k = 1 = r_1 = \sum_{k=1}^{1} r_k \binom{2}{1-k}$$

so the base case is true. Now, we assume the identity holds for $n \geq 1$, and show it holds for $n+1$ too. To this end, using (2.2) twice then shifting the index, we obtain

$$\sum_{k=1}^{n+1} r_k \binom{2(n+1)}{n+1-k} = \sum_{k=1}^{n+1} r_k \left(\binom{2n}{n-1-k} + 2\binom{2n}{n-k} + \binom{2n}{n+1-k} \right)$$

$$= \sum_{k=1}^{n-1} r_k \binom{2n}{n-(k+1)} + 2\sum_{k=1}^{n} r_k \binom{2n}{n-k} + \sum_{k=1}^{n+1} r_k \binom{2n}{n-(k-1)}$$

$$= \sum_{k=2}^{n} r_{k-1} \binom{2n}{n-k} + 2\sum_{k=1}^{n} r_k \binom{2n}{n-k} + \sum_{k=0}^{n} r_{k+1} \binom{2n}{n-k}$$

$$= 2\sum_{k=1}^{n} r_k \binom{2n}{n-k} + \sum_{k=1}^{n} (r_{k-1}+r_{k+1}) \binom{2n}{n-k} + r_1\binom{2n}{n} - r_0\binom{2n}{n-1}$$

$$= \sum_{k=1}^{n} r_k \binom{2n}{n-k} + \binom{2n}{n} - \binom{2n}{n-1}$$

where we have used $r_{k-1} + r_{k+1} = -r_k$, and $r_1 = r_0 = 1$ in the last equation. Finally, by the inductive hypothesis, we get

$$\sum_{k=1}^{n+1} r_k \binom{2(n+1)}{n+1-k} = \sum_{k=0}^{n-1} C_k + \frac{1}{n+1}\binom{2n}{n} = \sum_{k=0}^{n} C_k.$$

Thus, the desired identity is proved by induction. $\qquad\qquad\square$

Solution II.
To establish (2.3), we first find the generating function of $(r_k)_{k\geq 1}$ in closed form

$$\sum_{k=1}^{\infty} r_k x^k = \sum_{k=0}^{\infty} \left(x^{3k} + x^{3k+1} - 2x^{3k+2}\right) - 1 = \frac{x(1-x)}{1+x+x^2}.$$

Moreover, by the binomial theorem, $\sum_{i=0}^{2n} \binom{2n}{i} x^i = (1+x)^{2n}$, (2.3) is then transformed into

$$[x^n] \frac{x(1-x)(1+x)^{2n}}{1+x+x^2} = \sum_{k=0}^{n-1} C_k.$$

Since

$$\frac{x(1-x)(1+x)^{2n}}{1+x+x^2} = x(1-x)(1+x)^{2n-2} \frac{1}{1 - x/(1+x)^2}$$

$$= x(1-x)(1+x)^{2n-2} \sum_{i=0}^{\infty} \left(\frac{x}{(1+x)^2}\right)^i$$

$$= (1-x)\sum_{i=1}^{\infty} x^i (1+x)^{2n-2i}$$

$$= (1-x)\sum_{i=1}^{\infty} \left(\sum_{j=0}^{2n-2i} \binom{2n-2i}{j} x^{i+j}\right),$$

this implies that

$$[x^n]\left((1-x)\sum_{i=1}^{\infty}\left(\sum_{j=0}^{2n-2i}\binom{2n-2i}{j}x^{i+j}\right)\right) = \sum_{i=1}^{n}\left(\binom{2n-2i}{n-i} - \binom{2n-2i}{n-i-1}\right)$$

$$= \sum_{k=0}^{n-1}\left(\binom{2k}{k} - \binom{2k}{k-1}\right) = \sum_{k=0}^{n-1}C_k,$$

which confirms (2.3) as expected.   □

Remark. Catalan numbers $(C_n)_{n\geq 0} = 1, 1, 2, 5, 14, 42, \ldots$ can be found as the sequence $A000108$ in the OEIS (https://oeis.org/A000108). The wonderful book of R. P. Stanley [89] provides a comprehensive treatment of their properties and connections with modern combinatorics. Catalan numbers have a very large number of combinatorial interpretations. For example, Euler proved that C_n is the number of ways of cutting a convex polygon with $n+2$ sides into n triangles by connecting the labeled vertices with non-crossing diagonals. For $n = 4$ there are 14 possible ways as shown in Figure 2.1.

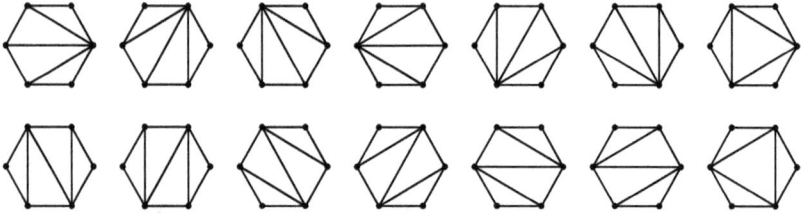

Figure 2.1: Triangulations of an hexagon

Given a $(n + 2)$-gon, we take two adjacent vertices. Now by picking one of the remaining n vertices, we form a triangle which splits the polygon into two subpolygons: one with $k+2$ vertices and the other with $n-k+1$ vertices for some $0 \leq k \leq n - 1$. These two polygons can be triangulated into C_k and C_{n-1-k} ways respectively. Hence, we find the recurrence

$$C_n = \sum_{k=0}^{n-1}C_kC_{n-1-k}$$

for $n \geq 1$, with $C_0 = 1$. From the above recurrence, it follows that the generating function $C(x) = \sum_{n\geq 0}C_nx^n$ satisfies the quadratic equation

$$C(x) = 1 + xC(x)^2$$

and so

$$C(x) = \frac{1 - \sqrt{1 - 4x}}{2x}.$$

Additional problems for practice.

1. Show that for $n \geq 1$,

$$\frac{1}{2}\sum_{k=0}^{n-1}\frac{C_k}{4^k} = 1 - \frac{\binom{2n}{n}}{4^n}.$$

2. Prove that for $n \geq 1$,

$$\sum_{k=1}^{n}(-1)^{k-1}\binom{n-k+1}{k}C_{n-k} = C_n.$$

3. Show that for $n \geq 0$,

$$\sum_{k=0}^{n}C_{2k}C_{2(n-k)} = 4^n C_n.$$

4. (Due to J. Touchard). Prove that for $n \geq 0$,

$$C_{n+1} = \sum_{k=0}^{\lfloor n/2 \rfloor}\binom{n}{2k}\cdot C_k \cdot 2^{n-2k}.$$

5. (Due to J. Cvijić). Show that for $n \geq 0$,

$$\sum_{k=0}^{2n}(-1)^k\binom{2n}{k}C_kC_{2n-k} = C_n\binom{2n}{n}.$$

6. **Problem 10264** (Proposed by L. Shapiro, 99(9), 1992). Let $C_n = \frac{1}{n+1}\binom{2n}{n}$ be the nth Catalan number and form the generating function $C(x) = \sum_{n\geq 0}C_n x^n$.

   Establish the identities

   (a) $\displaystyle\sum_{n\geq 0}(n+1)x^n C(x)^{2n+2} = \sum_{k\geq 0}(4x)^k.$

   (b) $\displaystyle\sum_{n\geq 0}(2n+1)x^n C(x)^{2n+1} = \sum_{k\geq 0}(4x)^k.$

7. **Problem 11832** (Proposed by D. E. Knuth, 122(4), 2015). Let $C(z)$ be the generating function of the Catalan numbers. Prove that

$$(\log(C(z)))^2 = \sum_{n=1}^{\infty}\binom{2n}{n}(H_{2n-1} - H_n)\frac{z^n}{n},$$

where $H_k = \sum_{j=1}^{k}1/j$ is the k-th harmonic number.

8. **Problem 11897** (Proposed by P. P. Dályay, 123(3), 2016). Prove for $n \geq 0$ that

$$\sum_{\substack{k+j=n \\ k,j \geq 0}} C_k \binom{2j+2}{j+1} = 2\binom{2n+2}{n}.$$

2.3   Perfect squares from an arithmetic function

Problem 11544 (Proposed by M. A. Alekseyev and F. Ruskey, 118(1), 2011).
Prove that if m is a positive integer, then

$$\sum_{k=0}^{m-1} \phi(2k+1)\left\lfloor \frac{m+k}{2k+1} \right\rfloor = m^2.$$

Here ϕ is the Euler totient function.

Discussion.
Let n be a positive integer. The Euler totient function (also often called the
Euler phi function) $\phi(n)$ is defined to be the number of positive integers not
exceeding n that are relative prime to n. It is well-known that

$$\sum_{d|n} \phi(d) = n, \qquad (2.4)$$

which states that n is the sum of values of the phi function at all the positive
divisors of n. One of the proofs of (2.4) is based on splitting the set of integers
from 1 to n into classes. Let

$$C_d = \{m : 1 \leq m \leq n \text{ and } \gcd(m,n) = d\},$$

and let $\mathrm{card}(C_d)$ be the number of integers $m \in C_d$. Each $m, 1 \leq m \leq n$, is
counted exactly once because $\gcd(m,n) = d$ is unique, so that

$$\sum_{d|n} \mathrm{card}(C_d) = n.$$

For example, when $n = 18$, we have

$$\begin{array}{ll}
C_1 = \{1,5,7,11,13,17\} & C_6 = \{6,12\} \\
C_2 = \{2,4,8,10,14,16\} & C_9 = \{9\} \\
C_3 = \{3,15\} & C_{18} = \{18\}.
\end{array}$$

On the other hand, $m \in C_d$, that is, $\gcd(m,n) = d$, if only if $\gcd(m/d, n/d) = 1$. Hence

$$\begin{aligned}
\phi(n/d) &= \mathrm{card}\{m : 1 \leq m \leq n/d \text{ and } \gcd(m, n/d) = 1\} \\
&= \mathrm{card}\{m : 1 \leq m \leq n \text{ and } \gcd(m,n) = d\}
\end{aligned}$$

and so

$$\sum_{d|n} \phi(d) = \sum_{d|n} \phi(n/d) = \sum_{d|n} \text{card}(C_d) = n.$$

By (2.4), we obtain

$$\sum_{k=0}^{m-1} \sum_{d|(2k+1)} \phi(d) = \sum_{k=0}^{m-1} (2k+1) = m^2. \qquad (2.5)$$

This transforms the desired problem into showing

$$\sum_{k=0}^{m-1} \sum_{d|(2k+1)} \phi(d) = \sum_{k=0}^{m-1} \phi(2k+1) \left\lfloor \frac{m+k}{2k+1} \right\rfloor.$$

Using the above splitting set approach, we will prove a more general identity in the following solution.

Solution.
We show the extension: Let $F(n) = \sum_{d|n} f(d)$. Then

$$\sum_{n=0}^{m-1} F(2n+1) = \sum_{k=0}^{m-1} f(2k+1) \left\lfloor \frac{m+k}{2k+1} \right\rfloor. \qquad (2.6)$$

To this end, replacing d by $2k+1$ and then interchanging the summation order, we obtain

$$\sum_{n=0}^{m-1} F(2n+1) = \sum_{n=0}^{m-1} \sum_{d|2n+1} f(d) = \sum_{n=0}^{m-1} \sum_{(2k+1)|(2n+1)} f(2k+1)$$

$$= \sum_{k=0}^{m-1} f(2k+1) \sum_{n=0,(2k+1)|(2n+1)}^{m-1} 1.$$

Here the last sum counts the number of odd multiples of $2k+1$ from 1 to $2m-1$, which is equal to the number of all the multiples of $2k+1$ minus that of the even multiples of $2k+1$. Therefore,

$$\sum_{n=0,(2k+1)|(2n+1)}^{m-1} 1 = \left\lfloor \frac{2m-1}{2k+1} \right\rfloor - \left\lfloor \frac{2m-1}{2(2k+1)} \right\rfloor = \left\lfloor \frac{m+k}{2k+1} \right\rfloor.$$

This proves (2.6). Setting $f(n) = \phi(n)$ in (2.6), together with (2.5), yields the desired identity. $\qquad \square$

Remark. After interchanging the order of summation in $\sum_{n=0}^{m-1} F(2n+1)$, there is another way to count the splitting set. Let α_d be the number of occurrences of $F(d)$ in the sum. Such terms occur when d is odd with $1 \leq d \leq 2m - 1$. Then

$$\sum_{n=0}^{m-1} \sum_{d|2n+1} f(d) = \sum_{\{1 \leq d \leq 2m-1, \, d \text{ is odd}\}} f(d) \cdot \alpha_d.$$

Since $f(d)$ appears once for each odd multiple of d not exceeding $2m - 1$, this yields $(2\alpha_d - 1)d \leq 2m - 1$, and so $\alpha_d = \lfloor \frac{2m+d-1}{2d} \rfloor$. In view of the fact that d is odd, we obtain

$$\alpha_d = \left\lfloor \frac{2m + d - 1}{2d} \right\rfloor = \left\lfloor \frac{2m + 2k + 1 - 1}{2(2k+1)} \right\rfloor = \left\lfloor \frac{m + k}{2k + 1} \right\rfloor,$$

which proves (2.6) again.

Similar identities have appeared in the literature. For example, Theorem 6.11 in [26, p. 199] states that if f and F are number-theoretic functions such that $F(n) = \sum_{d|n} f(d)$, then, for any positive integer N,

$$\sum_{n=1}^{N} F(n) = \sum_{k=1}^{N} f(k) \left\lfloor \frac{N}{k} \right\rfloor. \tag{2.7}$$

In addition, let $\tau(n)$ be the number of positive divisors of n and $\sigma_k(n)$ be the kth power sum of these divisors. As an application of (2.7), the following identities can be established:

$$\sum_{n=1}^{N} \tau(n) = \sum_{n=1}^{N} \left\lfloor \frac{N}{n} \right\rfloor,$$

$$\sum_{n=1}^{N} \sigma_k(n) = \sum_{n=1}^{N} n^k \left\lfloor \frac{N}{n} \right\rfloor,$$

$$\sum_{n=1}^{N} \phi(k) \left\lfloor \frac{N}{n} \right\rfloor = \binom{N + 1}{2}.$$

Additional problems for practice.

1. Prove the desired identity by induction. *Hint:* For any positive integers n and k,

$$\left\lfloor \frac{n + 1}{k} \right\rfloor - \left\lfloor \frac{n}{k} \right\rfloor = \begin{cases} 1, & \text{if } k \mid (n + 1), \\ 0, & \text{otherwise.} \end{cases}$$

2. If m is a positive integer, prove

$$\sum_{k=1}^{m} \phi(2k) \left\lfloor \frac{m+k}{2k} \right\rfloor = \frac{m(m+1)}{2},$$

$$\sum_{k=1}^{2m} \phi(k) \left\lfloor \frac{2m+k}{2k} \right\rfloor = \frac{m(3m+1)}{2},$$

$$\sum_{k=1}^{2m} \tau(k) = m + \sum_{k=1}^{m} \left\lfloor \frac{2m}{k} \right\rfloor.$$

3. Let $\mu(n)$ be the Möbius function. It is well-known that $\sum_{d|n} \mu(d) = 0$ for all positive integer $n > 1$. Prove that

$$\sum_{k=1}^{n} \mu(k) \left\lfloor \frac{n}{k} \right\rfloor = 1.$$

4. (Due to W. Chu) Show that

$$\sum_{n=0}^{m-1} \sigma_k(2n+1) = \sum_{n=0}^{m-1} (2n+1)^k \left\lfloor \frac{m+n}{2n+1} \right\rfloor.$$

5. **Problem 10829** (Proposed by W. Janous, 107(8), 2000). For a positive integer m, let

$$f(m) = \sum_{r=1}^{m} \frac{m}{\gcd(m,r)}.$$

Evaluate $f(m)$ in terms of the canonical factorization of m into a product of powers of distinct primes.

6. **Problem 11128** (Proposed by T. Apostol, 112(1), 2005). For positive integers m and n, let

$$S_m(n) = \sum_{k=1}^{n} \lfloor k^{1/m} \rfloor.$$

Find a closed-form expression in terms of *Bernoulli polynomials* for $S_m(n)$. (The generating function formula $\frac{te^{tx}}{e^t-1} = \sum_{j=0}^{\infty} B_j(x) \frac{t^j}{j!}$ determines the Bernoulli polynomials.)

2.4 Identity related to the parity constant

Problem 11685 (Proposed by D. E. Knuth, 120(1), 2013). Prove that

$$\prod_{k=0}^{\infty}\left(1+\frac{1}{2^{2^k}-1}\right)=\frac{1}{2}+\sum_{k=0}^{\infty}\frac{1}{\prod_{j=0}^{k-1}(2^{2^j}-1)}.$$

In other words, prove that

$$(1+1)\left(1+\frac{1}{3}\right)\left(1+\frac{1}{15}\right)\left(1+\frac{1}{255}\right)\cdots=\frac{1}{2}+1+1+\frac{1}{3}+\frac{1}{3\cdot15}+\frac{1}{3\cdot15\cdot255}+\cdots.$$

Discussion. First, we take advantage of the special structure of the left side of the desired identity to find the difference of consecutive nth partial products, and then telescope to yield the first solution.

On the other hand, the desired identity is a special case of the following identity

$$\prod_{n=0}^{\infty}\frac{1}{1-x^{2^n}}=1-x+2\sum_{k=0}^{\infty}\frac{x^{2^k}(1-x^{2^k})}{\prod_{j=0}^{k}(1-x^{2^j})}. \tag{2.8}$$

The infinite product, call it $F(x)$, in (2.8) can be viewed as a sub-product of $\prod_{n=1}^{\infty}1/(1-x^n)$ which gives the generating function of the integer partitions. Actually it is the generating function of the number of partitions of n into powers of 2 and it satisfies the equation

$$F(x^2)=\prod_{n=0}^{\infty}\frac{1}{1-x^{2^{n+1}}}=\prod_{n=1}^{\infty}\frac{1}{1-x^{2^n}}=(1-x)F(x).$$

This identity sheds light on our proof of (2.8).

Solution I.
Let $P_n=\prod_{k=0}^{n}\left(1+\frac{1}{2^{2^k}-1}\right)$, which is the nth partial product of the infinite product in the desired identity. Using

$$1+\frac{1}{2^{2^k}-1}=\frac{2^{2^k}}{2^{2^k}-1},$$

we find

$$P_n-P_{n-1}=P_{n-1}\left(1+\frac{1}{2^{2^n}-1}-1\right)=\prod_{k=0}^{n-1}\frac{2^{2^k}}{2^{2^k}-1}\left(\frac{1}{2^{2^n}-1}\right)$$

$$=\frac{\prod_{k=0}^{n-1}2^{2^k}}{\prod_{k=0}^{n}(2^{2^k}-1)}=\frac{2^{\sum_{k=0}^{n-1}2^k}}{\prod_{k=0}^{n}(2^{2^k}-1)}=\frac{2^{2^n-1}}{\prod_{k=0}^{n}(2^{2^k}-1)}$$

$$=\frac{1}{2}\frac{(2^{2^n}-1)+1}{\prod_{k=0}^{n}(2^{2^k}-1)}=\frac{1}{2}\left(\frac{1}{\prod_{j=0}^{n-1}(2^{2^j}-1)}+\frac{1}{\prod_{j=0}^{n}(2^{2^j}-1)}\right).$$

Note that the proposer defines $P_{-1} = 1$. Telescoping the above equation on n from 0 to N gives

$$P_N - 1 = \frac{1}{2} + \sum_{k=1}^{N} \frac{1}{\prod_{j=0}^{k-1} (2^{2^j} - 1)} + \frac{1}{2} \frac{1}{\prod_{j=0}^{N} (2^{2^j} - 1)}.$$

Letting $N \to \infty$ yields the desired identity. $\qquad\square$

Solution II.
Let

$$F(x) = \prod_{n=0}^{\infty} \frac{1}{1 - x^{2^n}} = \sum_{n=0}^{\infty} a_n x^n,$$

$$G(x) = 1 - x + 2 \sum_{k=0}^{\infty} \frac{x^{2^k} (1 - x^{2^k})}{\prod_{j=0}^{k}(1 - x^{2^j})} = \sum_{n=0}^{\infty} b_n x^n.$$

Clearly, the desired identity is equivalent to $F(1/2) = G(1/2)$. In the following, we prove that $a_n = b_n$ for all $n \geq 0$ by showing both a_n and b_n satisfy the same linear recursion with the same initial term.

The structure of F and G tells us a great deal about the coefficients of their power series. First, we have that equality $F(x^2) = (1 - x)F(x)$ implies

$$a_0 + \sum_{n=1}^{\infty} a_n x^{2n} = a_0 + \sum_{n=1}^{\infty} (a_n - a_{n-1}) x^n.$$

Matching the coefficients of like terms yields, for $n \geq 1$,

$$a_{2n+1} = a_{2n}, \quad a_{2n} = a_{2n-1} + a_n.$$

Next, we have

$$G(x^2) = 1 - x^2 + 2 \sum_{k=0}^{\infty} \frac{x^{2^{k+1}} (1 - x^{2^{k+1}})}{\prod_{j=0}^{k}(1 - x^{2^{j+1}})} = 1 - x^2 + 2 \sum_{k=0}^{\infty} \frac{x^{2^{k+1}} (1 - x^{2^{k+1}})}{\prod_{j=1}^{k+1}(1 - x^{2^j})}$$

$$= 1 - x^2 + 2(1 - x) \sum_{k=0}^{\infty} \frac{x^{2^{k+1}} (1 - x^{2^{k+1}})}{\prod_{j=0}^{k+1}(1 - x^{2^j})}$$

$$= 1 - x^2 + 2(1 - x) \sum_{k=1}^{\infty} \frac{x^{2^k} (1 - x^{2^k})}{\prod_{j=0}^{k}(1 - x^{2^j})}$$

$$= (1 - x) \left((1 + x) + 2 \left(\sum_{k=0}^{\infty} \frac{x^{2^k} (1 - x^{2^k})}{\prod_{j=0}^{k}(1 - x^{2^j})} - x \right) \right) = (1 - x)G(x).$$

Similarly, for $n \geq 1$, we obtain

$$b_{2n+1} = b_{2n}, \quad b_{2n} = b_{2n-1} + b_n.$$

Since $a_0 = F(0) = 1 = G(0) = b_0$, for all $n \geq 1$, $a_n = b_n$ follows from the uniqueness of the solution for the linear recursion. Therefore, $F(x) = G(x)$ for all $|x| < R$, where R is the radius of convergence of $F(x)$. Once we establish that $R \geq \sqrt{2}/2$, it follows that $F(1/2) = G(1/2)$ immediately. Since a_n is nondecreasing and $a_{2n-1} = a_{2(n-1)}$ for $n \geq 1$, we have

$$a_{2n} = a_{2n-1} + a_n = a_{2(n-1)} + a_n \leq 2a_{2(n-1)}.$$

By induction, we obtain that $1 \leq a_n \leq (\sqrt{2})^n$, form which $R \geq \sqrt{2}/2$ follows by the root test. $\qquad\square$

Remark. The reciprocal of the product in (2.8) is associated with the Thue-Morse sequence $\{t_n\}$ by

$$\prod_{n=0}^{\infty} \left(1 - x^{2^n} \right) = \sum_{n=0}^{\infty} (-1)^{t_n} x^n,$$

where $\{t_n\}$ is defined by the recursion:

$$t_0 = 0, \qquad t_{2n} = t_n, \qquad t_{2n+1} = 1 - t_n.$$

For each positive integer n, t_n gives the binary digit sum of n modulo 2. Let

$$\tau := \sum_{n=0}^{\infty} \frac{t_n}{2^{n+1}} = 0.4124540336\ldots,$$

which is called the *parity constant*. Hence

$$\prod_{k=0}^{\infty} \left(1 + \frac{1}{2^{2^k} - 1} \right) = F(1/2) = \frac{1}{2(1 - 2\tau)}.$$

In 1979, Woods and Robbins [104] proved the following surprising product:

$$\prod_{n=0}^{\infty} \left(\frac{2n+1}{2n+2} \right)^{(-1)^{t_n}} = \frac{\sqrt{2}}{2}.$$

Recently, by invoking the Dirichlet series associated with t_n, L. Tóth [99] derived closed forms for many sums in terms of the Riemann zeta function. We single out two elegant examples here:

$$\sum_{n=1}^{\infty} \frac{5t_{n-1} + 3t_n}{n^2} = 4\zeta(2),$$

$$\sum_{n=1}^{\infty} \frac{9t_{n-1} + 7t_n}{n^3} = 8\zeta(3).$$

This sequence has a rich history and plays important roles in many different areas of mathematics. For the interested reader, please refer to a survey article [5] and the details in [6].

Additional problems for practice.

1. Show that

$$\prod_{k=1}^{\infty} \frac{1}{1-x^{k^2}} = \sum_{k=0}^{\infty} \frac{x^{k^2}}{\prod_{j=0}^{k}(1-x^{j^2})}.$$

2. Let $n \geq 2$ be a positive integer, and let $x_1, \ldots, x_n$ be positive real numbers. Prove that

$$\left(\prod_{k=1}^{n}(1+x_k)\right)^{n-1} \geq \left(\prod_{1 \leq i < j \leq n}\left(1+\frac{2x_i x_j}{x_i + x_j}\right)\right)^2.$$

3. **Putnam 1990-B2.** Prove that for $|x| < 1, |z| > 1$,

$$1 + \sum_{j=1}^{\infty}(1+x^j)\frac{(1-z)(1-zx)(1-zx^2)\cdots(1-zx^{j-1})}{(z-x)(z-x^2)(z-x^3)\cdots(z-x^j)} = 0.$$

4. **Problem 10809** (Proposed by D. Beckwith, 107(6), 2000).

5. For $|x| < 1$, prove that

$$\sum_{n=0}^{\infty} \frac{x^{n(n+1)/2}}{1-x^n} = \sum_{n=0}^{\infty} \frac{x^n}{1-x^{2n}}.$$

6. **Problem 12022** (Proposed by M. Merca, 125(3), 2018). Let n be a positive integer, and let x be a real number not equal to -1 or 1. Prove

$$\sum_{k=0}^{n-1} \frac{(1-x^n)(1-x^{n-1})\cdots(1-x^{n-k})}{1-x^{k+1}} = n$$

and

$$\sum_{k=0}^{n-1}(-1)^k\frac{(1-x^n)(1-x^{n-1})\cdots(1-x^{n-k})}{1-x^{k+1}} \cdot x^{\binom{n+1-k}{2}} = nx^{\binom{n}{2}}.$$

7. **Problem 11210** (Proposed by M. Becker, 113(3), 2006). Show that

$$\prod_{n=0}^{\infty} \frac{(2n+1)^4}{(2n+1)^4 - (2/\pi)^4} = \frac{2e\sec(1)}{e^2+1}.$$

8. **Putnam 2004-B5.** Evaluate

$$\lim_{x \to 1^-} \prod_{n=0}^{\infty}\left(\frac{1+x^{n+1}}{1+x^n}\right)^{x^n}.$$

9. Let t_n be the nth term of the Thue-Morse sequence. Prove that there are no positive integers k and m such that

$$t_{k+j} = t_{k+m+j} = t_{k+2m+j},$$

for $0 \leq j \leq m - 1$.

10. **Open Problem.** Is the series $\sum_{n=1}^{\infty} t_n/n^z$ with $\mathrm{Re}(z) > 1$ an interesting series? i.e., does the series admit a closed form in terms of known constants?

2.5   Wrestle with Jacobi's triple product

Problem 12289 (Proposed by G. E. Andrews and M. Merca, 128(10), 2021). Prove

$$\sum_{n=0}^{\infty} 2\cos\left(\frac{(2n+1)\pi}{3}\right) q^{n(n+1)/2} = \prod_{n=1}^{\infty}(1 - q^n)(1 - q^{6n-1})(1 - q^{6n-5}),$$

when $|q| < 1$.

Discussion.
One of the most beautiful identities involving infinite products is perhaps *Jacobi's triple product identity*

$$\prod_{n=1}^{\infty}(1 + zq^{2n-1})(1 + z^{-1}q^{2n-1})(1 - q^{2n}) = \sum_{n=-\infty}^{\infty} z^n q^{n^2}. \qquad (2.9)$$

An elementary proof of it is due to the first proposer [10], who is viewed as the world's leading expert in the theory of q-series and integer partitions. He managed to show (2.9) by applying two identities that date back to Euler:

$$\prod_{k=1}^{\infty}(1 + zq^k) = \sum_{k=0}^{\infty} \frac{z^k q^{k(k+1)/2}}{(1 - q)(1 - q^2)\cdots(1 - q^k)}$$

and

$$\prod_{k=1}^{\infty} \frac{1}{1 - zq^k} = \sum_{k=0}^{\infty} \frac{z^k q^k}{(1 - q)(1 - q^2)\cdots(1 - q^k)}.$$

See also [11, Chapter 2]. Below, we illustrate how to wrestle (2.9) with substitutions to prove the desired identity.

Solution.
We first replace z by $z^2 q^{1/2}$ and q by $q^{1/2}$ in (2.9) to obtain

$$\prod_{n=1}^{\infty}(1 + z^2 q^n)(1 + z^{-2} q^{n-1})(1 - q^n) = \sum_{n=-\infty}^{\infty} z^{2n} q^{n(n+1)/2}.$$

For $n = 1$, separating the term $1 + z^{-2}q^{n-1} = 1 + z^{-2}$ from the product, then multiplying by z yields

$$(z + z^{-1}) \prod_{n=1}^{\infty} (1 + z^2 q^n)(1 + z^{-2}q^n)(1 - q^n) = \sum_{n=-\infty}^{\infty} z^{2n+1} q^{n(n+1)/2}.$$

Next, we substitute $z = e^{\pi i/3}$ on the right side to get

$$\sum_{n=-\infty}^{\infty} e^{(2n+1)\pi i/3} q^{n(n+1)/2}$$

$$= \sum_{n=0}^{\infty} e^{(2n+1)\pi i/3} q^{n(n+1)/2} + \sum_{n=-1}^{-\infty} e^{(2n+1)\pi i/3} q^{n(n+1)/2}$$

$$= \sum_{n=0}^{\infty} e^{(2n+1)\pi i/3} q^{n(n+1)/2} + \sum_{n=0}^{\infty} e^{-(2n+1)\pi i/3} q^{n(n+1)/2}$$

$$(\text{use } n \to -n, n - 1 \to n)$$

$$= \sum_{n=0}^{\infty} (e^{(2n+1)\pi i/3} + e^{-(2n+1)\pi i/3}) q^{n(n+1)/2}$$

$$= \sum_{n=0}^{\infty} 2 \cos\left(\frac{(2n+1)\pi}{3}\right) q^{n(n+1)/2}.$$

On the left side, correspondingly, we have

$$(z + z^{-1}) \prod_{n=1}^{\infty} (1 + z^2 q^n)(1 + z^{-2}q^n)(1 - q^n)$$

$$= \prod_{n=1}^{\infty} (1 + e^{2\pi i/3} q^n)(1 + e^{-2\pi i/3} q^n)(1 - q^n).$$

Since

$$(1 + e^{2\pi i/3} q^n)(1 + e^{-2\pi i/3} q^n) = 1 - q^n + q^{2n},$$

it suffices to prove

$$\prod_{n=1}^{\infty} (1 - q^n + q^{2n}) = \prod_{n=1}^{\infty} (1 - q^{6n-1})(1 - q^{6n-5}). \tag{2.10}$$

To this end, we rewrite (2.10) as

$$\prod_{n=1}^{\infty} (1 - q^n + q^{2n}) = \prod_{n=1}^{\infty} \frac{1 + q^{3n}}{1 + q^n} = \prod_{n=1}^{\infty} \frac{(1 - q^{6n})(1 - q^n)}{(1 - q^{3n})(1 - q^{2n})}. \tag{2.11}$$

Finally, since

$$\prod_{n=1}^{\infty}(1-q^{2n}) = \prod_{n=1}^{\infty}(1-q^{6n})(1-q^{6n-2})(1-q^{6n-4}),$$

$$\prod_{n=1}^{\infty}(1-q^{3n}) = \prod_{n=1}^{\infty}(1-q^{6n})(1-q^{6n-3}),$$

$$\prod_{n=1}^{\infty}(1-q^{n}) = \prod_{n=1}^{\infty}\prod_{j=0}^{5}(1-q^{6n-j}),$$

we obtain (2.10) by plugging these identities into (2.11) and removing the common factors. □

Remark. To see more applications of (2.9), we give a few interesting results that follow from this identity.

1. Letting $z = 1$ in (2.9) gives

$$\prod_{n=1}^{\infty}(1+q^{2n-1})^2(1-q^{2n}) = \sum_{n=-\infty}^{\infty} q^{n^2},$$

   which is the generating function for squares.

2. Letting $z = q$ in (2.9), then substituting q by $q^{1/2}$ yields

$$\prod_{n=1}^{\infty}(1+q^n)(1+q^{n-1})(1-q^n) = 2\prod_{n=1}^{\infty}(1+q^n)^2(1-q^n) = \sum_{n=-\infty}^{\infty} q^{n(n+1)/2}.$$

3. By replacing q by $q^{3/2}$ and then letting $z = -q^{1/2}$ in (2.9), leads to

$$\prod_{n=1}^{\infty}(1-q^{3n-1})(1-q^{3n-2})(1-q^{3n}) = \prod_{n=1}^{\infty}(1-q^n) = \sum_{n=-\infty}^{\infty}(-1)^n q^{n(3n+1)/2},$$

   which is the reciprocal of the generating function for $p(n)$, the number of partitions of n. By symmetry of n, replacing n by $-n$, the right hand side of the above series becomes

$$\sum_{n=-\infty}^{\infty}(-1)^n q^{n(3n-1)/2}.$$

   Since the integers in the form $n(3n-1)/2$ are known as pentagonal numbers, this presents a simple proof of the Euler's pentagonal number theorem.

Additional problems for practice.

1. Show that

$$\prod_{k=0}^{\infty} \frac{1}{1 - xq^k} = \sum_{k=1}^{\infty} \frac{x^k q^{k^2}}{(1 - q) \cdots (1 - q^k)(1 - xq) \cdots (1 - xq^k)}.$$

For a proof see [88, p. 65].

2. Prove that

$$\prod_{n=1}^{\infty} \frac{1 - q^n}{1 + q^n} = \sum_{n=-\infty}^{\infty} (-1)^n q^{n^2}.$$

3. **Quintuple product identity.** For $|q| < 1$, show that

$$\prod_{n=1}^{\infty} (1 - xq^n)(1 - x^{-1}q^{n-1})(1 - x^2 q^{2k-1})(1 - x^{-2} q^{2k-1})(1 - q^k)$$

$$= \sum_{n=-\infty}^{\infty} q^{(3n^2+n)/2}(x^{3n} - x^{-3n-1}).$$

4. **Problem 6562** (proposed by G. E. Andrews, 94(10), 1987). Let $Q(q) = \prod_{n=1}^{\infty}(1 - q^n)$ for $|q| < 1$. Euler's pentagonal number theorem asserts that

$$Q(q) = \sum_{n=0}^{\infty} (-1)^n q^{n(3n+1)/2}(1 - q^{2n+1}).$$

Jacobi showed that

$$Q(q)^3 = \sum_{n=0}^{\infty} (-1)^n (2n + 1) q^{n(n+1)/2}.$$

Prove that

$$Q(q)^2 = \sum_{n=0}^{\infty} (-1)^n q^{n(n+1)/2}(1 - q^{2n+2}) p_n(q),$$

where $p_n(q) = \sum_{r=0}^{n} q^{r(n-r)}$.

5. **Problem 11883** (proposed by H. Ohtsuka, 123(1), 2016). For $|q| > 1$, prove that

$$\sum_{k=0}^{\infty} \frac{1}{(q^{2^0} + q)(q^{2^1} + q) \cdots (q^{2^k} + q)} = \frac{1}{q - 1} \prod_{i=0}^{\infty} \frac{1}{q^{1-2^i} + 1}.$$

2.6   A Gaussian q-binomial identity

Problem 12327 (Proposed by M. Merca, 129(5), 2022). Let

$$\begin{bmatrix} n \\ k \end{bmatrix}_q = \frac{(1-q^n)(1-q^{n-1})\cdots(1-q^{n-k+1})}{(1-q^k)(1-q^{k-1})\cdots(1-q)}.$$

Prove

$$\sum_{k=0}^{n} \begin{bmatrix} n \\ k \end{bmatrix}_{q^2} q^k = \sum_{k=0}^{n} \begin{bmatrix} n \\ k \end{bmatrix}_{q^2} q^{k(k-1)+(n-k)^2-n(n-1)/2}$$

for $n \geq 0$.

Discussion.
Let L_n be the sum on the left side of the desired identity. To search for a pattern, we run *Mathematica* code `Sum[QBinomial[n,k,q^2]*q^k,{k,0,n}]` to get

$$L_2 = (1+q)(1+q^2) \qquad \text{and} \qquad L_3 = (1+q)(1+q^2)(1+q^3).$$

The emerged pattern prompts us to show that $L_n = \prod_{k=1}^{n}(1+q^k)$ in general.

Solution.
Let R_n be the sum on the right side of the desired identity. We show that $L_n = R_n = P(n) := \prod_{k=1}^{n}(1+q^k)$ by induction. We begin with proving that $L_n = P(n)$. For $n = 1$, we have

$$L_1 = 1 + \begin{bmatrix} 1 \\ 1 \end{bmatrix}_{q^2} q = 1 + \frac{1-q^2}{1-q^2} q = 1 + q = P(1).$$

By the well-known identity

$$\begin{bmatrix} n \\ k \end{bmatrix}_q = q^{n-k} \begin{bmatrix} n-1 \\ k-1 \end{bmatrix}_q + \begin{bmatrix} n-1 \\ k \end{bmatrix}_q = q^{n-k} \begin{bmatrix} n-1 \\ n-k \end{bmatrix}_q + \begin{bmatrix} n-1 \\ k \end{bmatrix}_q,$$

replacing q by q^2, and then using the induction hypothesis, we obtain

$$L_n = \sum_{k=1}^{n} q^{2(n-k)} \begin{bmatrix} n-1 \\ n-k \end{bmatrix}_{q^2} q^k + \sum_{k=0}^{n-1} \begin{bmatrix} n-1 \\ k \end{bmatrix}_{q^2} q^k$$

$$= q^n \sum_{k=1}^{n} \begin{bmatrix} n-1 \\ n-k \end{bmatrix}_{q^2} q^{n-k} + \prod_{k=1}^{n-1}(1+q^k)$$

$$= q^n \prod_{k=1}^{n-1}(1+q^k) + \prod_{k=1}^{n-1}(1+q^k) = P(n).$$

On the other hand, to prove $R_n = P(n)$, let

$$a(n,k) = k(k-1) + (n-k)^2 - \frac{n(n-1)}{2}.$$

It is easy to verify that

$$a(n,k) = a(n-1, n-k) \quad \text{and} \quad a(n,k) + 2k = a(n-1, k) + n.$$

For $n = 1$, we have $R_1 = q + 1 = P(1)$. For $n > 1$, by another well-known identity

$$\begin{bmatrix} n \\ k \end{bmatrix}_q = \begin{bmatrix} n-1 \\ k-1 \end{bmatrix}_q + q^k \begin{bmatrix} n-1 \\ k \end{bmatrix}_q = \begin{bmatrix} n-1 \\ n-k \end{bmatrix}_q + q^k \begin{bmatrix} n-1 \\ k \end{bmatrix}_q,$$

replacing q by q^2, and then using the induction hypothesis, we obtain

$$R_n = \sum_{k=1}^{n} \begin{bmatrix} n-1 \\ n-k \end{bmatrix}_{q^2} q^{a(n,k)} + \sum_{k=0}^{n-1} q^{2k} \begin{bmatrix} n-1 \\ k \end{bmatrix}_{q^2} q^{a(n,k)}$$

$$= \sum_{k=1}^{n} \begin{bmatrix} n-1 \\ n-k \end{bmatrix}_{q^2} q^{a(n-1,n-k)} + q^n \sum_{k=0}^{n-1} \begin{bmatrix} n-1 \\ k \end{bmatrix}_{q^2} q^{a(n-1,k)}$$

$$= \prod_{k=1}^{n-1} (1 + q^k) + q^n \prod_{k=1}^{n-1} (1 + q^k) = P(n).$$

Thus, the desired identity is proved by induction. $\quad\square$

Remark. The q-binomial coefficient $\begin{bmatrix} n \\ k \end{bmatrix}_q$ is the q-analog of the binomial number $\binom{n}{k}$, which means that we retrieve the classic one as $q \to 1$. The q-binomial coefficient is not just a rational function of q, it is actually a polynomial. Notice that $\begin{bmatrix} n+m \\ m \end{bmatrix}_q$ is the generating function for integer partitions whose Ferrers diagrams fit into a rectangle $n \times m$, that is the partitions with at most n parts, all at most m . The following identity is a q-analog of the binomial theorem,

$$\prod_{k=1}^{n} (1 + tq^k) = \sum_{k=0}^{n} \begin{bmatrix} n \\ k \end{bmatrix}_q q^{k(k+1)/2} t^k. \qquad (2.12)$$

A combinatorial proof is immediately obtained. The left-hand side is

$$\prod_{k=1}^{n} (1 + tq^k) = \sum_{i=0}^{n} \sum_{j=0}^{n} a(i,j) t^i q^j$$

where $a(i,j)$ is the number of partitions of j with i distinct parts each less or equal to n. On the other hand, in the right-hand side, the terms $\begin{bmatrix} n \\ k \end{bmatrix}_q$ counts the partitions with at most k parts all at most $n-k$. We leave it to the reader to find a simple bijection between these two kinds of partitions (see [13, p. 70]).

Additional problems for practice.

1. Show that

$$(1 - q^{n+1}) \begin{bmatrix} n \\ k \end{bmatrix}_q = (1 - q^{n+1-k}) \begin{bmatrix} n+1 \\ k \end{bmatrix}_q .$$

Use this identity to prove that

$$(1 - q^{2n+2}) L_n = (1 - q^{n+1}) L_{n+1}.$$

This leads to $L_{n+1} = (1 + q^{n+1}) L_n$ and yields another proof of $L_n = P(n)$.

2. Let n and k be positive integers. Prove that

$$\sum_{k=0}^{n} (-1)^k (1 - q^n) \cdots (1 - q^{n-k+1}) q^{\binom{k}{2}} \begin{bmatrix} n \\ k \end{bmatrix}_q = q^{n^2}.$$

3. Let n be a positive integer. Show that

$$\prod_{k=1}^{n} (1 + xq^k)(1 + x^{-1} q^{k-1}) = \sum_{k=-n}^{n} \begin{bmatrix} 2n \\ n+k \end{bmatrix}_q q^{k(k+1)/2} x^k.$$

4. Prove that for any integer $n \geq 0$,

$$\sum_{k=0}^{n} \begin{bmatrix} n \\ k \end{bmatrix}_q^2 q^{k^2} = \begin{bmatrix} 2n \\ n \end{bmatrix}_q ,$$

and then deduce

$$\sum_{k=0}^{\infty} \frac{q^{k^2}}{(1-q)^2 (1-q^2)^2 \cdots (1-q^k)^2} = \prod_{n=1}^{\infty} \frac{1}{1-q^n}.$$

5. **Problem 12078** (Proposed by H. Ohtsuka, 125(12), 2018). Let $\begin{bmatrix} n+1 \\ k \end{bmatrix}_q$ be the q-binomial coefficient defined by

$$\begin{bmatrix} n \\ k \end{bmatrix}_q = \prod_{i=0}^{k-1} \frac{1 - q^{n-i}}{1 - q^{k-i}}.$$

For a positive integer s and for $0 < q < 1$, prove

$$\sum_{n=1}^{\infty} \frac{q^{sn}}{\begin{bmatrix} s+n \\ s+1 \end{bmatrix}_q} = \frac{q^s(1 - q^{s+1})}{1 - q^s}.$$

6. **Problem 12183** (Proposed by H. Ohtsuka, 127(5), 2020). For integers m, n, and r with $m \geq 1$ and $n \geq r \geq 0$, prove

$$\sum_{k=0}^{n} \frac{(-1)^k q^{\binom{k+1}{2} - rk}}{1 - q^{k+m}} \begin{bmatrix} n \\ k \end{bmatrix}_q = \frac{q^{rm}}{1 - q^m} \begin{bmatrix} m+n \\ m \end{bmatrix}_q^{-1}.$$

2.7 A rational function identity

Problem 11828 (Proposed by R. Tauraso, 122(3), 2015). Let n be a positive integer, and let z be a complex number that is not a kth root of unity for any k with $1 \leq k \leq n$. Let S be the set of all lists $(a_1, \ldots, a_n)$ of n nonnegative integers such that $\sum_{k=1}^{n} k a_k = n$. Prove that

$$\sum_{a \in S} \prod_{k=1}^{n} \frac{1}{a_k! k^{a_k} (1 - z^k)^{a_k}} = \prod_{k=1}^{n} \frac{1}{1 - z^k}.$$

Discussion.

Let $P(n)$ be the proposition which states the desired identity holds for the positive integer n. Here, induction is hard to apply because there is no clear insight how to proceed from the truth of $P(n)$ to the truth of $P(n+1)$. We will handle this problem in two steps. First, we find the generating function of the left-hand side sequence in closed form. Then, we take advantage of the symmetry of the generating function and find the coefficients of the power series by way of a functional equation.

Solution.

Let

$$G(z, x) = \sum_{n=0}^{\infty} \left(\sum_{a_1 + \cdots + k a_k = n} \prod_{k=1}^{n} \frac{1}{a_k! k^{a_k} (1 - z^k)^{a_k}} \right) x^n.$$

Then

$$G(z, x) = \prod_{k=1}^{\infty} \left(\sum_{a_k \geq 0} \frac{x^{k a_k}}{a_k! k^{a_k} (1 - z^k)^{a_k}} \right) = \prod_{k=1}^{\infty} \exp\left(\frac{x^k}{k(1 - z^k)} \right)$$

$$= \exp\left(\sum_{k=1}^{\infty} \frac{x^k}{k(1 - z^k)} \right) = \exp\left(\sum_{k=1}^{\infty} \sum_{j=0}^{\infty} \frac{x^k (z^k)^j}{k} \right)$$

$$= \exp\left(\sum_{j=0}^{\infty} \sum_{k=1}^{\infty} \frac{(x z^j)^k}{k} \right) = \prod_{j=0}^{\infty} \exp\left(\sum_{k=1}^{\infty} \frac{(x z^j)^k}{k} \right)$$

$$= \prod_{j=0}^{\infty} \exp\left(\ln\left(\frac{1}{1 - x z^j} \right) \right) = \prod_{j=0}^{\infty} \frac{1}{1 - x z^j}.$$

On the other hand, let

$$G(z, x) = \prod_{j=0}^{\infty} \frac{1}{1 - x z^j} = \sum_{n=0}^{\infty} b_n(z) x^n.$$

It suffices to show that

$$b_n(z) = \prod_{k=1}^{n} \frac{1}{1 - z^k}.$$

To this end, it is easy to verify that $G(z, xz) = (1 - x)G(z, x)$, i.e.

$$\sum_{n=0}^{\infty} b_n(z) z^n x^n = (1 - x) \left(\sum_{n=0}^{\infty} b_n(z) x^n \right).$$

This yields $b_n(z) - b_{n-1}(z) = z^n b_n(z)$ for all $n \geq 1$. Since $b_0(z) = 1$, for $|z| < 1$, we find

$$b_n(z) = \frac{1}{1 - z^n} b_{n-1}(z) = \frac{1}{1 - z^n} \cdot \frac{1}{1 - z^{n-1}} b_{n-2}(z) = \cdots = \prod_{k=1}^{n} \frac{1}{1 - z^k},$$

as claimed. The validity of this formula for every complex number z other than a kth root of unity for any k with $1 \leq k \leq n$ will follow by analytic continuation. $\qquad \square$

Remark. The first part of the above solution gives also proof of this identity

$$\sum_{n=0}^{\infty} Z_n(x_1, x_2, \ldots, x_n) y^n = \exp \left(\sum_{k=1}^{\infty} \frac{x_k y^k}{k} \right) \qquad (2.13)$$

where $Z_n(x_1, x_2, \ldots, x_n)$ is the *cycle index* of the symmetric group S_n. The cycle index of S_n is a polynomial in n variables given by the formula

$$Z_n(x_1, x_2, \ldots, x_n) = \sum_{\substack{a_1 + 2a_2 + \cdots + na_n = n \\ a_1 \geq 0, a_2 \geq 0, \ldots, a_n \geq 0}} \frac{1}{\prod_{k=1}^{n} (a_k! \, k^{a_k})} x_1^{a_1} x_2^{a_2} \cdots x_n^{a_n}.$$

Notice that if $a_1 + 2a_2 + \cdots + na_n = n$ then

$$\frac{n!}{\prod_{k=1}^{n} (a_k! \, k^{a_k})}$$

is the number of permutations in S_n which have precisely a_k cycles of length k for $k = 1, \ldots, n$.

An easy application of (2.13) is that if we set $x_1 = 0$ and $x_2 = x_3 = \cdots = 1$ then we obtain the exponential generating function of the number of permutations without fixed points (derangements):

$$D(y) = \sum_{n=0}^{\infty} \frac{D_n y^n}{n!} = \exp \left(\sum_{k=2}^{\infty} \frac{y^k}{k} \right) = e^{-y} \exp(-\ln(1 - y)) = \frac{e^{-y}}{1 - y}.$$

It follows that

$$D_n = n! \sum_{k=0}^{\infty} \frac{(-1)^k}{k!}.$$

Similarly, by setting $x_1 = x_2 = 1$ and $x_3 = x_4 = \cdots = 0$ we get the exponential generating function of the number of self-inverse permutations (involutions):

$$\exp \left(y + \frac{y^2}{2} \right) = 1 + y + 2 \frac{y^2}{2!} + 4 \frac{y^3}{3!} + 10 \frac{y^4}{4!} + 26 \frac{y^5}{5!} + 76 \frac{y^6}{6!} + \cdots.$$

See entry *A000085* in OEIS (https://oeis.org/A000085).

Additional problems for practice.

1. Let $a_0 = a_1 = 1$ and for $k \geq 2$,

$$a_k = \frac{(z+2)(z+4)\cdots(z+2(k-1))}{(z+1)(z+2)\cdots(z+k-1)}.$$

   Find $\sum_{k=0}^{n}(-1)^k \binom{n}{k} a_k$ in closed from.

2. Show that for any integer $n \geq 0$,

$$\frac{(z+1)\cdots(z+n)}{z(z-1)\cdots(z-n)} = \sum_{k=0}^{n} \binom{n}{k}\binom{n+k}{k}\frac{(-1)^{n-k}}{z-k}.$$

3. (Due to H. Prodinger) Show that for any integer $n > 0$,

$$\sum_{k=1}^{n} \binom{n}{k}\frac{(-1)^{k-1}}{\binom{z+k}{k}}\sum_{j=1}^{k}\frac{1}{z+j} = \frac{n}{(z+n)^2}.$$

4. Let $f(z) = \sum_{k=1}^{N} a_k/(z-r_k)^{\alpha_k}$ with complex numbers a_k and r_k and positive integers α_k. Show that the coefficient of z^n is given by

$$[z^n]f(z) = \sum_{k=1}^{N}\frac{a_k}{(-r_k)^{\alpha_k}}\frac{1}{r_k^n}\binom{\alpha_k+n-1}{\alpha_k-1}.$$

2.8   An identity associated with the zeros of $z^n + 1$

Problem 11947 (Proposed by G. Stoica, 123(10), 2016). Let n be a positive integer, and let $z_1, z_2, \ldots, z_n$ be the zeros in $\mathbb{C}$ of $z^n + 1$. For $a > 0$, prove

$$\frac{1}{n}\sum_{k=1}^{n}\frac{1}{|z_k - a|^2} = \frac{\sum_{k=0}^{n-1} a^{2k}}{(1+a^n)^2}.$$

Discussion.
We will present two proofs to the desired identity. First, recall that if $p(z)$ is an nth degree polynomial with distinct zeros z_k with $1 \leq k \leq n$, then

$$\frac{p'(z)}{p(z)} = \sum_{k=1}^{n}\frac{1}{z - z_k}. \tag{2.14}$$

Letting $p(z) = z^n + 1$ will yield the first proof. Second, in view of the right-hand side of the desired identity, we introduce the rational function

$$f(z) = \frac{\sum_{k=0}^{n-1} z^{2k}}{(1+z^n)^2} = \frac{z^n - 1}{(z^2 - 1)(1 + z^n)}.$$

Finding the partial fraction decomposition of $f(z)$ by the residue formula, and applying $z = a$ will lead to the second proof.

Solution I.

We prove the desired identity indeed holds for any complex number $a \neq z_k$ $(1 \le k \le n)$. Let $p(z) = z^n + 1$. By the assumption and (2.14), we obtain

$$\frac{nz^{n-1}}{z^n + 1} = \sum_{k=1}^{n} \frac{1}{z - z_k}.$$

Multiplying this identity by $2a$, then setting $z = a$ yields

$$\frac{2na^n}{a^n + 1} = \sum_{k=1}^{n} \frac{2a}{a - z_k} = \sum_{k=1}^{n} \frac{(a + z_k) + (a - z_k)}{a - z_k} = n + \sum_{k=1}^{n} \frac{a + z_k}{a - z_k}.$$

Dividing by n gives

$$\frac{1}{n} \sum_{k=1}^{n} \frac{a + z_k}{a - z_k} = \frac{2a^n}{a^n + 1} - 1 = \frac{a^n - 1}{a^n + 1}. \tag{2.15}$$

Using $|z|^2 = z \cdot \bar{z}$, where $\bar{z}$ denotes the complex conjugate of z, we have

$$\frac{a + z_k}{a - z_k} = \frac{(a + z_k)(\bar{a} - \overline{z_k})}{(a - z_k)(\bar{a} - \overline{z_k})} = \frac{(|a|^2 - |z_k|^2) + (\bar{a}z_k - a\overline{z_k})}{|a - z_k|^2}.$$

Here $\mathrm{Re}((\bar{a}z_k - a\overline{z_k})) = 0$ because $\overline{\bar{a}z_k - a\overline{z_k}} = -(\bar{a}z_k - a\overline{z_k})$. Since $|z_k| = 1$, we obtain

$$\mathrm{Re}\left(\frac{a + z_k}{a - z_k}\right) = \frac{|a|^2 - |z_k|^2}{|a - z_k|^2} = \frac{|a|^2 - 1}{|a - z_k|^2}.$$

Similarly,

$$\mathrm{Re}\left(\frac{a^n - 1}{a^n + 1}\right) = \frac{|a^n|^2 - 1}{|a^n + 1|^2}.$$

Comparing the real part on each side of (2.15) yields

$$\frac{1}{n} \sum_{k=1}^{n} \frac{|a|^2 - 1}{|a - z_k|^2} = \frac{|a^n|^2 - 1}{|a^n + 1|^2},$$

and so

$$\frac{1}{n} \sum_{k=1}^{n} \frac{1}{|a - z_k|^2} = \frac{1 + |a|^2 + \cdots + (|a|^2)^{n-1}}{|a^n + 1|^2},$$

where the factorization

$$\left(|a|^2\right)^n - 1 = (|a|^2 - 1)\left(1 + |a|^2 + \cdots + \left(|a|^2\right)^{n-1}\right)$$

is used. Now, the special case $a > 0$ gives the desired identity. □

Solution II.
Let

$$f(z) = \frac{\sum_{k=0}^{n-1} z^{2k}}{(1 + z^n)^2} = \frac{z^n - 1}{(z^2 - 1)(1 + z^n)}.$$

Since $z^n = -1 = e^{\pi i}$, by De Moivre's formula, we have $z_k = \exp((2k-1)\pi i/n)$ for $k = 1, \ldots, n$, which are simple poles of $f(z)$. Hence,

$$\text{Res}(f(z), z_k) = \text{Res}\left(\frac{(z^n - 1)/(z^2 - 1)}{1 + z^n}, z_k\right)$$

$$= \lim_{z \to z_k} \frac{(z^n - 1)/(z^2 - 1)}{nz^{n-1}} = \frac{2}{n(z_k - \overline{z_k})}.$$

For the parity of n, we consider two cases:
(1) If $n = 2m$, then $z_k \neq -1$ for all $1 \leq k \leq n$ and z_k are only zeros of the denominator of $f(z)$. Let the partial fraction decomposition of $f(z) = \sum_{k=1}^{n} \alpha_k/(z - z_k)$. Then

$$\alpha_k = \lim_{z \to z_k} (z - z_k)f(z) = \text{Res}(f(z), z_k) = \frac{2}{n(z_k - \overline{z_k})}.$$

Since $\overline{z_k} = z_{n+1-k}$, it implies that $\alpha_k = -\alpha_{n+1-k}$. For $a > 0$, summing by pair, we find

$$f(a) = \sum_{k=1}^{m} \frac{\alpha_k}{a - z_k} + \sum_{k=1}^{m} \frac{\alpha_{n+1-k}}{a - z_{n+1-k}} = \sum_{k=1}^{m} \alpha_k \left(\frac{1}{a - z_k} - \frac{1}{a - \overline{z_k}}\right)$$

$$= \sum_{k=1}^{m} \frac{\alpha_k(z_k - \overline{z_k})}{|a - z_k|^2} = \frac{1}{n} \sum_{k=1}^{m} \frac{2}{|z_k - a|^2} = \frac{1}{n} \sum_{k=1}^{n} \frac{1}{|z_k - a|^2}.$$

(2) If $n = 2m + 1$, then $z_k \neq -1$ for $k \neq m + 1$ and $z_{m+1} = -1$. Now, the corresponding partial fraction decomposition of f becomes

$$f(z) = \sum_{k=1}^{m} \frac{\alpha_k}{z - z_k} + \frac{1}{n(z + 1)^2} + \sum_{k=m+2}^{2m+1} \frac{\alpha_k}{z - z_k}, \quad \text{with } \alpha_k = \frac{2}{n(z_k - \overline{z_k})}.$$

Following the same lines as in Case (1), for $a > 0$, we have

$$f(a) = \sum_{k=1}^{m} \frac{\alpha_k}{a - z_k} + \frac{1}{n(a+1)^2} + \sum_{k=1}^{m} \frac{\alpha_{n+1-k}}{a - z_{n+1-k}}$$

$$= \frac{1}{n(a+1)^2} + \sum_{k=1}^{m} \alpha_k \left(\frac{1}{a - z_k} - \frac{1}{a - \overline{z_k}} \right)$$

$$= \frac{1}{n((-1) - a)^2} + \sum_{k=1}^{m} \frac{\alpha_k(z_k - \overline{z_k})}{|z_k - a|^2} = \frac{1}{n} \sum_{k=1}^{n} \frac{1}{|z_k - a|^2}.$$

$\square$

Remark. The identity (2.14) displays an elegant connection between the logarithmic derivative of a polynomial and its roots. One should note that there is another relation between the coefficients of a polynomial and functions of its roots. Let $p(z) = \prod_{k=1}^{n}(z - z_k)$ with $\mathbf{z} = (z_1, z_2, \ldots, z_n)$. Then

$$p(z) = z^n - e_1(\mathbf{z})z^{n-1} + \cdots + (-1)^k e_k(\mathbf{z})z^{n-k} + \cdots + (-1)^n e_n(\mathbf{z}),$$

where e_k is the kth elementary symmetric function of the n variables defined by

$$e_k(x_1, x_2, \ldots, x_n) = \sum_{1 \le i_1 < i_2 \cdots < i_k \le n} x_{i_1} x_{i_2} \cdots x_{i_k}.$$

These functions form a fundamental basis in the study of symmetric functions, serving as the building blocks for other symmetric polynomials and providing crucial insights into combinatorial identities and relationships. They are particularly useful in solving problems related to polynomial roots and generating functions.

Additional problems for practice.

1. Let the complex numbers $z_1, z_2, \ldots, z_n \in D = \{z : |z| \le r < 1\}$. Show there exists a $z_0 \in D$ such that

$$\sqrt[n]{\prod_{k=1}^{n}(1 + z_k)} = 1 + z_0.$$

   This can be viewed as a complex mean product theorem.

2. Show that

$$\prod_{k=1}^{n}(1 + z_k z) = \sum_{k=0}^{n} e_k(z_1, z_2, \ldots, z_n)z^k.$$

3. **Problem 10697** (Proposed by J. L. Díaz-Barrero, 105(10), 1998). Given n distinct nonzero complex numbers $z_1, z_2, \ldots, z_n$, show that

$$\sum_{k=1}^{n} \frac{1}{z_k} \prod_{j=1, j \ne k}^{n} \frac{1}{z_k - z_j} = \frac{(-1)^{n+1}}{z_1 z_2 \cdots z_n}.$$

4. **Problem 11008** (Proposed by J. L. Díaz-Barrero and J. J. Egozcue, 110(4), 2003). Let $A(z) = \sum_{k=0}^{n} a_k z^k$ be a monic polynomial with complex coefficients and with zeros $z_1, z_2, \ldots, z_n$. Prove that

$$\frac{1}{n} \sum_{k=1}^{n} |z_k|^2 < 1 + \max_{1 \le k \le n} |a_{n-k}|^2.$$

5. **Problem 11319** (Proposed by D. Beckwith, 114(9), 2007). Let q be an integer greater than 1. For $n \ge 1$, let Φ_n be the polynomial function on the complex numbers given by $\Phi_n(z) = \sum_{j=0}^{n-1} z^j$. Let $S(k)$ denote the sum of the digits in the base q representation of k. Show that for $|z| < 1$,

$$\prod_{n=1}^{\infty} \Phi_q(z^{\Phi_n(q)}) = \sum_{m=0}^{\infty} z^{(qm - S(m))/(q-1)}.$$

6. **Problem 11736** (Proposed by M. Merca, 120(9), 2013). For $n \ge 1$, let f be the symmetric polynomial in variables $x_1, x_2, \ldots, x_n$ given by

$$f(x_1, x_2, \ldots, x_n) = \sum_{k=0}^{n-1} (-1)^{k+1} e_k(x_1 + x_1^2, x_2 + x_2^2, \cdots, x_n + x_n^2),$$

where e_k is the kth elementary polynomial in n variables. Also, let ω be a primitive nth root of unity. Prove that

$$f(1, \omega, \omega^2, \ldots, \omega^{n-1}) = L_n - L_0,$$

where L_k is the kth Lucas number.

7. Let $E(k, n) := e_k(1, 1/2, \ldots, 1/n)$. In 1946, Erdös and Niven proved that there are only finitely many positive integers n for which one or more $E(n, k)$ are integers. For example $E(1, 2) = 3/2, E(2, 2) = 1/2, E(2, 3) = 1$. Show that if $n \ge 4$ and $1 \le k \le n$, then $E(k, n)$ is not an integer.

8. Let $r_k(n)$ be Ramanujan's sum defined by

$$r_k(n) = \sum_{\{m=1,\, \gcd(m,k)=1\}}^{n} \xi_n^{mk}, \quad (\text{where } \xi_n = e^{2\pi i/n}),$$

and let $\Phi_n(x)$ be the nth cyclotomic polynomial defined by

$$\Phi_n(x) = \prod_{\{k=1,\, \gcd(k,n)=1\}}^{n} (x - \xi_n^k).$$

Prove that, for $|x| < 1$ and $n > 1$,

$$\frac{\Phi_n'(x)}{\Phi_n(x)} = \frac{1}{x^n - 1} \sum_{k=1}^{n} r_k(n) x^{k-1}.$$

2.9 An identity struck by the elementary symmetric functions

Problem 12298 (Proposed by G. Stoica, 129(1), 2022). Let n be a positive integer, S_n be the group of all permutations of $\{1, 2, \ldots, n\}$, and z be a primitive complex nth root of unity. Prove

$$\sum_{\tau \in S_n} \prod_{j=1}^{n} (1 - x_j z^{\tau(j)}) = n! \left(1 - \prod_{i=1}^{n} x_i\right)$$

for any $x_1, x_2, \ldots, x_n \in \mathbb{C}$.

Discussion.

To get a feel for this problem, we examine the case $n = 3$. Expanding the products, collecting the like terms, and then using $1 + z + z^2 = 0$, we have

$$\sum_{\tau \in S_3} \prod_{j=1}^{3} (1 - x_j z^{\tau(j)}) = 6 - 2(1 + z + z^2)(x_1 + x_2 + x_3)$$

$$+ 2(1 + z + z^2)(x_1 x_2 + x_1 x_3 + x_2 x_3) - 6x_1 x_2 x_3$$
$$= 6(1 - x_1 x_2 x_3).$$

Recall the kth elementary symmetric function of the n variables defined by

$$e_k(x_1, x_2, \ldots, x_n) = \sum_{1 \le i_1 < i_2 \cdots < i_k \le n} x_{i_1} x_{i_2} \cdots x_{i_k}. \tag{2.16}$$

In particular, we have $e_0(x_1, x_2, x_3) = 1$, $e_1(x_1, x_2, x_3) = x_1 + x_2 + x_3$,

$$e_2(x_1, x_2, x_3) = x_1 x_2 + x_1 x_3 + x_2 x_3, \quad \text{and} \quad e_3(x_1, x_2, x_3) = x_1 x_2 x_3.$$

This special case suggests how to proceed in the general case:

1. Expand $\sum_{\tau \in S_n} \prod_{j=1}^{n} (1 - x_j z^{\tau(j)})$ in terms of $e_k(x_1, \ldots, x_n)$ for $0 \le k \le n$.

2. Show that all coefficients of e_k are zero for $1 \le k \le n - 1$ and the coefficients of e_0 and e_n are $n!$.

Solution.

As we mentioned in the remark of Section 2.5, if $p(z) = \prod_{k=1}^{n} (z - z_k)$ with $\mathbf{z} = (z_1, z_2, \ldots, z_n)$, then

$$p(z) = z^n - e_1(\mathbf{z}) z^{n-1} + \cdots + (-1)^k e_k(\mathbf{z}) z^{n-k} + \cdots + (-1)^n e_n(\mathbf{z}),$$

where e_k is defined by (2.16). Applying this identity to $p(z) = z^n - 1$ leads to

$$z^n - 1 = \prod_{j=1}^{n} (z - z^j) = z^n + \sum_{k=1}^{n} (-1)^k e_k(z, z^2, \ldots, z^n) z^{n-k}.$$

Comparing the coefficients of like terms in each side yields

$$e_k(z, z^2, \ldots, z^n) = \sum_{1 \le i_1 < \cdots < i_k \le n} z^{i_1 + \cdots + i_k} = 0, \quad \text{for } 1 \le k \le n-1,$$

and

$$e_n(z, z^2, \ldots, z^n) = z^{1+2+\cdots+n} = (-1)^{n-1}.$$

Thus, for any nonempty set $J \subset \{1, 2, \ldots, n\}$,

$$\sum_{\tau \in S_n} \prod_{j \in J} z^{\tau(j)} = \sum_{1 \le i_1 < \cdots < i_{|J|} \le n} \sum_{\tau(J)=(i_1,\ldots,i_{|J|})} z^{i_1+\cdots+i_{|J|}}$$

$$= |J|!(n-|J|)! \sum_{1 \le i_1 < \cdots < i_{|J|} \le n} z^{i_1 + \cdots + i_{|J|}} = |J|!(n-|J|)! e_{|J|}$$

$$= \begin{cases} (-1)^{n-1} n!, & \text{if } J = \{1, \ldots, n\}, \\ 0, & \text{otherwise.} \end{cases} \tag{2.17}$$

Using the elementary symmetric functions $e_k(x_1, \ldots, x_n)$, we finally arrive at

$$\sum_{\tau \in S_n} \prod_{j=1}^{n} (1 - x_j z^{\tau(j)}) = n! - \left(\sum_{\tau \in S_n} z^{\tau(i_1)} \right) e_1(x_1, \ldots, x_n)$$

$$+ \left(\sum_{\tau \in S_n} z^{\tau(i_1)+\tau(i_2)} \right) e_2(x_1, \ldots, x_n)$$

$$- \cdots + (-1)^n \left(\sum_{\tau \in S_n} z^{\tau(i_1)+\cdots+\tau(i_n)} \right) e_n(x_1, \ldots, x_n)$$

$$= n! + (-1)^{2n-1} n! e_n(x_1, \ldots, x_n) = n! \left(1 - e_n(x_1, \ldots, x_n)\right)$$

$$= n! \left(1 - \prod_{i=1}^{n} x_i \right).$$

$\square$

Remark. There is another way to establish (2.17). By the q-binomial theorem (2.12), we have

$$\sum_{\tau \in S_n} \prod_{j \in J} z^{\tau(j)} = |J|!(n-|J|!) \cdot [t^k] \prod_{m=1}^{n} (1 + tz^m)$$

$$= |J|!(n-|J|!) z^{(k(k+1)/2} \begin{bmatrix} n \\ k \end{bmatrix}_z.$$

If z is the primitive complex nth root of unity, then (2.17) follows from

$$\begin{bmatrix} n \\ k \end{bmatrix}_z = \begin{cases} 1, & \text{when } k = n, \\ 0, & \text{otherwise.} \end{cases}$$

Additional problems for practice.

1. Show that for any integer $n > 0$,

$$\sum_{\sigma \in S_n} \prod_{k=1}^{n} \frac{1}{\sum_{j=1}^{k} x_{\sigma(j)}} = \frac{1}{\prod_{k=1}^{n} x_k}.$$

2. Prove that if n and k are positive integers then

$$\frac{1}{n!} \sum_{\sigma \in S_n} k^{c(\sigma)} = \binom{n+k-1}{n},$$

   where $c(\sigma)$ is the number of cycles of the permutation σ.

3. Let $t(n, k)$ be the number of permutations $\sigma \in S_n$ which have k *inversions*, i.e., the pairs (i, j) such that $i < j$ and $\sigma(i) > \sigma(j)$. Then

$$\sum_{k=0}^{\binom{n}{2}} t(n, k) x^k = (1 + x)(1 + x + x^2) \cdots (1 + x + \cdots + x^{n-1}).$$

4. **Problem 11713** (Proposed by M. Bencze, 120(6), 2013). Let $x_1, \ldots, x_n$ be nonnegative real numbers. Prove that

$$\prod_{k=1}^{n} (1 + x_k) \le 1 + \sum_{k=1}^{n} \frac{1}{k!} \left(1 - \frac{k}{2n}\right)^{k-1} \left(\sum_{k=1}^{n} x_k\right)^k.$$

   Hint. Note that $\prod_{k=1}^{n}(1+x_k) = 1 + \sum_{k=1}^{n} e_k(x_1, \ldots, x_n)$. Then use the symmetric mean inequality.

5. **Problem 11802** (Proposed by I. Mezö, 121(10), 2014). Let $H_n(2) = \sum_{k=1}^{n} 1/k^2$ and let $D_n = n! \sum_{k=0}^{n} (-1)^k/k!$. D_n is the derangement number of n, that is, the number of permutations of $\{1, \ldots, n\}$ that fix no element. Prove that

$$\sum_{n=1}^{\infty} \frac{(-1)^n H_n(2)}{n!} = \frac{\pi^2}{6e} - \sum_{n=0}^{\infty} \frac{D_n}{n!(n+1)^2}.$$

2.10 A sum of products of truncated binomial expansions

Problem 12293 (Proposed by H. Ohtsuka and R. Tauraso, 129(1), 2022). Let n be a positive integer and r be a positive real number. Prove

$$\sum_{k=0}^{n} (-1)^k \left(\sum_{j=0}^{k} r^j \binom{n}{j} \right) \left(\sum_{j=0}^{k} (-r)^j \binom{n}{j} \right) = \left(\frac{(r+1)^n + (r-1)^n}{2} \right)^2.$$

Discussion.
Since the binomial expansions that appear in the summands of the desired identity are truncated, we can't apply the binomial theorem to simplify them. However, on the right-hand side of the desired identity, by the binomial theorem, we have

$$\frac{(r+1)^n + (r-1)^n}{2} = \begin{cases} \binom{n}{0} + \binom{n}{2} r^2 + \cdots + \binom{n}{n} r^n, & \text{if } n \text{ is even,} \\ \binom{n}{1} r + \binom{n}{3} r^3 + \cdots + \binom{n}{n} r^n, & \text{if } n \text{ is odd.} \end{cases}$$
$$(2.18)$$

This motivates us to retreat the left side by some appropriate substitutions. Surprisingly, by taking advantage of the symmetric structure of the left-hand side, we can establish the desired identity by telescoping.

Solution.
Motivated by the above discussion, we introduce $a_0 = 0$ and for $0 < k \le n+1$,

$$a_k = \begin{cases} \binom{n}{0} + \binom{n}{2} r^2 + \cdots + \binom{n}{k-1} r^{k-1}, & \text{if } k \text{ is odd,} \\ \binom{n}{1} r + \binom{n}{3} r^3 + \cdots + \binom{n}{k-1} r^{k-1}, & \text{if } k \text{ is even.} \end{cases}$$

Then

$$\sum_{j=0}^{k} r^j \binom{n}{j} = a_{k+1} + a_k, \qquad (-1)^k \sum_{j=0}^{k} (-r)^j \binom{n}{j} = a_{k+1} - a_k,$$

and so

$$\sum_{k=0}^{n} (-1)^k \left(\sum_{j=0}^{k} r^j \binom{n}{j} \right) \left(\sum_{j=0}^{k} (-r)^j \binom{n}{j} \right) = \sum_{k=0}^{n} (a_{k+1} + a_k)(a_{k+1} - a_k)$$

$$= \sum_{k=0}^{n} \left(a_{k+1}^2 - a_k^2 \right) = a_{n+1}^2.$$

In view of the definition of a_k and (2.18), this proves the desired identity. □

Remark. Along the same lines used in the above solution, we can prove the following generalization: for any complex numbers $z_0, z_1, \ldots, z_n$,

$$\sum_{k=0}^{n} (-1)^k \left(\sum_{j=0}^{k} z_j \right) \left(\sum_{j=0}^{k} (-1)^j z_j \right) = \left(\sum_{m=0}^{\lceil n/2 \rceil} z_{n-2m} \right)^2.$$

The desired identity is the special case of $z_j = r^j \binom{n}{j}$.

Additional problems for practice.

1. **Putnam 1987-B2.** Let r, s, and t be integers with $0 \le r, 0 \le s$, and $r + s \le t$. Prove that

$$\sum_{k=0}^{s} \frac{\binom{s}{k}}{\binom{t}{r+k}} = \frac{t+1}{(t+1-s)\binom{t-s}{r}}.$$

2. **Problem 11509** (Proposed by W. Stanford, 117(7), 2010). Let m be a positive integer. Prove that

$$\sum_{k=m}^{m^2-m+1} \frac{\binom{m^2-2m+1}{k-m}}{k\binom{m^2}{k}} = \frac{1}{m\binom{2m-1}{m}}.$$

3. **Problem 11026** (Proposed by J. Sondow, 110(7), 2003). Let H_n denote the nth harmonic number $\sum_{k=1}^{n} 1/k$. Let $H_0 = 0$. Prove that for positive integers n and k with $k \le n$,

$$\sum_{i=0}^{k-1} \sum_{j=k}^{n} (-1)^{i+j-1} \binom{n}{i} \binom{n}{j} \frac{1}{j-i} = \sum_{i=0}^{k-1} \binom{n}{i}^2 (H_{n-i} - H_i).$$

Hint. Show that for $j \ne i$

$$\sum_{j=0}^{n} \binom{n}{j} \frac{(-1)^j}{j-i} = (-1)^{i-1} \binom{n}{i} (H_{n-i} - H_i).$$

4. **Problem 11711** (Proposed by J. A. Grzesik, 120(5), 2013). Show, for integers n and k with $n \ge 2$ and $1 \le k \le n$, that

$$(-1)^{n-k} \binom{n}{k} k \sum_{j=1, j \ne k}^{n} \frac{1}{k-j} = -\sum_{j=1, j \ne k}^{n} (-1)^{n-j} \binom{n}{j} \frac{j}{k-j}.$$

5. **Problem 12049** (Proposed by Z. K. Silagadze, 125(6), 2018). For all nonnegative integers m and n with $m \leq n$, prove

$$\sum_{k=m}^{n} \frac{(-1)^{k+m}}{2k+1} \binom{n+k}{n-k} \binom{2k}{k-m} = \frac{1}{2n+1}.$$

Hint. Apply the Wilf-Zeilberger algorithm to

$$F(n,k) = \frac{(-1)^{k+m}(2n+1)}{2k+1} \binom{n+k}{n-k} \binom{2k}{k-m}.$$

6. **Problem 11212** (Proposed by D. Beckwith, 113(3), 2006). Show that for an arbitrary integer $n > 0$

$$\sum_{k=0}^{n} (-1)^k \binom{n}{k} \binom{2n-2k}{n-1} = 0.$$

7. **Problem 11914** (Proposed by R. Chapman and R. Tauraso, 123(5), 2016). Show that for all positive integers m and n,

$$\sum_{k=1}^{n} (-4)^{-k} \binom{n-k}{k-1} \sum_{j=1}^{3m} (-2)^{-j} \binom{n+1-2k}{j-1} \binom{m-k}{3m-j} = 0.$$

Here $\binom{x}{k} = \frac{1}{k!} \prod_{i=0}^{k-1}(x-i)$ for $x \in \mathbb{R}$.

8. **Problem 11844** (Proposed by H. Ohtsuka and R. Tauraso, 122(5), 2015). For nonnegative integers m and n, prove

$$\sum_{k=0}^{n} (m-2k) \binom{m}{k}^3 = (m-n) \binom{m}{n} \sum_{j=0}^{m-1} \binom{j}{n} \binom{j}{m-n-1}.$$

9. **Problem 11928** (Proposed by H. Ohtsuka, 123(7), 2016). For positive integers n and m and for a sequence $\{a_i\}_{i \geq 1}$, prove

$$\sum_{i=0}^{n} \sum_{j=0}^{m} \binom{n}{i} \binom{m}{j} a_{i+j} = \sum_{k=0}^{n+m} \binom{n+m}{k} a_k$$

and

$$\sum_{0 \leq i < j \leq n} \binom{n}{i} \binom{n}{j} \binom{i+j}{n} = \sum_{0 \leq i < j \leq n} \binom{n}{i} \binom{n}{j}^2.$$

10. **Problem 12304** (Proposed by M. Bataille, 129(2), 2022). Let m and n be positive integers with $m < n$. Prove

$$\left(\sum_{k=0}^{m} \binom{m}{k} \frac{(-1)^k}{n-k} \right) \left(\sum_{k=0}^{m} \binom{n}{k} \frac{(-1)^k}{k+1} \right) = \sum_{k=0}^{m} \binom{m}{k} \frac{(-1)^k}{(n-k)(k+1)}.$$

Hint. Find the closed forms for $\sum_{k=0}^{m} \binom{m}{k} \frac{(-1)^k}{n-k}$ and $\sum_{k=0}^{m} \binom{n}{k} \frac{(-1)^k}{k+1}$.

2.11 An identity with alternating weighted binomial coefficients

Problem 12477 (Proposed by O. Kouba, 131(6), 2024). Let n be a nonnegative integer. Prove

$$\sum_{j=0}^{n}(-1)^j \binom{n}{j} \frac{2^{n-j}}{2n+2j-1} = \sum_{j=0}^{n}(-1)^{j-1} \binom{n}{j} \frac{4^{2j}}{\binom{4j}{2j}}.$$

Discussion.
When $n = 0$, each side of the desired identity equals -1, so the identity is true for $n = 0$. For $n \geq 1$, we will proceed to establish this identity in two ways. First, since

$$\frac{1}{2n+2j-1} = \int_0^1 x^{2n+2j-2}\, dx,$$

by the binomial theorem, we have

$$I_L := \sum_{j=0}^{n}(-1)^j \binom{n}{j} \frac{2^{n-j}}{2n+2j-1} = \sum_{j=0}^{n}(-1)^j \binom{n}{j} 2^{n-j} \int_0^1 x^{2n+2j-2}\, dx$$

$$= \int_0^1 \left(x^{2n-2} \sum_{j=0}^{n} \binom{n}{j} 2^{n-j}(-x^2)^j \right)\, dx = \int_0^1 x^{2n-2}(2-x^2)^n\, dx.$$

Thus, to prove the validity of the identity, we just need to show that

$$I_R := \sum_{j=0}^{n}(-1)^{j-1} \binom{n}{j} \frac{4^{2j}}{\binom{4j}{2j}} = \int_0^1 x^{2n-2}(2-x^2)^n\, dx. \qquad (2.19)$$

Second, we will present a proof based on the following two combinatorial identities:

$$\sum_{k=0}^{n}(-1)^k \binom{n}{k} \frac{x}{x+k} = \binom{x+n}{n}^{-1} \qquad (2.20)$$

and

$$\sum_{k=0}^{n}(-1)^k \binom{n}{k}\binom{2k}{j} = (-1)^n \binom{n}{i-n} 2^{2n-j}. \qquad (2.21)$$

They are (1.41) and (3.64) in [44], respectively.

Solution I.
To prove (2.19), by Wallis' integral formula

$$\int_0^{\pi/2} \sin^{4j+1}\theta\, d\theta = \frac{4^{2j}}{(4j+1)\binom{4j}{2j}},$$

we have

$$I_R = \sum_{j=0}^{n} (-1)^{j-1} \binom{n}{j} (4j+1) \int_0^{\pi/2} \sin^{4j+1} \theta \, d\theta.$$

Integrating by parts gives

$$I_R = \sum_{j=0}^{n} (-1)^{j-1} \binom{n}{j} \int_0^{\pi/2} \tan \theta \, d(\sin^{4j+1} \theta - 1)$$

$$= \sum_{j=0}^{n} (-1)^{j-1} \binom{n}{j} \left([\tan \theta (\sin^{4j+1} \theta - 1)]_0^{\pi/2} - \int_0^{\pi/2} (\sin^{4j+1} \theta - 1) \sec^2 \theta \, d\theta \right)$$

$$= \int_0^{\pi/2} \left(\sum_{j=0}^{n} (-1)^j \binom{n}{j} (\sin^{4j+1} \theta - 1) \sec^2 \theta \right) d\theta$$

$$= \int_0^{\pi/2} \left(\sum_{j=0}^{n} (-1)^j \binom{n}{j} \sin^{4j+1} \theta \sec^2 \theta \right) d\theta,$$

In the last equation, we have used $\sum_{j=0}^{n} (-1)^j \binom{n}{j} = 0$. Next, applying the binomial theorem yields

$$I_R = \int_0^{\pi/2} (1 - \sin^4 \theta)^n \sin \theta \sec^2 \theta \, d\theta = \int_0^{\pi/2} [\cos^2 \theta (2 - \cos^2 \theta)]^n \sin \theta \sec^2 \theta \, d\theta$$

$$= \int_0^{1} x^{2n-2} (2 - x^2)^n \, dx \quad (\text{use } x = \cos \theta).$$

This shows that $I_L = I_R$ as claimed. $\qquad\square$

Solution II.
Notice that

$$\binom{2m}{m} = \frac{(2m)!! \cdot (2m-1)!!}{m! \cdot m!} = \frac{2^m m! (2m-1)!!}{m! \cdot m!}$$

$$= \frac{2^{2m} (1/2)(3/2) \cdots ((2m-1)/2)}{m!} = 2^{2m} \binom{m-1/2}{m}.$$

Now we rewrite the right-hand side I_R and then apply (2.20) with $x = -1/2$ to obtain

$$I_R = \sum_{k=0}^{n} (-1)^{k-1} \binom{n}{k} \binom{-1/2 + 2k}{2k}^{-1}$$

$$= \sum_{k=0}^{n} (-1)^{k-1} \binom{n}{k} \left(\sum_{j=0}^{2k} (-1)^j \binom{2k}{j} \frac{-1}{2j-1} \right).$$

Interchanging the order of summation and then applying (2.21) yields

$$
\begin{aligned}
I_R &= \sum_{j=0}^{2n} \frac{(-1)^j}{2j-1} \left(\sum_{j/2 \le k \le n} (-1)^k \binom{n}{k} \binom{2k}{j} \right) \\
&= \sum_{j=0}^{2n} \frac{(-1)^j}{2j-1} \left(\sum_{k=0}^{n} (-1)^k \binom{n}{k} \binom{2k}{j} \right) \\
&= \sum_{j=0}^{2n} \frac{(-1)^{j-n}}{2j-1} \binom{n}{j-n} 2^{2n-j} = \sum_{j=n}^{2n} \frac{(-1)^{j-n}}{2j-1} \binom{n}{j-n} 2^{2n-j} \\
&= \sum_{k=0}^{n} (-1)^k \binom{n}{k} \frac{2^{n-k}}{2n+2k-1},
\end{aligned}
$$

where the substitution $k = j - n$ is used in the last equation. This proves the desired identity. $\qquad\square$

Remark. Identity (2.20) appears in Gould's collection [44, (1.41)]. Here we present an elementary proof of (2.20) by using partial fractions. To this end, notice that

$$
\binom{x+n}{n} = \frac{(x+n)(x+n-1)\cdots(x+1)}{n!}.
$$

Let the partial fraction decomposition be

$$
\frac{1}{x\binom{x+n}{n}} = \sum_{k=0}^{n} \frac{A_k}{x+k}.
$$

Then

$$
\begin{aligned}
A_k &= \lim_{x \to -k} (x+k) \frac{n!}{(x+n)(x+n-1)\cdots(x+1)} \\
&= \frac{n!}{(-1)^k k! \cdot (n-k)!} = (-1)^k \binom{n}{k},
\end{aligned}
$$

and so

$$
\frac{1}{x\binom{x+n}{n}} = \sum_{k=0}^{n} (-1)^k \binom{n}{k} \frac{1}{x+k}.
$$

Multiplying this equation by x yields (2.20).

Additional problems for practice.

1. Let n and m be positive integers with $m < n$. Show that

$$
\sum_{k=0}^{n} (-1)^k \binom{n}{k} \frac{k^m}{x+k} = (-1)^m x^{m-1} \binom{x+n}{n}^{-1}.
$$

2. (Due to T. B. Staver) Show that if n is a positive integer then

$$\sum_{k=1}^{n} \frac{1}{k}\binom{2k}{k} = \frac{2n+1}{3n^2}\binom{2n}{n}\sum_{k=1}^{n}\frac{1}{\binom{n-1}{k-1}^2}.$$

3. Let H_n be nth harmonic number. Show that

$$\sum_{k=0}^{n}(-1)^k\binom{n}{k}\frac{4^k}{\binom{2k}{k}} = \frac{1}{1-2n} \quad \text{and} \quad \sum_{k=1}^{n}(-1)^k\binom{n}{k}\frac{4^k}{k\binom{2k}{k}} = H_n - 2H_{2n}.$$

4. Show that if n and m are positive integers then

$$\sum_{k=1}^{n}(-1)^{k-1}\binom{n}{k}\sum_{1\leq i_1\leq\cdots\leq i_m\leq k}\frac{1}{i_1\cdots i_m} = \frac{1}{n^m}.$$

5. **Problem 12106** (Proposed by H. Ohtsuka, 126(4), 2019). For any positive integer n, prove

$$\sum_{k=1}^{n}\binom{n}{k}\sum_{1\leq i\leq j\leq k}\frac{1}{ij} = \sum_{1\leq i\leq j\leq n}\frac{2^n - 2^{n-i}}{ij}.$$

6. **Problem 11103** (Proposed by G. Galperin and H. Gauchman, 111(8), 2004). Prove that for every positive integer n,

$$\sum_{k=1}^{n}\frac{1}{k\binom{n}{k}} = \frac{1}{2^{n-1}}\sum_{\substack{k=1\\k\ \text{odd}}}^{n}\binom{n}{k}k.$$

7. **Problem 12415** (Proposed by R. Tauraso, 130(8), 2023). For any nonnegative integer n, evaluate

$$\sum_{j=0}^{2n}\sum_{k=\lceil j/2\rceil}^{j}\binom{2n+2}{2k+1}\binom{n+1}{2k-j}.$$

8. **Problem 11916** (Proposed by H. Ohtsuka and R. Tauraso, 123(6), 2016). Show that if n, r, and s are positive integers, then

$$\binom{n+r}{n}\sum_{k=0}^{s-1}\binom{r+k}{r-1}\binom{n+k}{n} = \binom{n+s}{n}\sum_{k=0}^{r-1}\binom{s+k}{s-1}\binom{n+k}{n}.$$

9. **Problem 12016** (Proposed by H. Ohtsuka and R. Tauraso, 125(1), 2018). For nonnegative integers m, n, r, and s, prove that

$$\sum_{k=0}^{s}\binom{m+r}{n-k}\binom{r+k}{k}\binom{s}{k} = \sum_{k=0}^{r}\binom{m+s}{n-k}\binom{s+k}{k}\binom{r}{k}.$$

10. **Problem 12156** (Proposed by H. Ohtsuka, 127(1), 2020). For positive integers m and n and nonnegative integers r and s, prove

$$\sum_{0 \le j_1 \le \cdots \le j_m \le r} \frac{\binom{n+s}{n}\binom{n+j_1}{n}\binom{s+j_1}{n}}{\prod_{i=1}^{m}(n+j_i)} = \sum_{0 \le j_1 \le \cdots \le j_m \le s} \frac{\binom{n+r}{n}\binom{n+j_1}{n}\binom{r+j_1}{n}}{\prod_{i=1}^{m}(n+j_i)}.$$

Hint. For nonnegative integer k, first show that

$$\sum_{0 \le j_1 \le \cdots \le j_m \le s} \frac{\binom{n+j_1}{n}\binom{j_1}{k}}{\prod_{i=1}^{m}(n+j_i)} = \frac{\binom{n+s}{n}\binom{s}{k}}{(n+k)^m}.$$

2.12 An identity with the generalized binomial coefficients

Problem 12364 (Proposed by R. Tauraso, 130(1), 2023). Let n be a positive integer, and let z be a complex number not in $\{-1, \ldots, -n\}$. Prove

$$\sum_{1 \le j \le k \le n} \frac{(-1)^{k-1}}{(z+j)k^2}\binom{z+n}{n-k} = \binom{z+n}{n} \sum_{1 \le j \le k \le n} \frac{1}{(z+j)^2 k},$$

where $\binom{\alpha}{k} = (1/k!) \prod_{i=0}^{k-1}(\alpha - i)$.

Discussion.
First, we observe that the approaches used in previous sections shed no insight on how to proceed with this identity directly. We therefore try to reformulate the problem. Let

$$F(z) := \frac{1}{\binom{z+n}{n}} \sum_{1 \le j \le k \le n} \frac{(-1)^{k-1}}{(z+j)k^2}\binom{z+n}{n-k}.$$

Then we have to show that

$$F(z) = \sum_{1 \le j \le k \le n} \frac{1}{(z+j)^2 k}.$$

We now try to rewrite the function F in a more suitable way. Notice that

$$\frac{\binom{z+n}{n-k}}{\binom{z+n}{n}} = \frac{n!}{(n-k)!(z+1)(z+2)\cdots(z+k)}.$$

Moreover, since $d/dz((z+j)^{-1}) = -(z+j)^{-2}$, we find

$$\frac{d}{dz}\left(\frac{1}{(z+1)(z+2)\cdots(z+k)}\right) = -\sum_{j=1}^{k} \frac{1}{(z+j)(z+1)(z+2)\cdots(z+k)}.$$

Thus

$$F(z) = \frac{d}{dz}\left(\sum_{k=1}^{n} \frac{(-1)^k n!}{(n-k)!k^2} \cdot \frac{1}{(z+1)(z+2)\cdots(z+k)}\right).$$

By applying partial fractions and some well-known combinatorial identities, we are able to prove the desired identity.

Solution.
Using partial fractions, we obtain

$$\frac{1}{(z+1)(z+2)\cdots(z+k)} = \sum_{j=1}^{k} \frac{(-1)^{j-1}}{(j-1)!(k-j)!(z+j)}$$

and so

$$F(z) = \frac{d}{dz}\left(\sum_{k=1}^{n} \frac{(-1)^k n!}{(n-k)!k^2} \sum_{j=1}^{k} \frac{(-1)^{j-1}}{(j-1)!(k-j)!(z+j)}\right)$$

$$= -\frac{d}{dz}\left(\sum_{j=1}^{n}\sum_{k=j}^{n} \binom{n}{k}\binom{k-1}{j-1} \frac{(-1)^{k-j}}{k(z+j)}\right).$$

To simplify the coefficient of $1/(z+j)$, using the *hockey stick identity*,

$$\binom{n}{k} = \sum_{i=k}^{n} \binom{i-1}{k-1},$$

we find

$$\sum_{k=j}^{n} \binom{n}{k}\binom{k-1}{j-1}\frac{(-1)^{k-j}}{k} = \sum_{k=j}^{n}\left(\sum_{i=k}^{n}\binom{i-1}{k-1}\right)\binom{k-1}{j-1}\frac{(-1)^{k-j}}{k}$$

$$= \sum_{i=j}^{n}\frac{1}{i}\sum_{k=j}^{i}\binom{i}{k}(-1)^{k-j}\binom{k-1}{j-1}$$

$$= \sum_{i=j}^{n}\frac{1}{i}\sum_{k=j}^{i}\binom{i}{i-k}\binom{-j}{k-j}.$$

Due to the Vandermonde convolution formula, we have

$$\sum_{k=j}^{i}\binom{i}{i-k}\binom{-j}{k-j} = \binom{i-j}{i-j} = 1,$$

and we finally arrive at

$$F(z) = -\frac{d}{dz}\left(\sum_{j=1}^{n}\sum_{i=j}^{n}\frac{1}{i(z+j)}\right) = -\sum_{j=1}^{n}\sum_{i=j}^{n}\frac{1}{i}\frac{d}{dz}\left(\frac{1}{z+j}\right)$$

$$= \sum_{1\leq j\leq i\leq n}\frac{1}{i(z+j)^2}.$$

□

Remark. Notice that

$$\frac{d}{dx}\binom{x+n}{m} = \binom{x+n}{m}\sum_{k=1}^{m}\frac{1}{1+x+n-k}.$$

Define the generalized harmonic numbers by

$$H_0(x) = 0 \quad \text{and} \quad H_n(x) = \sum_{k=1}^{n}\frac{1}{x+k} \quad \text{for } n = 1, 2, \dots.$$

Then

$$\frac{d}{dx}\binom{x+n}{m} = \binom{x+n}{m}(H_n(x) - H_{n-m}(x)).$$

In particular, for $0 \leq m \leq n$,

$$\frac{d}{dx}\binom{x+n}{m}\bigg|_{x=0} = \binom{x+n}{m}(H_n - H_{n-m}).$$

Applying differential operators to known binomial identities enables us to discover new identities. For example, differentiating the binomial convolution identity

$$\sum_{k=0}^{n}\binom{n}{k}\binom{x+pn+n}{n-k} = \binom{x+pn+2n}{n}$$

yields

$$\sum_{k=0}^{n}\binom{n}{k}\binom{n+pn}{n-k}(H_{pn+n} - H_{pn+k}) = \binom{pn+2n}{n}(H_{pn+2n} - H_{pn+n})$$

or

$$\sum_{k=0}^{n}\binom{n}{k}\binom{n+pn}{n-k}H_{pn+k} = \binom{pn+2n}{n}(2H_{pn+n} - H_{pn+2n})$$

Letting $p = 0$ immediately produces the well-known identity

$$\sum_{k=0}^{n}\binom{n}{k}^2 H_k = \binom{2n}{n}(2H_n - H_{2n}).$$

Moreover, if the shifted factorial is defined by $(x)_0 = 1$ and $(x)_n = x(x + 1) \cdots (x + n - 1)$ for $n \geq 1$, by applying differential operators to the Gauss hypergeometric summation formula

$$\sum_{n=1}^{\infty} \frac{(a)_n (b)_n}{n!(c)_n} H_n(c - 1) = \frac{\Gamma(c)\Gamma(c - a - b)}{\Gamma(c - a)\Gamma(c - b)}$$

$$\cdot \left(\psi(c - a) + \psi(c - b) - \psi(c) - \psi(c - a - b) \right),$$

where ψ is the polygamma function, H. Chen [32] derived many interesting series involving nonlinear binomial coefficients like

$$\sum_{n=1}^{\infty} \frac{1}{16^n (2n - 1)^2} \binom{2n}{n}^2 H_n = \frac{12 - 16 \ln 2}{\pi}.$$

The interested reader is encouraged to pursue new results by differential operators.

Additional problems for practice.

1. Show that if $\sum_{k=1}^{n} \binom{z+n}{n-k} x_k = \binom{z+n}{n} y_n$ for any positive integer n, then

$$\sum_{k=1}^{n} \binom{z + n}{n - k} \frac{x_k}{k} = \binom{z + n}{n} \sum_{k=1}^{n} \frac{y_k}{k}.$$

Using this result prove the proposed identity.

2. If two sequences are related by $\sum_{k=1}^{n} (-1)^{k-1} \binom{n}{k} a_k = b_n$ then, for all positive integers n and t,

$$\sum_{k=1}^{n} \frac{(-1)^{k-1} \binom{n}{k} a_k}{(z + k)^t} = \frac{1}{\binom{z+n}{n}} \sum_{1 \leq k_1 \leq \cdots \leq k_t \leq n} \frac{b_{k_1} \binom{z+k_1}{k_1}}{\prod_{j=1}^{t} (z + k_j)}.$$

For a proof see [9, Lemma 3.1].

3. Show that

$$\sum_{k=0}^{n} (-1)^k 2^{-n-k} \binom{n + k}{k} \binom{2x}{n - k} \frac{n - k}{n + k} = \binom{x}{n}.$$

4. Let $P_n(x)$ be the Legendre polynomial defined by

$$P_n(x) = \frac{1}{2^n n!} \frac{d^n}{dx^n} (x^2 - 1)^n.$$

Show that

$$P_n(2x - 1) = (-1)^n \sum_{k=0}^{n} \binom{n}{k} \binom{n + k}{k} (-x)^k.$$

5. **Problem 11798** (Proposed by F. Holland, 121(8), 2014). For positive integers n, let f_n be the polynomial given by

$$f_n(x) = \sum_{k=0}^{n} \binom{n}{k} x^{\lfloor k/2 \rfloor}.$$

(a) Prove that if $n + 1$ is prime, then f_n is irreducible over $\mathbb{Q}$.

(b) Prove for all n,

$$f_n(1 + x) = \sum_{k=0}^{\lfloor n/2 \rfloor} \binom{n-k}{k} 2^{n-2k} x^k.$$

6. **Problem 11403** (Proposed by Y. Yu, 115 (10), 2008). Let n be an integer greater than 1, and let f_n be the polynomial given by

$$f_n = \sum_{i=0}^{n} \binom{n}{i} (-x)^{n-i} \prod_{j=0}^{i-1} (x + j).$$

Find the degree of f_n.

7. **Problem 11909** (Proposed by H. Ohtsuka, 123(5), 2016). Prove that for every positive integer m there exists a polynomial P_m in two variables, with integer coefficients, such that for all integers n and r with $0 \le r \le n$,

$$\sum_{k=-r}^{r} \binom{n}{r+k} \binom{n}{r-k} k^{2m} = \frac{P_m(n,r)}{\prod_{j=1}^{m}(2n - 2j + 1)} \binom{2n}{2r}.$$

Chapter 3

Geometry

Here the reader finds 10 problems exploring different results in Euclidean geometry, such as Anne's theorem, Napoleon's theorem, Heron's formula, Brianchon's theorem, and Newton's quadrilateral theorem. Also, several notable points of a triangle, including the Nagel point and Fermat point, show some unexpected properties. For most solutions, we decided to choose analytical techniques such as vectors, complex numbers or barycentric coordinates. While commenting on our work, we included references to facilitate the understanding of certain aspects of the chosen approach.

3.1 A line perpendicular to the Newton line

Problem 12147 (Proposed by L. Gonzalez and T. Q. Hung, 126(10), 2019). Let $ABCD$ be a quadrilateral that is not a parallelogram. The Newton line of $ABCD$ is the line that connects the midpoints of the diagonals AC and BD. Let X be the intersection of the perpendicular bisectors of AB and CD, and let Y be the intersection of the perpendicular bisectors of BC and DA. Prove that XY is perpendicular to the Newton line of $ABCD$.

Discussion.
The parallelogram is excluded because in that case, both midpoints of the diagonals coincide with the point of intersection of the two diagonals and therefore the Newton line remains undefined. We will adopt an approach through vectors.

Since X is the intersection of the perpendicular bisectors of AB and CD, we have

$$\left\langle X - \frac{A+B}{2}, A - B \right\rangle = 0 \quad \text{and} \quad \left\langle X - \frac{C+D}{2}, C - D \right\rangle = 0, \quad (3.1)$$

DOI: 10.1201/9781003607809-3

where the $\langle \cdot, \cdot \rangle$ denotes the scalar product. Similarly, since Y is the intersection of the perpendicular bisectors of BC and DA, we have

$$\left\langle Y - \frac{B+C}{2}, B - C \right\rangle = 0 \quad \text{and} \quad \left\langle Y - \frac{D+A}{2}, D - A \right\rangle = 0. \qquad (3.2)$$

These four equations provide us with the necessary information about points X and Y to solve the problem.

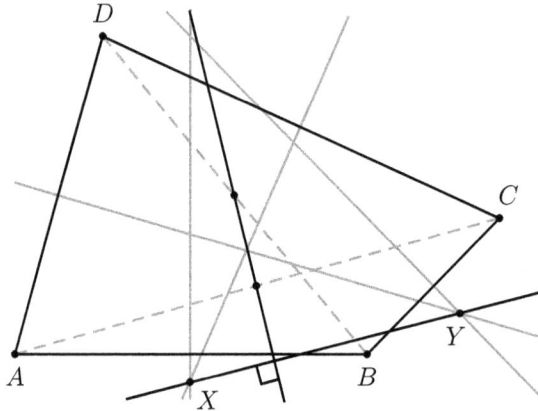

Figure 3.1: Problem 12147

Solution.
By taking the sum of the above four equations, we find

$$\left\langle X - \frac{A+B}{2}, A - B \right\rangle + \left\langle X - \frac{C+D}{2}, C - D \right\rangle$$
$$+ \left\langle Y - \frac{B+C}{2}, B - C \right\rangle + \left\langle Y - \frac{D+A}{2}, D - A \right\rangle = 0$$

which is reduced to

$$2\left\langle X - Y, \frac{A+C}{2} - \frac{B+D}{2} \right\rangle = 0.$$

This equality implies that the vector $X - Y$ is perpendicular to the line that connects the midpoint of the diagonal AC and the midpoint of the diagonal BD, i.e., the Newton line of $ABCD$.

Notice that if the quadrilateral $ABCD$ is concyclic then $X - Y$ is the null vector because both X and Y coincide with the center of the circumscribing circle. $\qquad \square$

Remark. Another noteworthy property of the Newton line is due to Pierre-Léon Anne: Let $ABCD$ be a convex quadrilateral that is not a parallelogram.

A point P satisfies the equality

$$\text{Area}(PAB) + \text{Area}(PCD) = \text{Area}(PBC) + \text{Area}(PDA)$$

if and only if P lies on the Newton line of $ABCD$.

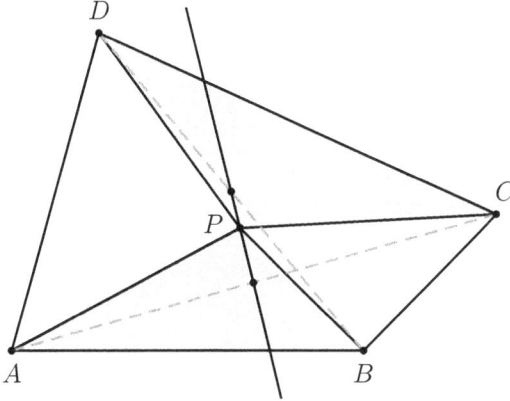

Figure 3.2: Anne's theorem

For its proof see, for example, [7, p. 116].

From Anne's theorem we immediately deduce Newton's theorem: the center of the circle inscribed in a tangential quadrilateral lies on the Newton line of the quadrilateral. Indeed if we denote the center of the incircle by I and its radius by r then

$$\text{Area}(IAB) + \text{Area}(ICD) = r(|AB| + |CD|) = r(|BC| + |DA|)$$
$$= \text{Area}(IBC) + \text{Area}(IDA).$$

Therefore, the point I lies on the Newton line by Anne's theorem.

Additional problems for practice.

1. **Problem E3299** (Proposed by J. Yamouta, 95(10), 1988). Suppose $ABCD$ is a plane quadrilateral with no two sides parallel. Put $E = AB \cap CD$ and $F = AD \cap BC$. If M, N, P are the midpoints of AC, BD, EF, respectively, and $AE = aAB$, $AF = bAD$, where a and b are nonzero real numbers, prove that $MP = abMN$.

2. **Problem 11572** (Proposed by S. Sakmar, 118(5), 2011). Given a circle C and two points A and B outside C, give a Euclidean construction to find a point P on C such that if Q and S are the second intersections with C of AP and BP respectively, then QS is perpendicular to AB.

3. **Problem 11737** (Proposed by N. T. Binh, 120(9), 2013). Given an acute triangle ABC, let O be its circumcenter, let M be the intersection of lines AO and BC, and let D be the other intersection of AO with the circumcircle of ABC. Let E be that point on AD such that M is the midpoint of ED. Let F be the point at which the perpendicular to AD at M meets AC. Prove that EF is perpendicular to AB.

4. **Problem 11841** (Proposed by L. Giugiuc, 122(5), 2015). Let $ABCD$ be a convex quadrilateral. Let E be the midpoint of AC, and let F be the midpoint of BD. Show that

$$AB + BC + CD + DA \geq AC + BD + 2EF.$$

5. **Problem 12007** (K. Altintas and L. Giugiuc, 124(10), 2017). Let G be the centroid of triangle ABC, and let M be an interior point of ABC. Let D, E, and F be the centroids of subtriangles CMB, AMC, and BMA, respectively.

(a) Prove that the lines AD, BE, and CF are concurrent.

(b) Suppose that $M \neq G$ and that P is the point of concurrency in part (a). Prove that G, P, and M are collinear, with P between G and M, and $PM = 3PG$.

6. **Problem 12452** (T. Q. Hung, 131(3), 2024). A *symmedian line* in a triangle is the reflection of a median across the corresponding angle bisector. The *symmedian point* of a triangle is the intersection of the three symmedian lines. Let $\triangle ABC$ be a nonequilateral triangle with circumcircle ω, and let a, b, c be the lengths of sides BC, CA, AB, respectively. Suppose P is a point on ω that minimizes or maximizes $a^4 PA^4 + b^4 PB^4 + c^4 PC^4$ over all points on ω. Prove that P, the center of ω, and the symmedian point of $\triangle ABC$ are collinear.

3.2   About two cyclic quadrilaterals

Problem 12294 (Proposed by T. Q. Hung, 129(1), 2022). Let $A_1 A_2 A_3 A_4$ be a quadrilateral inscribed in a circle with center O. Let $B_1 B_2 B_3 B_4$ be the quadrilateral that contains $A_1 A_2 A_3 A_4$ in its interior such that, for $1 \leq k \leq 4$ and with subscripts taken cyclically, $B_k B_{k+1}$ is parallel to $A_k A_{k+1}$ and at distance $|A_k A_{k+1}|$ from it. Because $B_1 B_2 B_3 B_4$ has the same angles as $A_1 A_2 A_3 A_4$, there is a circle in which it is inscribed. Let P be the center of that circle. Show that $A_1 A_3$, $A_2 A_4$, and OP are concurrent.

Discussion.
We will show a more general result: the concurrency holds when $B_k B_{k+1}$ is at distance $r|A_k A_{k+1}|$ from $A_k A_{k+1}$ for $1 \leq k \leq 4$ and for any $r > 0$.

To this end, we will use complex numbers. Without loss of generality, we may assume that the circumcircle of $A_1A_2A_3A_4$ is the unit circle $|z| = 1$. Moreover, since the point B_3 is the circumcenter of the triangle with the vertices A_3, $A_3 + 2r(A_2 - A_3)i$, and $A_3 - 2r(A_4 - A_3)i$, we can determine its position. Infact, the circumcenter of the triangle with vertices $u, v, w \in \mathbb{C}$ is given by the ratio of determinants

$$C(u, v, w) = \begin{vmatrix} u & |u|^2 & 1 \\ v & |v|^2 & 1 \\ w & |w|^2 & 1 \end{vmatrix} \bigg/ \begin{vmatrix} u & \bar{u} & 1 \\ v & \bar{v} & 1 \\ w & \bar{w} & 1 \end{vmatrix}.$$

A similar argument can be done for the other points B_1, B_2, and B_4. After finding the center P, we will be able to prove the required concurrency.

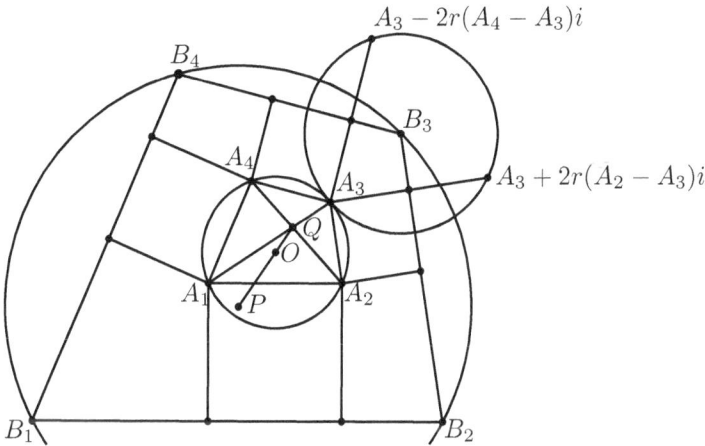

Figure 3.3: Problem 12294

Solution.
Since the point A_1 lies on the unit circle, let

$$A_1 = \cos(\theta_1) + i\sin(\theta_1) = \frac{1 - a^2}{1 + a^2} + \frac{2ai}{1 + a^2}$$

where $a = \tan(\theta_1/2)$. Similarly, we have

$$A_2 = \frac{1 - b^2}{1 + b^2} + \frac{2bi}{1 + b^2}, \quad A_3 = \frac{1 - c^2}{1 + c^2} + \frac{2ci}{1 + c^2}, \quad A_4 = \frac{1 - d^2}{1 + d^2} + \frac{2di}{1 + d^2}$$

with $a < b < c < d$. Moreover, by the previous Discussion, we find that

$$B_1 = A_1 + 2ri\, C(0, A_1 - A_2, A_4 - A_1),$$
$$B_2 = A_2 + 2ri\, C(0, A_2 - A_3, A_1 - A_2),$$
$$B_3 = A_3 + 2ri\, C(0, A_3 - A_4, A_2 - A_3),$$
$$B_4 = A_4 + 2ri\, C(0, A_4 - A_1, A_3 - A_4),$$

and so

$$P = C(B_1, B_2, B_3) = 2\lambda r \cdot Q$$

where

$$\lambda = \frac{(ab+1)(c-d) + (cd+1)(a-b)}{(a-c)(b-d)}$$

and

$$Q = \frac{A_1 A_3 (A_2 + A_4) - A_2 A_4 (A_1 + A_3)}{A_1 A_3 - A_2 A_4}$$
$$= \frac{2(a+c-b-d) + 2(ac-bd)i}{(ab+1)(c-d) + (cd+1)(a-b)} - 1$$

is the intersection of $A_1 A_3$ and $A_2 A_4$. Since $O = 0$ and $2\lambda r$ is a real number, we conclude that the points Q, O, and P are collinear, which means that $A_1 A_3$, $A_2 A_4$, and OP are concurrent. $\qquad\qquad\square$

Remark. In general, if u_1, u_2, v_1 and v_2 lie on the unit circle $|z| = 1$ with $u_1 u_2 \neq v_1 v_2$ then, by solving

$$\begin{cases} z + u_1 u_2 \bar{z} = u_1 + u_2 \\ z + v_1 v_2 \bar{z} = v_1 + v_2 \end{cases},$$

we find that the intersection of the two chords $[u_1, u_2]$ and $[v_1, v_2]$ is the complex number

$$z = \frac{u_1 u_2 (v_1 + v_2) - v_1 v_2 (u_1 + u_2)}{u_1 u_2 - v_1 v_2}.$$

Another fascinating theorem of plane geometry that can be elegantly proven using complex numbers is Napoleon's theorem (although there is no concrete evidence that he was the one who discovered it). It states that if three equilateral triangles $A'BC$, $AB'C$, and ABC' are constructed externally (or internally) on the sides of a triangle ABC, then their respective centers G_1, G_2, G_3 form an equilateral triangle.

Here is a proof. We first note that a triangle PQR oriented positively is equilateral if and only if $e^{i\pi/3}(Q - P) = R - P$, that is $P + \omega Q + \omega^2 R = 0$ with $\omega = e^{2\pi i/3}$. Hence, if we construct externally the equilateral triangles

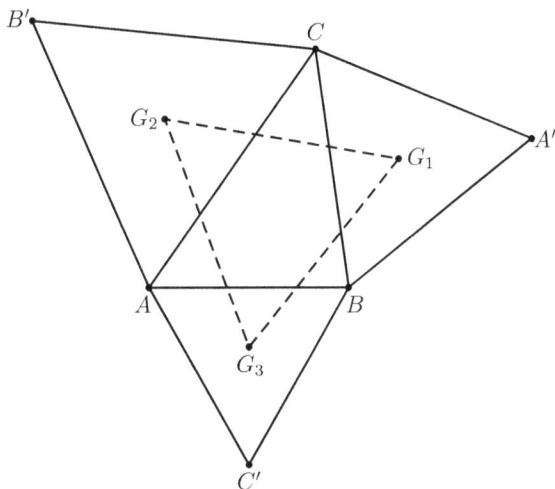

Figure 3.4: Napoleon's theorem

$A'BC$, $AB'C$, and ABC', we find that

$$\begin{cases} A' + \omega C + \omega^2 B = 0 \\ B' + \omega A + \omega^2 C = 0 \\ C' + \omega B + \omega^2 A = 0 \end{cases}.$$

Moreover, $G_1 = (A'+C+B)/3$, $G_2 = (B'+A+C)/3$, and $G_3 = (C'+B+A)/3$. In order to show that $G_1 G_2 G_3$ is an equilateral triangle, it remains to evaluate

$$3(G_1 + \omega G_2 + \omega^2 G_3) = A' + C + B + \omega(B' + A + C) + \omega^2(C' + B + A)$$
$$= (A' + \omega C + \omega^2 B) + \omega(B' + \omega A + \omega^2 C) + \omega^2(C' + \omega B + \omega^2 A) = 0,$$

as we expected. How would we adjust this proof to cover the case when the equilateral triangles are erected internally? For a different proof of see [34, p. 63]).

For more applications of complex numbers in solving geometry problems we refer to [28, Chapter 6]. Another way to show that complex numbers and geometry can be combined beautifully is the book of L.-S. Hahn [54].

Additional problems for practice.

1. **Problem 10415** (Proposed by E. Kitchen, 101(9), 1994). Let T be a triangle whose centroid is at the origin. Choose $\lambda \in \mathbb{R}$, $\lambda > 1$, and dilate one of the *Napoleon triangles* of T by a factor of $-\lambda$ and the other by a factor of $\lambda/(1 - \lambda)$. Prove that T is (simultaneously) perspective with both dilated triangles.

2. **Problem 10483** (Proposed by S. Rabinowitz, 102(9), 1995). Given an odd positive integer n, let $A_1, A_2, \ldots, A_n$ be a regular n-gon with circumcircle Γ. A circle O_i with radius r is drawn externally tangent to Γ at A_i for $i = 1, 2, \ldots, n$. Let P be any point on Γ between A_n and A_1. A circle C (with any radius) is drawn externally tangent to Γ at P. Let t_i be the length of the common external tangent between the circles C and O_i. Prove that

$$\sum_{i=1}^{n}(-1)^i t_i = 0.$$

3. (Due to Van Aubel) A square is drawn externally on each edge of a quadrilateral $ABCD$. Prove that the lines PR and QS which connect the center P, Q, R, S of these squares have equal length and that they intersect at right angles.

4. **Problem 10533** (Proposed by A. Flores, 103(6), 1996). On a parallelogram P construct exterior squares on the sides. The centers of these squares form a square Q_E. On the same parallelogram construct the interior squares on the sides. The centers of these squares form another square Q_I.

   (a) Show that $\text{Area}(Q_E) - \text{Area}(Q_I) = 2\text{Area}(P)$.

   (b) Is there a generalization when P is replaced by an arbitrary convex quadrilateral?

5. **Problem 10783** (Proposed by W. W. Chao, 107(2), 2000). Let $ABCD$ be a cyclic quadrilateral such that AD is not parallel to BC. Given points E and F on the line CD, let G and H be the circumcenters of BCE and ADF. Prove that the lines AB, CD, and GH are concurrent or parallel if and only if there is a circle through A, B, E, and F.

6. (Japanese Theorem) Let $ABCD$ be a convex quadrilateral inscribed in a circle. Let r_a and I_a be the inradius and the incenter of BCD respectively. Define r_b, I_b, r_c, I_c, r_d and I_d similarly. Show that $I_a I_b I_c I_d$ form a rectangle and $r_a + r_c = r_b + r_d$.

7. **Problem 11714** (Proposed by N. Minculete and C. Barbu, 120(6), 2013). Let $ABCD$ be a cyclic quadrilateral. Let $e = |AC|$ and $f = |BD|$. Let r_a be the inradius of BCD, and define r_b, r_c, and r_d similarly. Prove that $e r_a r_c = f r_b r_d$.

8. **Problem 12343** (Proposed by T. Q. Hung, 129(8), 2022). Let $ABCD$ be a convex quadrilateral with $AB = a$, $BC = b$, $CD = c$, $DA = d$, $AC = e$, and $BD = f$. Prove that $ABCD$ is a cyclic quadrilateral if and only if

$$\frac{f^2 - e^2}{ac + bd} = \frac{(a^2 - c^2)(b^2 - d^2)}{(ab + cd)(ad + bc)}.$$

3.3 Concurrency of three lines

Problem 12431 (Proposed by T. Q. Hung, 130(10), 2023). Let $ABCD$ be a tangential quadrilateral. Let I be the center of the incircle, and let P be the intersection of the diagonals of $ABCD$. Suppose that the incircle touches the sides AB, BC, CD, and DA at points E, F, G, and H, respectively. Let W, X, Y, and Z be the centroids of triangles ECD, FDA, GAB, and HBC, respectively. Prove that the lines XZ, YW, and PI are concurrent.

Discussion.
Let the vertices $A, B, C,$ and D of the quadrilateral be four complex numbers. Let

$$a = |AH| = |AE|, \quad b = |BF| = |BE|, \quad c = |CF| = |CG|, \quad d = |DG| = |DH|.$$

Then we have

$$E = \frac{aB + bA}{a + b}, \quad F = \frac{bC + cB}{b + c}, \quad G = \frac{cD + dC}{c + d}, \quad H = \frac{dA + aD}{d + a}$$

and

$$W = \frac{E + C + D}{3}, \quad X = \frac{F + D + A}{3}, \quad Y = \frac{G + A + B}{3}, \quad Z = \frac{H + B + C}{3}.$$

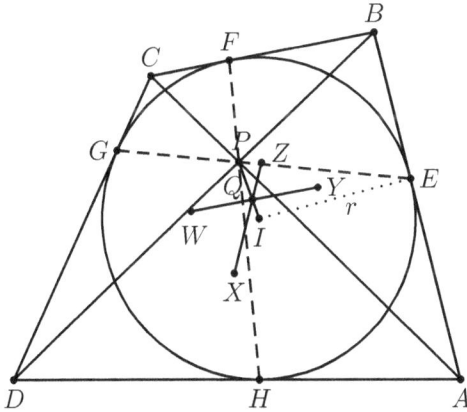

Figure 3.5: Problem 12431

Furthermore, let Q be the intersection point of YW and XZ. It is easily to verify that

$$Q = tY + (1 - t)W = sX + (1 - s)Z$$

where

$$t = \frac{c + d}{a + b + c + d} \quad \text{and} \quad s = \frac{b + c}{a + b + c + d}.$$

Now it suffices to determine the positions of points I and P and then prove that Q lies on the line PI.

Solution.
Recall that in a tangential quadrilateral $ABCD$, the two diagonals AC and BD and the tangency chords GE and FH are concurrent. Therefore,

$$P = t'F + (1-t')H = s'G + (1-s')E = \frac{aC + cA + dB + bD}{a+b+c+d}$$

where

$$t' = \frac{b+c}{a+b+c+d} \quad \text{and} \quad s' = \frac{c+d}{a+b+c+d}.$$

Moreover, going from E to I along the inradius of length r, we find that

$$I = E + ir\frac{B-E}{|BE|} = \frac{aB+bA}{a+b} + i\frac{r}{b}\left(B - \frac{aB+bA}{a+b}\right)$$
$$= \frac{(a+ir)B + (b-ir)A}{a+b}.$$

In a similar way, we also get

$$I = \frac{(b+ir)C + (c-ir)B}{b+c} = \frac{(c+ir)D + (d-ir)C}{c+d} = \frac{(d+ir)A + (a-ir)D}{d+a}.$$

Hence

$$I = \frac{(a+b)I + (b+c)I + (c+d)I + (d+a)I}{2(a+b+c+d)}$$
$$= \frac{b+d}{a+b+c+d} \cdot \frac{A+C}{2} + \frac{a+c}{a+b+c+d} \cdot \frac{B+D}{2}. \qquad (3.3)$$

Now it is easy to show that Q lies on the line PI. Indeed, we have

$$Q = \frac{P+2I}{3}.$$

$\square$

Remark. Notice that if the quadrilateral $ABCD$ is a rhombus then $P = I$. The fact that in the tangential quadrilateral $ABCD$, the two diagonals AC and BD and the tangency chords GE and FH are concurrent can be seen as a limiting case of Brianchon's theorem, which states that if a hexagon is circumscribed around a conic section then the lines joining opposite vertices are concurrent (see for example [3, p. 64]).

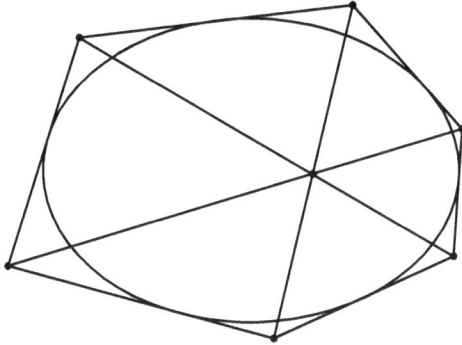

Figure 3.6: Brianchon's theorem

Indeed, if we apply Brianchon's theorem to the hexagon $AEBCGD$ concerning the incircle then we find that AC, BD and GE are concurrent. Now we repeat the same argument for the hexagon $ABFCDH$ and it follows that AC, BD and FH are concurrent. So the desired property is verified.

A further note: the equality (3.3) confirms Newton's quadrilateral theorem: the center of the incircle I lies on the Newton line, the line connecting the midpoints of the diagonals AC and BD.

Additional problems for practice.

1. **Problem 10405** (Proposed by H. Gülicher, 101(8), 1994).
   Let $A_1A_2A_3A_4A_5A_6$ be a hexagon *circumscribed* about a conic, and form the intersections $P_i = A_iA_{i+2} \cap A_{i+1}A_{i+3}$ ($i = 1, \ldots, 6$ all indices mod 6). Show that the P_i are the vertices of a hexagon *inscribed* in a conic.

2. **Problem 10678** (Proposed by C. Kimberling, 105(7), 1998). Let C be the incircle of a triangle $A_1A_2A_3$. Suppose that whatever $\{i, j, k\} = \{1, 2, 3\}$, there is a circle through A_j and A_k meeting C in a single point B_i. Prove that the lines A_1B_1, A_2B_2, A_3B_3 are concurrent.

3. **Problem 11723** (Proposed by L. R. King, 120(7), 2013). Let A, B, and C be three points in the plane, and let D, E, and F be points lying on BC, CA, and AB, respectively. Show that there exists a conic tangent to BC, CA, and AB at D, E, and F, respectively, if and only if AD, BE, and CF are concurrent.

4. **Problem 11880** (Proposed by D. Andrica, 123(1), 2016). Let $ABCD$ be any plane quadrilateral (not necessarily convex or even simple). Let a parallelogram be created by constructing through the ends of each diagonal of $ABCD$ lines parallel to the other diagonal. Show that each

diagonal of this parallelogram passes through the intersection point of
a pair of opposite sides of $ABCD$.

5. **Problem 12076** (Proposed by T. Beke, 125(10), 2018). From each of the
   three feet of the altitudes of an arbitrary triangle, produce two points by
   projecting this foot onto the other two sides. Show that the six points
   produced in this way are concyclic.

3.4 A triangle inscribed in another triangle

Problem 12269 (Proposed by M. Şahin and A. C. Güllü, 128(7), 2021). Let
ABC be an acute triangle. Suppose that D, E, and F are points on sides
BC, CA, and AB, respectively, such that FD is perpendicular to BC, DE is
perpendicular to CA, and EF is perpendicular to AB. Prove

$$\frac{AF}{AB} + \frac{BD}{BC} + \frac{CE}{CA} = 1.$$

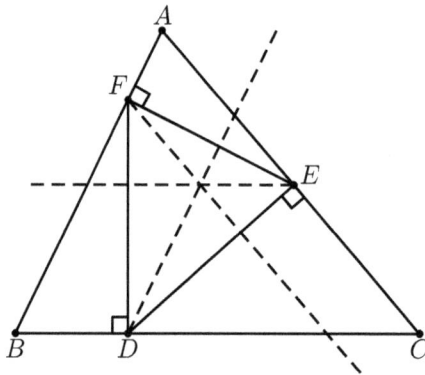

Figure 3.7: Problem 12269

Discussion.
Let $a = BC$, $b = CA$, $c = AB$, $x = BD$, $y = CE$, and $z = AF$. So we have
to show that

$$\frac{x}{a} + \frac{y}{b} + \frac{z}{c} = 1.$$

By the law of cosines and the condition of perpendicularity, we have

$$\frac{z}{b-y} = \frac{AF}{AE} = \cos(A) = \frac{b^2 + c^2 - a^2}{2bc}$$

which immediately implies

$$\frac{z}{c} = \frac{b^2 + c^2 - a^2}{2c^2}\left(1 - \frac{y}{b}\right).$$

By working similarly with the other two angles, we can find two more equations and then try to express all the ratios x/a, y/b, z/c in terms of the sides.

Solution.
By solving the following system of the three linear equations

$$\begin{cases} \dfrac{x}{a} = \dfrac{c^2 + a^2 - b^2}{2a^2}\left(1 - \dfrac{z}{c}\right) \\ \dfrac{y}{b} = \dfrac{a^2 + b^2 - c^2}{2b^2}\left(1 - \dfrac{x}{a}\right) \\ \dfrac{z}{c} = \dfrac{b^2 + c^2 - a^2}{2c^2}\left(1 - \dfrac{y}{b}\right) \end{cases}$$

with respect to x/a, y/b, z/c, we obtain

$$\frac{x}{a} = \frac{a^2 + c^2 - b^2}{a^2 + b^2 + c^2}, \quad \frac{y}{b} = \frac{a^2 + b^2 - c^2}{a^2 + b^2 + c^2}, \quad \frac{z}{c} = \frac{b^2 + c^2 - a^2}{a^2 + b^2 + c^2}. \tag{3.4}$$

Therefore

$$\frac{x}{a} + \frac{y}{b} + \frac{z}{c} = \frac{(a^2 + c^2 - b^2) + (a^2 + b^2 - c^2) + (b^2 + c^2 - a^2)}{a^2 + b^2 + c^2} = 1.$$

□

Remark. The formulas given in (3.4) show that any acute triangle ABC admits a triangle DEF and such a triangle is unique. Moreover, the following three lines are concurrent: the parallel to AB through D, the parallel to BC through E, and the parallel to CA through F. Indeed, since AB is perpendicular to FE, it follows that the parallel to AB through D is also perpendicular to FE which means that it is the altitude from D of the triangle DEF. Similarly, we find the other two altitudes of the triangle DEF. We may conclude that such lines meet at the orthocenter of the triangle DEF.

Additional problems for practice.

1. **Problem 11554** (Proposed by Z. Yun, 118(2), 2011). In triangle ABC let I be the incenter, and let A', B', C' be the reflections of I through sides BC, CA, AB, respectively. Prove that the lines AA', BB', and CC' are concurrent.

2. **Problem 11942** (Proposed by F. Parvanescu, 123(9), 2016). In acute triangle ABC, let D be the foot of the altitude from A, let E be the foot of the perpendicular from D to AC, and let F be a point on segment DE. Prove that AF is perpendicular to BE if and only if $|FE|/|FD| = |BD|/|CD|$.

3. **Problem 12027** (Proposed by A. Hannan, 125(3), 2018). Let ABC be a triangle with circumradius R and inradius r. Let D, E, and F be the points where the incircle of ABC touches BC, CA, and AB, respectively, and let X, Y, and Z be the second points of intersection between the incircle of ABC and AD, BE, and CF, respectively. Prove

$$\frac{AX}{XD} + \frac{BY}{YE} + \frac{CZ}{ZF} = \frac{R}{r} - \frac{1}{2}.$$

4. **Problem 12154** (Proposed by M. Lukarevski, 127(1), 2020). Let r_a, r_b, and r_c be the exradii of a triangle with circumradius R and inradius r. Prove

$$\frac{r_a}{r_b + r_c} + \frac{r_b}{r_c + r_a} + \frac{r_c}{r_a + r_b} \geq 2 - \frac{r}{R}.$$

5. **Problem 12357** (Proposed by Van Khea and D. S. Marinescu, 129(10), 2022). Suppose that triangles ABC and DEF have the same centroid, where D, E, and F are on the segments BC, CA, and AB, respectively. Let I be the incenter of triangle ABC. Prove

$$\frac{AI}{AD} + \frac{BI}{BE} + \frac{CI}{CF} \leq 2.$$

3.5   A surprising bisection

Problem 12165 (Proposed by T. Q. Hung and N. M. Ha, 127(2), 2020). Let $MNPQ$ be a square with center K inscribed in triangle ABC with N and P lying on sides AB and AC, respectively, while M and Q lie on side BC. Let the incircle of BMN touch side BM at S and side BN at F, and let the incircle of CQP touch side CQ at T and side CP at E. Let L be the point of intersection of lines FS and ET. Prove that KL bisects the segment ST.
Discussion.
We set up a coordinate system with the x-axis along the side BC such that $A = (0, h)$. Let $B = (-a_1, 0)$ and $C = (a_2, 0)$ with $a_1 + a_2 = a = |BC|$, $b = |AC|$ and $c = |AB|$ and let l be the side of the square. Since the triangle ABC is similar to the triangle ANP, we find that $\frac{h}{a} = \frac{h-l}{l}$ and so $l = \frac{ah}{a+h}$. Thus, the four vertices of the square and its center are

$$M = \frac{l}{a}(-a_1, 0), \ Q = \frac{l}{a}(a_2, 0), \ N = \frac{l}{a}(-a_1, a), \ P = \frac{l}{a}(a_2, a)$$

respectively. Its center is

$$K = \frac{l}{2a}(a_2 - a_1, a)$$

For this setup, once we find the coordinates of the points $J = \frac{S+T}{2}$ and L, proving the alignment of the three points K, J and L will be straightforward.

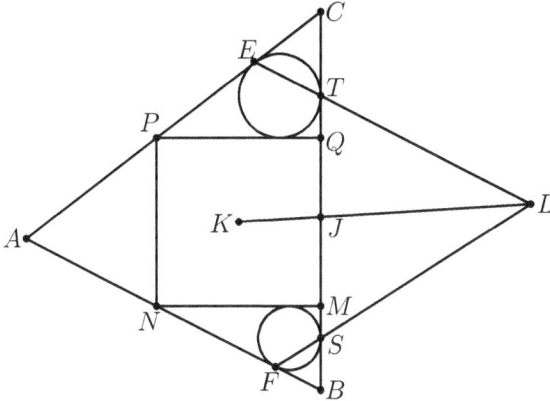

Figure 3.8: Problem 12165

Solution.
From the discussion above, we find the radii of the incircles of the right triangles BMN and CQP are

$$r_1 = \frac{la_1}{a_1 + h + c} \quad \text{and} \quad r_2 = \frac{la_2}{a_2 + h + b},$$

respectively. Therefore, $S = (x_M - r_1, 0)$ and $T = (x_Q + r_2, 0)$, which implies

$$J = \frac{S + T}{2} = \frac{l}{2a}\left(a_2 - a_1 + \frac{a_2 a}{a_2 + h + b} - \frac{a_1 a}{a_1 + h + c}, 0\right).$$

Moreover, the lines FS and ET are respectively

$$y = -\frac{x - x_S}{\tan\left(\frac{\angle B}{2}\right)} = -\frac{(a_1 + c)(x - x_S)}{h}, \quad y = \frac{x - x_T}{\tan\left(\frac{\angle C}{2}\right)} = \frac{(a_2 + b)(x - x_T)}{h}.$$

Consequently, their intersection point is

$$L = -\frac{l}{2a}\left(a_1 - a_2 + 2c - 2b, \frac{2(a_2 + b)(a_1 + c)}{h} - 2h - a\right).$$

Finally, it is straightforward to verify that $(1 - t)K + tL = J$ for

$$t = \frac{ah}{2((a_2 + b)(a_1 + c) - h^2)}.$$

This confirms the alignment of the three points K, J and L and completes the proof. $\qquad\square$

Remark. By using the same approach, it can be shown that KL bisects the segment ST even if $MNPQ$ is a rectangle inscribed in the triangle ABC.

Additional problems for practice.

1. **Problem 10317** (Proposed by J. B. Romero Márquez, 100(6), 1993). Let $\triangle ABC$ be inscribed in the circle Γ and let A', B', C' be the midpoints of the arcs BC, CA, AB respectively.

   a) Prove that the incenter of $\triangle ABC$ is the orthocentre of $\triangle A'B'C'$.

   b) Prove that the pedal triangle $\triangle A'B'C'$ is homothetic to $\triangle ABC$.

2. **Problem 11046** (Proposed by C. Solana, 110(9), 2003). Let ABC be a triangle, let I be the incircle of ABC, and let r be the radius of I. Let K_1, K_2, and K_3 be the three circles outside I and tangent to I and to two of the three edges of ABC. Let r_i, be the radius of K_i, $1 \leq i \leq 3$. Show that
$$r = \sqrt{r_1 r_2} + \sqrt{r_2 r_3} + \sqrt{r_3 r_1}.$$

3. **Problem 11171** (Proposed by C. Cheng, 112(7), 2005). Let A_1, A_2, A_3, and A_4 list in order the vertices of a convex quadrilateral Q. Treat all indices as integers modulo 4. Let L_k denote the line through A_k and A_{k+1}, and let C_k be the circle tangent to L_{k-1}, L_k, and L_{k+1} outside Q. Let M_k be the line through the points where C_k is tangent to L_{k-1} and to L_{k+1}. Let E_k be the intersection of M_k with M_{k+1}. Prove that $E_2 E_4$ bisects $A_2 A_4$.

4. **Problem 12393** (Proposed by F. Masroor, 130(5), 2023). Let E and F be points in the interior of $\triangle ABC$ such that $\triangle ABF$ and $\triangle ACE$ are similar (with similarity mapping A to A, B to C, and F to E). Prove that EF is parallel to AC if and only if AE bisects BC.

5. **Problem 12253** (Proposed by A. Girban and B. D. Suceava, 128(5), 2021). Let ABC be a triangle, and let D and E be the contact points of the incircle of ABC with the segments BC and CA, respectively. Let M be the intersection of the line DE and the line through A parallel to BC. Prove that the bisector of $\angle ABC$ passes through the midpoint of DM.

3.6 A tangent to a circle

Problem 12245 (Proposed by J. Chen, 128(4), 2021). Suppose that two circles α and β, with centers P and Q, respectively, intersect orthogonally at A and B. Let CD be a diameter of β that is exterior to α. Let E and F be points on α such that CE and DF are tangent to α, with C and E on one side of PQ and D and F on the other side of PQ. Let S be the intersection of CF and QA, and let T be the intersection of DE and QB. Prove that ST is parallel to CD and is tangent to α.

Discussion.
Without loss of generality we may assume that β is the unit circle centered at $Q = (0,0)$ with

$$A = (\cos(u), \sin(u)) \quad \text{and} \quad C = (\cos(v), \sin(v))$$

such that $u, v \in (0, \pi)$, $u \neq \pi/2$ and $u \neq v$. Moreover, let

$$B = (\cos(u), -\sin(u)) \quad \text{and} \quad D = (-\cos(v), -\sin(v)).$$

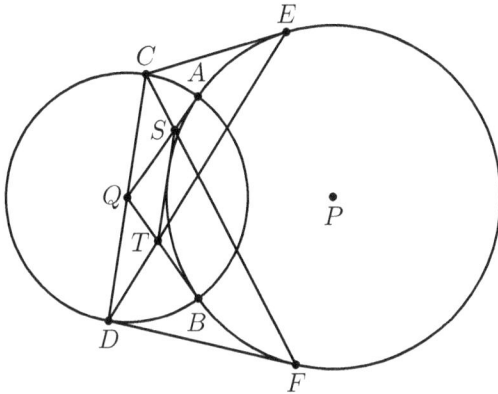

Figure 3.9: Problem 12245

Then the lines QA and QB are

$$\sin(u)x - \cos(u)y = 0 \quad \text{and} \quad \sin(u)x + \cos(u)y = 0,$$

respectively. This also implies that $P = (1/\cos(u), 0)$ and R, the radius of the circle α, is given by

$$R^2 = |PA|^2 = \left(\cos(u) - \frac{1}{\cos(u)}\right)^2 + \sin^2(u) = \tan^2(u).$$

Now, it remains to find the lines DE and CF, the points T and S and finally verify that the line ST is parallel to CD and tangent to the circle α.

Solution.
The polar of C with respect to the circle α is

$$\left(\cos(v) - \frac{1}{\cos(u)}\right)x + \sin(v)y - \frac{\cos(v)}{\cos(u)} + 1 = 0.$$

Since the above equation is satisfied by D, it follows that this polar is the line DE.

Similarly, the polar of D with respect to the circle α is

$$\left(\cos(v) + \frac{1}{\cos(u)}\right)x + \sin(v)y - \frac{\cos(v)}{\cos(u)} - 1 = 0.$$

Since the above equation is satisfied by C, it follows that this polar is the line CF.

Hence, the intersection of DE and QB is

$$T = \left(\frac{\cos(u) - \cos(v)}{1 - \cos(u+v)}, -\tan(u)\frac{\cos(u) - \cos(v)}{1 - \cos(u+v)}\right)$$

whereas the intersection of the CF and QA is

$$S = \left(\frac{\cos(u) + \cos(v)}{1 + \cos(u-v)}, \tan(u)\frac{\cos(u) + \cos(v)}{1 + \cos(u-v)}\right).$$

Therefore, the equation of the line ST becomes

$$\sin(v)x - \cos(v)y + \frac{\sin(u) - \sin(v)}{\cos(u)} = 0$$

which implies that ST is parallel to the line CD whose equation is

$$\sin(v)x - \cos(v)y = 0.$$

Finally, the squared distance of the center $P = (P_x, P_y) = (1/\cos(u), 0)$ from the line ST is

$$\frac{\left(\sin(v)P_x - \cos(v)P_y + \frac{\sin(u)-\sin(v)}{\cos(u)}\right)^2}{(\sin(v))^2 + (-\cos(v))^2} = \tan^2(u) = R^2$$

which proves that the line ST is tangent to the circle α.   □

Remark. In general, for any conic section α of equation

$$Ax^2 + Bxy + Cy^2 + Dx + Ey + F = 0,$$

we define the polar of a point (x_0, y_0) with respect to α as the line $ax+by+c = 0$ where

$$\begin{bmatrix} A & B/2 & D/2 \\ B/2 & C & E/2 \\ D/2 & E/2 & F \end{bmatrix}\begin{bmatrix} x_0 \\ y_0 \\ 1 \end{bmatrix} = \begin{bmatrix} a \\ b \\ c \end{bmatrix}.$$

Polars have several useful properties. If a point lies on the conic section then its polar is the tangent line through this point to the conic section. If two tangent lines can be drawn from a point to the conic section, then its polar passes through both tangent points.

Additional problems for practice.

1. **Problem E3293** (Proposed by J. Keane, 95(9), 1988). Suppose that the distinct circles C_1 and C_2 intersect at P and Q. Suppose that the tangent to C_1 at P intersects C_2 again at A, the tangent to C_2 at P intersects C_1 again at B, and the line AB separates P and Q. Let C_3 be the circle externally tangent to C_1, externally tangent to C_2, tangent to line AB, and lying on the same side of AB as Q. Prove that the circles C_1 and C_2 intercept equal segments on one of the tangents to C_3 through P.

2. **Problem 10755** (Proposed by J. Fukuta, 106(8), 1999). An arbitrary circle O is drawn through vertices B and D of a convex quadrilateral $ABCD$. Let O_1 be the circle tangent to lines AB and AD and tangent to O internally at a point of O on the opposite side of line BD from A. Let O_2 be the circle tangent to lines CB and CD and tangent to O internally at a point of O on the opposite side of line BD from C. Let R_1 and R_2 be the radii of circles O_1 and O_2, respectively, and let r_1 and r_2 be the radii of the incircles of triangles ABD and CBD, respectively. Prove that the quadrilateral $ABCD$ is inscribable in a circle if and only if
$$\frac{r_1}{R_1} + \frac{r_2}{R_2} = 1.$$

3. **Problem 11717** (Proposed by N. T. Binh, 120(6), 2013). Given a circle c and line segment AB tangent to c at a point E that lies strictly between A and B, provide a compass and straightedge construction of the circle through A and B to which c is internally tangent.

4. **Problem 12325** (Proposed by D. Luu, 129(5), 2022). Let $ABCD$ be a quadrilateral with a circumscribed circle ω and an inscribed circle γ. Prove that there are two circles α and β with the following property: for any triangle MEF with (1) M on ω, (2) E and F on the line AB, and (3) the lines ME and MF tangent to γ, the circumcircle of $\triangle MEF$ is tangent to α and β.

5. **Problem 12438** (Proposed by R. Foote and G. Galperin, 131(1), 2024). Let $ABCD$ be an isosceles trapezoid inscribed in a circle γ with AB parallel to CD. Let F bisect AB and G bisect CD. Let E be the ellipse with minor axis FG and major axis of length AC. Prove that E is internally tangent to γ at two points.

3.7 Collinearity of three points

Problem 12371 (Proposed by L. Dong (Vietnam), 130(2) 2023). Let ABC be a triangle with incenter I. Let D be the point where the incircle of ABC

touches BC, let H be the perpendicular projection of A on BC, and let K be the midpoint of AH. Suppose that the lines IB and IC meet AH at the points M and N, respectively. Prove that the orthocenter of $\triangle IMN$ lies on the line KD.

Discussion.
Let $A = (0, h)$, $B = (x_B, 0)$, $C = (x_C, 0)$ with $h > 0$ and $x_B < x_C$. Then $H = (0, 0)$ and $K = (0, \frac{h}{2})$. Let $I = (x_I, y_I)$ be the incenter of $\triangle ABC$, then $y_I = r$, where r is the inradius. Moreover, we find that

$$D = (x_I, 0), \quad M = \left(0, \frac{r x_B}{x_B - x_I}\right), \quad N = \left(0, \frac{r x_C}{x_C - x_I}\right).$$

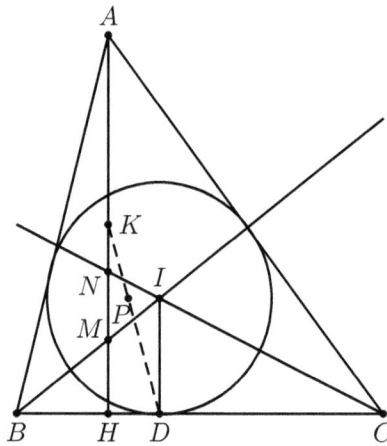

Figure 3.10: Problem 12371

Once we determine the orthocenter $P = (x_P, y_P)$ of the triangle IMN, we can show that P, K, and D are collinear, namely, $P = tK + (1 - t)D$ for some real number t.

Solution.
Notice that $y_P = y_I = r$ because PI is orthogonal to AH. By imposing the orthogonality condition between NP and BM,

$$\langle P - N, M - B \rangle = \left\langle \left(x_P, r - \frac{r x_C}{x_C - x_I}\right), \left(-x_B, \frac{r x_B}{x_B - x_I}\right)\right\rangle$$

$$= -x_P x_B + \frac{r^2 x_I x_B}{(x_I - x_B)(x_C - x_I)} = 0,$$

we obtain

$$x_P = \frac{r^2 x_I}{(x_I - x_B)(x_C - x_I)}.$$

We now verify that

$$P = tK + (1 - t)D \qquad \text{for} \quad t = \frac{2a}{a+b+c}$$

with $a = |BC|$, $b = |CA|$, $c = |AB|$, which means that the orthocenter P lies on the line KD.

To this end, the equality for the y-coordinate is immediate:

$$r = t\frac{h}{2} = \frac{ah}{a+b+c},$$

which is equivalent to

$$r(a + b + c) = ah.$$

This holds because both sides are equal to 2 times the area of $\triangle ABC$.

As regards the x-coordinate, we have to check that

$$\frac{r^2 x_I}{(x_I - x_B)(x_C - x_I)} = (1 - t)x_I = \frac{b+c-a}{a+b+c}x_I.$$

Indeed, since $x_C - x_I = |CD| = (a+b-c)/2$ and $x_I - x_B = |BD| = (a+c-b)/2$, the above equality is equivalent to

$$4r^2(a + b + c)^2 = (a + b + c)(b + c - a)(a + b - c)(a + c - b).$$

This holds, because, by Heron's formula, both sides are equal to 16 times the squared area of $\triangle ABC$. □

Remark. Notice that the triangle inequality $b + c > a$ implies that $t = 2a/(a + b + c) \in (0, 1)$. Hence, more precisely, the orthocenter P lies on the segment KD.

Actually, it is well known that the orthocenter of a triangle ABC lies along another special line, i. e. the *Euler line* which joins the circumcenter O and the centroid G of ABC. An easy proof of this property can be carried out by using complex numbers. Indeed, we may assume that the circumcircle of ABC is the unit circle centered at $O = 0$. The centroid is $G = (A+B+C)/3$. We claim that the orthocenter H is

$$H = O + 3G = A + B + C$$

which shows the collinearity. All that is left is to verify that $H - A$ is orthogonal to $B - C$:

$$\text{Re}\left((H - A)\overline{(B - C)}\right) = \text{Re}\left((B + C)(\overline{B} - \overline{C})\right)$$
$$= \text{Re}\left(|B|^2 + C\overline{B} - B\overline{C} - |C|^2\right) = 1 + 0 - 1 = 0$$

and we are done.

Additional problems for practice.

1. **Problem 10308** (Proposed by R. Connelly and J. H. Hubbard, 100(5), 1993). Suppose that P_1, P_2, P_3, Q_1, Q_2, Q_3 are six points in the plane and that the distance between P_i and Q_j $(i, j = 1, 2, 3)$ is $i + j$. Show that the six points are collinear.

2. **Problem 10673** (Proposed by M. Mazur, 105(6), 1998). Let C be the circle inscribed in the triangle $A_1 A_2 A_3$ and let $P_i \in A_{i+1} A_{i+2}$ (subscripts taken modulo 3) be such that the lines $P_i A_i$ are concurrent. Let t_i be the second tangent from P_i to C, the first being $A_{i+1} A_{i+2}$. Prove that Q_1, Q_2, Q_3 defined by $Q_i = t_i \cap P_{i+1} P_{i+2}$ are collinear.

3. **Problem 12073** (Proposed by H. Karakus, 125(9), 2018). Given a scalene triangle ABC, let G denote its centroid and H denote its orthocenter. Let P_A be the second point of intersection of the two circles through A that are tangent to BC at B and at C. Similarly define P_B and P_C. Prove that G, H, P_A, P_B, and P_C are concyclic.

4. **Problem 12224** (Proposed by C. Perng, 128(1), 2021). Let ABC be a triangle, with D and E on AB and AC, respectively. For a point F in the plane, let DF intersect BC at G and let EF intersect BC at H. Furthermore, let AF intersect BC at I, let DH intersect EG at J, and let BE intersect CD at K. Prove that I, J, and K are collinear.

5. **Problem 12238** (Proposed by T. Q. Hung, 128(3), 2021). Let $ABCD$ be a convex quadrilateral with $AD = BC$. Let P be the intersection of the diagonals AC and BD, and let K and L be the circumcenters of triangles PAD and PBC, respectively. Show that the midpoints of segments AB, CD, and KL are collinear.

6. **Problem 12483** (Proposed by L. Dong, 131(8), 2024). Given $\triangle ABC$, let D and E be the intersections of the incircle ω with BC and AC, respectively, and let K be the reflection of D in A. Let J be the excenter opposite B and let H denote the second intersection of the circumcircle of JEC with ω. Prove that K, E, and H are collinear.

3.8 Another property of the Nagel point

Problem 12291 (Proposed by L. Giugiuc and P. Braica, 128(10), 2021). The *Nagel point* of a triangle is the point common to the three segments that join a vertex of the triangle to the point at which an excircle touches the opposite side. Let ABC be a triangle with incenter I and Nagel point J. Prove that AJ is perpendicular to the line through the orthocenters of triangles IAB and IAC.

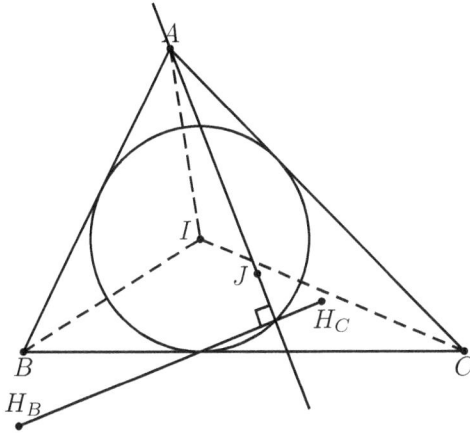

Figure 3.11: Problem 12291

Discussion.
We solve this problem by using barycentric coordinates: each point P in the plane is assigned an ordered triple of real numbers $(x : y : z)$ such that $x+y+z = 1$ and $P = xA+yB+zC$. Notice that $A = (1 : 0 : 0)$, $B = (0 : 1 : 0)$, $C = (0 : 0 : 1)$. Let $a = |BC|$, $b = |CA|$, $c = |AB|$ and $s = (a+b+c)/2$. Then

$$I = \left(\frac{a}{2s} : \frac{b}{2s} : \frac{c}{2s} \right) \quad \text{and} \quad J = \left(\frac{s-a}{s} : \frac{s-b}{s} : \frac{s-c}{s} \right),$$

which provide a computational approach for geometry problems. With this approach, here the harder part is to establish the barycentric coordinates of H_B and H_C, the orthocenters of triangles IAC and IAB, respectively. Once this is accomplished, the orthogonality will be a straightforward verification.

Solution.
The orthocenter H_B of the triangle IAC has (not normalized) barycentric coordinates with respect to IAC:

$$\left(\frac{1}{b'^2 + c'^2 - a'^2} : \frac{1}{c'^2 + a'^2 - b'^2} : \frac{1}{a'^2 + b'^2 - c'^2} \right)$$

where $a' = |AC| = b$, $b' = |IC| = \sqrt{\frac{ab(a+b-c)}{a+b+c}}$, $c' = |IA| = \sqrt{\frac{bc(b+c-a)}{a+b+c}}$.
After converting the normalized barycentric coordinates of H_B with respect to the triangle IAC to the barycentric coordinates with respect to the triangle ABC, we find

$$H_B = \left(\frac{c-b}{a+c-b} : \frac{b}{a+c-b} : \frac{a-b}{a+c-b} \right).$$

A similar computation can be done for the orthocenter H_C of the triangle IAB,

$$H_C = \left(\frac{b-c}{a+b-c} : \frac{a-c}{a+b-c} : \frac{c}{a+b-c} \right).$$

Finally, it just remains to show that AJ is perpendicular to $H_B H_C$: we verify that

$$\langle A - J, H_B - H_C \rangle = 0$$

which is a tedious, but fairly straightforward task.   □

Remark. We recall that, in general, if $[p : q : r]$ are the barycentric coordinates of a point P with respect to a triangle DEF, such that

$$D = (u_1 : v_1 : w_1), \quad E = (u_2 : v_2 : w_2), \quad F = (u_3 : v_3 : w_3)$$

are the barycentric coordinates with respect to the triangle ABC, then the barycentric coordinates of P with respect to triangle ABC are as follows,

$$(u_1 p + u_2 q + u_3 r : v_1 p + v_2 q + v_3 r : w_1 p + w_2 q + w_3 r).$$

For more applications of barycentric coordinates in solving geometry problems, we refer to [28, Chapter 7].

Additional problems for practice.

1. **Problem 10763** (Proposed by J. Anglesio, 106(9), 1999). Let ABC be a triangle; let O be its circumcenter, H its orthocenter, I its incenter, N its Nagel point, and X, Y, Z its excenters. Let S be defined so that O is the midpoint of HS, and let T denote the midpoint of SN. It is known that the orthocenter rand the nine-point center of triangle XYZ are I and $=$, respectively. Prove

   (a) the circumcenter of triangle XYZ is T; and

   (b) the centroid of triangle XYZ is the centroid of SIN.

2. **Problem 11683** (Proposed by R. Struble, 119(10), 2012). Given a triangle ABC, let FC be the foot of the altitude from the incenter to AB. Define F_B and F_C similarly. Let C_A be the circle with center A that passes through F_B and F_C, and define C_B and C_C similarly. The *Gergonne point* of a triangle is the point at which segments AF_A, BF_B, and CF_C meet. Determine, up to similarity, all isosceles triangles such that the Gergonne point of the triangle lies on one of the circles C_A, C_B, or C_C.

3. **Problem 11881** (Proposed by R. Bagby, 123(1), 2016). Given a triangle T, let T' be the triangle whose vertices are the points where the three excircles of T are tangent to the sides of T.

   (a) Prove that the ratio of the area of T' to the area of T is $r/(2R)$, where r and R are the inradius and circumradius of T.

(b) Show also that the triangle formed by the points of tangency of the incircle likewise has area $r/(2R)$ times that of the original triangle.

4. **Problem 12092** (Proposed by M. Diao and A. Wu, 126(2), 2019). Let ABC be a triangle, and let P be a point in the plane of the triangle satisfying $\angle BAP = \angle CAP$. Let Q and R be diametrically opposite P on the circumcircles of $\triangle ABP$ and $\triangle ACP$, respectively. Let X be the point of concurrency of line BR and line CQ. Prove that XP and BC are perpendicular.

5. **Problem 12144** (Proposed by N. M. Ha and T. Q. Hung, 126(9), 2019). Let $MNPQ$ be a square inscribed in quadrilateral $ABCD$ with M, N, P, and Q lying on sides AB, BC, CD, DA, respectively. Let W, X, Y, and Z be the points where the incircles of triangles AQM, BMN, CNP, and DPQ touch QM, MN, NP, and PQ, respectively. Prove that $ABCD$ has an inscribed circle if and only if WY is perpendicular to XZ.

3.9 An inequality involving the Fermat point of a triangle

Problem 12319 (Proposed by M. Bencze, 129(4), 2022). Let ABC be a triangle with all angles less than $120°$, and let F be the Fermat point of ABC (the point in the interior that minimizes the sum of the distances to A, B, and C). Prove

$$\frac{FA^4}{AB^2} + \frac{FB^4}{BC^2} + \frac{FC^4}{CA^2} \geq \frac{FA^3 + FB^3 + FC^3}{FA + FB + FC}.$$

Discussion.
Since all angles of ABC are less than $120°$, F is an interior point of the triangle from which each side subtends an angle of $120°$. Hence, by the law of cosines,

$$AB^2 = x^2 + xy + y^2, \quad BC^2 = y^2 + yz + z^2, \quad CA^2 = z^2 + zx + x^2$$

where $x = FA$, $y = FB$, $z = FC$. We now see that the required inequality is equivalent to that for any positive real numbers x, y, z,

$$\sum_{\text{cyc}} x \cdot \sum_{\text{cyc}} \frac{x^4}{x^2 + xy + y^2} \geq \sum_{\text{cyc}} x^3 \tag{3.5}$$

where $\sum_{\text{cyc}} t(x, y, z) = t(x, y, z) + t(y, z, x) + t(z, x, y)$.

Solution.
We begin with

$$\sum_{\text{cyc}} x \cdot \sum_{\text{cyc}} \frac{x^4}{x^2 + xy + y^2} = \sum_{\text{cyc}} x \cdot \sum_{\text{cyc}} \frac{x(x^3 - y^3) + xy^3}{x^2 + xy + y^2}$$

$$= \sum_{\text{cyc}} x \cdot \sum_{\text{cyc}} x(x - y) + \sum_{\text{cyc}} x \cdot \sum_{\text{cyc}} \frac{xy^3}{x^2 + xy + y^2}$$

$$= \sum_{\text{cyc}} x^3 - 3xyz + \sum_{\text{cyc}} x \cdot \sum_{\text{cyc}} \frac{xy^3}{x^2 + xy + y^2}$$

$$= \sum_{\text{cyc}} x^3 + xyz \left(\sum_{\text{cyc}} x \cdot \sum_{\text{cyc}} \frac{y^2}{z(x^2 + xy + y^2)} - 3 \right)$$

$$\geq \sum_{\text{cyc}} x^3.$$

The last step is concluded based on the following fact: by the Cauchy-Schwarz inequality,

$$\sum_{\text{cyc}} z(x^2 + xy + y^2) \cdot \sum_{\text{cyc}} \frac{y^2}{z(x^2 + xy + y^2)} \geq \left(\sum_{\text{cyc}} y \right)^2$$

and, after noting that $\sum_{\text{cyc}} z(x^2 + xy + y^2) = \sum_{\text{cyc}} x \cdot \sum_{\text{cyc}} xy$, we find that

$$\sum_{\text{cyc}} x \cdot \sum_{\text{cyc}} \frac{y^2}{z(x^2 + xy + y^2)} \geq \frac{\left(\sum_{\text{cyc}} y \right)^2}{\sum_{\text{cyc}} xy} = \frac{\sum_{\text{cyc}} y^2}{\sum_{\text{cyc}} xy} + 2 \geq 1 + 2 = 3.$$

□

Remark. The cyclic inequality (3.5) has two nice variants of lower degree: for $x, y, z > 0$,

$$\sum_{\text{cyc}} \frac{x^2}{x^2 + xy + y^2} \geq 1 \qquad\qquad (3.6)$$

and

$$\sum_{\text{cyc}} \frac{x^3}{x^2 + xy + y^2} \geq \frac{1}{3} \sum_{\text{cyc}} x. \qquad\qquad (3.7)$$

Indeed, since $\frac{x}{y} + \frac{y}{x} \geq 2$, it follows

$$\sum_{\text{cyc}} \frac{x^3}{x^2 + xy + y^2} = \sum_{\text{cyc}} \left(x - \frac{xy(x + y)}{x^2 + xy + y^2} \right) = \sum_{\text{cyc}} \left(x - \frac{x + y}{\frac{x}{y} + 1 + \frac{y}{x}} \right)$$

$$\geq \sum_{\text{cyc}} \left(x - \frac{x + y}{3} \right) = \frac{x + y + z}{3}.$$

and (3.7) is verified. The interested reader is invited to prove the first inequality (3.6).

Additional problems for practice.

1. **Problem 10282** (Proposed by P. Erdős, 107(8), 1993). Let A, B, C be the vertices of a triangle inscribed in a unit circle, and le P be a point in the interior of the triangle ABC. Show that

$$|PA| \cdot |PB| \cdot |PC| < \frac{32}{27}.$$

2. **Problem 11435** (Proposed by P. Ligouras, 116(5), 2009). In a triangle T, let a, b, and c be the lengths of the sides, r the inradius, and R the circumradius. Show that

$$\frac{a^2bc}{(a+b)(a+c)} + \frac{b^2ca}{(b+c)(b+a)} + \frac{c^2ab}{(c+a)(c+b)} \le \frac{9}{2}rR.$$

3. **Problem 11951** (Proposed by O. Kouba, 124(1), 2017). Let ABC be a triangle that is not obtuse. Denote by a, b, and c the lengths of the sides opposite A, B, and C, respectively, and denote by h_a, h_b, and h_c the lengths of the altitudes dropped from A, B, and C, respectively. Prove that

$$\frac{a^2}{h_b^2 + h_c^2} + \frac{b^2}{h_c^2 + h_a^2} + \frac{c^2}{h_a^2 + h_b^2} < \frac{5}{2}.$$

Show also that $5/2$ is the smallest possible constant in this inequality.

4. **Problem 12056** (Proposed by L. Giugiuc, K. Altintas, and F. Stanescu, 125(7), 2018). Let $ABCD$ be a rectangle inscribed in a circle S of radius R, and let P be a point inside S. The lines AP, BP, CP, and DP intersect S a second time at K, L, M, and N, respectively. Prove

$$AK^2 + BL^2 + CM^2 + DN^2 \ge \frac{16R^4}{R^2 + OP^2}.$$

5. **Problem 12175** (Proposed by G. Fera, 127(4), 2020). Let I be the incenter and G be the centroid of a triangle ABC. Prove

$$2 < \frac{IA^2}{GA^2} + \frac{IB^2}{GB^2} + \frac{IC^2}{GC^2} \le 3.$$

6. **Problem 12217** (Proposed by G. Fera, 127(10), 2020). Let I be the incenter and G be the centroid of a triangle ABC. Prove

$$\frac{3}{2} < \frac{AI}{AG} + \frac{BI}{BG} + \frac{CI}{CG} \le 3.$$

7. **Problem 12303** (Proposed by G. Apostolopoulos, 129(2), 2022). Let R and r be the circumradius and inradius, respectively, of triangle ABC. Let D, E, and F be chosen on sides BC, CA, and AB so that AD, BE, and CF bisect the angles of ABC. Prove

$$\frac{FD}{AB+BC} + \frac{DE}{BC+CA} + \frac{EF}{CA+AB} \leq \frac{3}{8}\left(1 + \frac{R}{2r}\right).$$

8. **Problem 12491** (Proposed by T. Q. Hung, 131(9), 2024). Let P be any point in the plane of triangle ABC. Let r be the inradius of ABC, let h_a, h_b, h_c be the lengths of the altitudes from A, B, C, respectively. Prove

$$PA + PB + PC \geq h_a + h_b + h_c - 3r.$$

3.10 Supplementary pairs of Heronian triangles

Problem 12259 (Proposed by G. Fera, 128(6), 2021). A triangle is *Heronian* if it has integer sides and integer area. A pair of noncongruent Heronian triangles is called a *supplementary* pair if the triangles have the same perimeter and the same area and some interior angle of one is the supplement of some interior angle of the other. Prove that there are infinitely many supplementary pairs of Heronian triangles.

Discussion.
There are several ways to generate all Heronian triangles. The following parametrization approach is due to Euler: every Heronian triangle has its sides proportional to the integers of the form

$$a := mr(p^2 + q^2), \quad b := pq(m^2 + r^2), \quad c := (mq + rp)(mp - rq)$$

where m, p, q, and r are positive integers such that $mp > rq$. The perimeter and the area of the triangle $T = (a, b, c)$ are

$$\text{perimeter}(T) = 2mp(mq + rp) \quad \text{and} \quad \text{area}(T) = mpqr(mq + rp)(mp - rq),$$

respectively. By this approach, here we only need to choose the appropriate positive integers m, p, q, and r which satisfy all the required properties.

Solution.
Let $n \geq 2$. We generate the Heronian triangle T_1 by taking $m = n(n^2 + 1)$, $p = 1$, $q = n$, $r = 1$ and then dividing each side by the common factor n to get

$$(a_1, b_1, c_1) = ((n^2 + 1)^2, n^2(n^2 + 1)^2 + 1, n^2(n^4 + n^2 + 1)).$$

Similarly, we generate the Heronian triangle T_2 by taking $m = n^2 + 1$, $p = n$, $q = 1$, $r = n^3$ and then dividing each side by the common factor n again to get

$$(a_2, b_2, c_2) = (n^2(n^2 + 1)^2, (n^2 + 1)^2 + n^6, n^4 + n^2 + 1).$$

It is easy to check that T_1 and T_2 are non-congruent triangles, but they have the same perimeter and the same area:

$$\text{perimeter}(T_1) = \text{perimeter}(T_2) = 2(n^2 + 1)(n^4 + n^2 + 1)$$

and

$$\text{area}(T_1) = \text{area}(T_2) = n^3(n^2 + 1)(n^4 + n^2 + 1).$$

Moreover, let s_i be the half-perimeter of T_i for $i = 1, 2$. Then

$$\tan\left(\frac{B_1}{2}\right) = \sqrt{\frac{(s_1 - a_1)(s_1 - c_1)}{s_1(s_1 - b_1)}} = n$$

and

$$\tan\left(\frac{B_2}{2}\right) = \sqrt{\frac{(s_2 - a_2)(s_2 - c_2)}{s_2(s_2 - b_2)}} = \frac{1}{n}$$

which implies that $B_1/2 = \pi/2 - B_2/2$ and so the angle B_1 is the supplement of B_2. Since here the integer $n \geq 2$ is arbitrary, we have generated an infinite family of supplementary pairs (T_1, T_2) of noncongruent Heronian triangles. $\square$

Remark. Among the various properties of Heronian triangles, we would like to mention the following: every Heronian triangle can be realized as a lattice triangle (which are those which can be drawn with vertices at integral coordinates). A proof can be found in [106].

Additional problems for practice.

1. **Problem 6628** (Proposed by R. A. Melter, 97(4), 1990). Call a triangle a Heron Triangle if it has integer sides and integer area. Fermat showed that there does not exist a Heron right triangle whose area is a perfect square. However, the triangle with sides 9, 10, 17 has area 36. Prove that there are infinitely many Heron triangles whose sides have no common factor and whose area is a perfect square.

2. **Problem 10900** (Proposed by G. Rice, 108(9), 2001). It is clear from the law of cosines that every angle that occurs in a triangle with integer sides has a rational cosine. Is the converse true? Does every angle between 0 and π with a rational cosine occur in some triangle with integer sides?

3. **Problem 10924** (Proposed by A. J. Sasane, 109(2), 2002). A regular polygon of 2001 sides is inscribed in a circle of unit radius. Prove that its side and all its diagonals have irrational lengths.

4. **Problem 11134** (Proposed by K. D. Boklan, 112(2), 2005). Fix primes p and q. Prove that there are at most six integers x such that the area of the triangle with side-lengths p, q, and x is a positive integer.

5. **Problem 11979** (Proposed by Z. Franco, 124(5), 2017). Let O and I denote the circumcenter and incenter of a triangle ABC. Are there infinitely many nonsimilar scalene triangles ABC for which the lengths AB, BC, CA, and OI are all integers?

Chapter 4

Combinatorics

In this chapter, we gathered 15 problems spanning topics from integer partitions and tilings to applications of the pigeonhole principle. Various methods have been used to unravel the interactions between combinatorics and other branches of mathematics, especially number theory. A particular focus is placed on the following subjects: recursions, generating functions, enumeration, arithmetic properties of combinatorial quantities, permutations and graph theory. Many of the problems have links to current research topics, and references have been provided to support readers who wish to explore them further.

4.1   An equality for integer partitions

Problem 11237 (Proposed E. Deutsch, 113(7), 2006). Prove that the number of 2s occurring in all partitions of n is equal to the number of singletons occurring in all partitions of $n - 1$, where a *singleton* in a partition is a part occurring once. For example, partitions of 5 yield four 2s: one from $3 + 2$, two from $2 + 2 + 1$, and one from $2 + 1 + 1 + 1$; partitions of 4 yield four singletons: one from 4, two from $3 + 1$, and one from $2 + 1 + 1$.

Discussion.
Let $p(n)$ be the number of integer partitions of n. It is well-known that the generating function of this sequence is

$$P(x) = \sum_{n=0}^{\infty} p(n)x^n = \prod_{n=1}^{\infty} \frac{1}{1 - x^n}$$
$$= 1 + x + 2x^2 + 3x^3 + 5x^4 + 7x^5 + 11x^6 + 15x^7 + o(x^7)$$

(see the sequence $A000041$ in the OEIS (https://oeis.org/A000041)). Our plan is first to find the generating function $T(x)$ which counts the 2s and then

DOI: 10.1201/9781003607809-4

the generating function $S(x)$ which counts the singletons. Thus, the required problem becomes to show that $T(x) = xS(x)$.

Solution.

To compute the number of 2s, we introduce a parameter t. Since

$$\frac{1-x^2}{1-tx^2} \cdot P(x) = 1 + x + (1+t)x^2 + (2+t)x^3 + (3+t+t^2)x^4 + o(x^4),$$

the generating function which counts the 2s is

$$T(x) = \sum_{n=0}^{\infty} t(n)x^n = \frac{d}{dt}\left(\frac{1-x^2}{1-tx^2} \cdot P(x)\right)\Big|_{t=1}$$

$$= \frac{x^2}{1-x^2} \cdot P(x) = x^2 + x^3 + 3x^4 + 4x^5 + 8x^6 + 11x^7 + o(x^7).$$

On the other hand, since the generating function which counts the singletons that are equal to $n \geq 1$ is

$$x^n(1-x^n) \cdot P(x),$$

the generating function which counts all the singletons is

$$S(x) = \sum_{n=0}^{\infty} s(n)x^n = \left(\sum_{n=1}^{\infty} x^n(1-x^n)\right) \cdot P(x)$$

$$= \left(\frac{x}{1-x} - \frac{x^2}{1-x^2}\right) \cdot P(x) = \frac{x}{1-x^2} \cdot P(x)$$

Therefore, $T(x) = xS(x)$. Comparing the coefficient of x^n in $T(x)$ and $xS(x)$ yields

$$t(n) = s(n-1)$$

as required.   □

Remark. The generating function which counts the number of occurrences of the integer k in all partitions of n, can be obtained in a similar way

$$T_k(x) = \frac{d}{dt}\left(\frac{1-x^k}{1-tx^k} \cdot P(x)\right)\Big|_{t=1} = \frac{x^k}{1-x^k} \cdot P(x).$$

A remarkable result, commonly known as the Elder's theorem, which involves these kind of numbers is the following: The number of occurrences of an integer k in all partitions of n is equal to the number of parts occurring at least k times in all the partitions of n. For example $p(5) = 7$:

$$5 = 4+1 = 3+2 = 3+1+1 = 2+2+1 = 2+1+1+1 = 1+1+1+1+1.$$

Thus, for $k = 3$, the number of 3s is $0 + 0 + 1 + 1 + 0 + 0 + 0 = 2$ and the number of parts occurring at least thrice is $0 + 0 + 0 + 0 + 0 + 1 + 1 = 2$.

Here is a quick proof. Since the number of partitions of n where an integer j occurs at least k time is $p(n - kj)$, it follows that their sum is

$$\sum_{j=1}^{n} p(n - kj) = [x^n]\left(\sum_{j=1}^{n} x^{kj}\right) \cdot P(x) = \frac{x^k}{1 - x^k} \cdot P(x) = [x^n]T_k(x)$$

and the proof is completed.

There are many beautiful identities involving integer partitions. One of the most extraordinary is the following (see [18, p. 34]):

$$\sum_{n=0}^{\infty} p(5n + 4)q^n = 5\frac{(q^5; q^5)_{\infty}^5}{(q; q)_{\infty}^6},$$

where $(a; q) = \prod_{n=0}^{\infty}(1 - aq^n)$ for $|q| < 1$. This identity immediately implies one of the congruences discovered by S. Ramanujan in 1919, that is, for any integer $n \geq 0$,

$$p(5n + 4) \equiv 0 \pmod{5}.$$

Additional problems for practice.

1. Prove that the number of partitions of n in which no part appears exactly once is equal to the number of partitions of n into parts not congruent to $\pm 1 \pmod 6$.

2. **Problem 10969** (Proposed by R. P. Stanley, 109(8), 2002). Given a partition λ of n, let λ' denote the conjugate partition. Let $p(n)$ denote the total number of partitions of n, and let $t(n)$ denote the number of them satisfying

$$\mathcal{O}(\lambda) \equiv \mathcal{O}(\lambda') \pmod 4,$$

where $\mathcal{O}(\lambda)$ denotes the number of odd parts of the partition λ. Show that $t(n) = (p(n) + f(n))/2$, where

$$\sum_{n=0}^{\infty} f(n)x^n = \prod_{i=1}^{\infty} \frac{1 + x^{2i-1}}{(1 - x^{4i})(1 - x^{4i-2})^2}.$$

3. **Problem 11772** (Proposed by M. Merca, 121(4), 2014). Let n be a positive integer. Prove that the number of integer partitions of $2n + 1$ that do not contain 1 as a part is less than or equal to the number of integer partitions of $2n$ that contain at least one odd part.

4. **Problem 11675** (Proposed by M. Merca, 119(9), 2012). Let p be the partition function. Let $p(0) = 1$, and let $p(n) = 0$ for $n < 0$. Prove that for $n \geq 0$ with $n \neq 3$,

$$p(n) - 4p(n - 3) + 4p(n - 5) - p(n - 8) > 0.$$

4.2   A recurrence involving integer partitions

Problem 11730 (Proposed M. Merca, 120(8), 2013). Let p be the partition function (counting the ways to write n as a sum of positive integers), extended so that $p(0) = 1$ and $p(n) = 0$ for $n < 0$. Prove that

$$\sum_{k=0}^{\infty} \sum_{j=0}^{2k} (-1)^k p\left(n - \frac{k(3k+1)}{2} - j\right) = 1.$$

Discussion. Let $P(x)$ be the generating function of the partition. By Euler's pentagonal number theorem

$$\prod_{n=1}^{\infty}(1 - x^n) = \sum_{k=-\infty}^{\infty} (-1)^k x^{\frac{k(3k+1)}{2}}, \tag{4.1}$$

in view of the left hand side is $1/P(x)$, it follows that

$$\left(\sum_{n=0}^{\infty} p(n)x^n\right) \cdot \left(\sum_{k=-\infty}^{\infty} (-1)^k x^{\frac{k(3k+1)}{2}}\right) = 1.$$

Extracting the coefficient of x^m from the above product, we find

$$\sum_{k=-\infty}^{\infty} (-1)^k p\left(m - \frac{k(3k+1)}{2}\right) = \sum_{n+\frac{k(3k+1)}{2}=m} (-1)^k p(n) = \begin{cases} 1 & \text{if } m = 0, \\ 0 & \text{otherwise.} \end{cases}$$
$$\tag{4.2}$$

We will use this identity as a guide to solving the problem.

Solution.
By induction, we prove that $S(n) = 1$ for all non-negative integers n, where

$$S(n) = \sum_{k=0}^{\infty} \sum_{j=0}^{2k} (-1)^k p\left(n - \frac{k(3k+1)}{2} - j\right).$$

Then

$$S(n-1) = \sum_{k=0}^{\infty} \sum_{j=0}^{2k} (-1)^k p\left(n - 1 - \frac{k(3k+1)}{2} - j\right)$$

$$= \sum_{k=0}^{\infty} \sum_{j=1}^{2k+1} (-1)^k p\left(n - \frac{k(3k+1)}{2} - j\right),$$

Hence, recalling that $S(0) = p(0) = 1$, for $n > 0$, we have

$$S(n) - S(n-1)$$
$$= \sum_{k=0}^{\infty} (-1)^k \left(p\left(n - \frac{k(3k+1)}{2} \right) - p\left(n - \frac{k(3k+1)}{2} - 2k - 1 \right) \right)$$
$$= \sum_{k=0}^{\infty} (-1)^k p\left(n - \frac{k(3k+1)}{2} \right) + \sum_{k=0}^{\infty} (-1)^{k+1} p\left(n - \frac{k(3k+2)}{2} \right)$$
$$= \sum_{k=0}^{\infty} (-1)^k p\left(n - \frac{k(3k+1)}{2} \right) - \sum_{j=-\infty}^{-1} (-1)^j p\left(n - \frac{j(3j+1)}{2} \right)$$
$$= \sum_{k=-\infty}^{\infty} (-1)^k p\left(n - \frac{k(3k+1)}{2} \right) = 0,$$

where we have applied (4.2) in the last step. $\qquad \square$

Remark. Notice that the identity (4.2) provides an efficient algorithm for computing $p(n)$:

$$p(n) = \sum_{k\geq 1} (-1)^{k-1} \left(p\left(n - \frac{k(3k+1)}{2} \right) + p\left(n - \frac{k(3k-1)}{2} \right) \right).$$

For instance,

$$p(12) = p(11) + p(10) - p(7) - p(5) + p(0) = 56 + 42 - 15 - 7 + 1 = 77.$$

Moreover, if we rewrite (4.1) as

$$\frac{1}{(q;q)_\infty} \sum_{n=0}^{\infty} (-1)^n q^{\frac{n(3n+1)}{2}} (1 - q^{2n+1}) = 1$$

then it becomes evident that

$$\frac{1}{(q;q)_\infty} \sum_{n=0}^{m-1} (-1)^n q^{\frac{n(3n+1)}{2}} (1-q^{2n+1}) = 1 + (-1)^{k-1} \sum_{n=k}^{\infty} \frac{q^{\frac{k(k-1)}{2}+(k+1)n}}{(q;q)_n} \begin{bmatrix} n-1 \\ k-1 \end{bmatrix}_q,$$

is a truncated version of the Euler's pentagonal number theorem. This result is due to the proposers and a proof can be found in [14].

Additional problems for practice.

1. Let $b(n)$ be the number of *binary partitions* of n, that is the partitions all of whose parts are powers of 2. For example $b(7) = 6$ because 7 can be written as

$$1+1+1+1+1+1+1 = 2+1+1+1+1+1 = 2+2+1+1+1$$
$$= 2+2+2+1 = 4+1+1+1 = 4+2+1.$$

Show that for any integer $n \geq 0$,

$$b(2n + 1) = b(2n) \quad \text{and} \quad b(2n) = b(2n - 1) + b(n).$$

2. Show that for any positive integer

$$np(n) = \sum_{k=0}^{n-1} p(k)\sigma(n - k)$$

where $\sigma(n)$ denotes the sum of the divisors of n.
Hint: First take the logarithm of both sides of $\sum_{n=1}^{\infty} p(n)x^n = 1/\prod_{n=1}^{\infty}(1 - x^n)$ and then differentiate.

3. **Problem 10628** (Proposed by G. E. Andrews, 104(10), 1997). Let $p_{a,b}(n)$ denote the number of partitions of n that contain no parts of size a or b. For $n > 0$, prove that

$$\sum_{k=1}^{\infty} (-1)^k p_{k,2k}\left(n - \frac{3k(k - 1)}{2}\right) = 0.$$

4. **Problem 11787** (Proposed by M. Merca, 121(6), 2014). Prove that

$$\sum_{k=1}^{\infty} (-1)^{k-1} k p_k \left(n - \frac{k(k + 1)}{2}\right) = \sum_{k=-\infty}^{\infty} (-1)^k \tau \left(n - \frac{k(3k - 1)}{2}\right).$$

Here, $p_k(n)$ denotes the number of partitions of n in which the greatest part is less than or equal to k (with $p_k(0) = 1$ and $p_k(n) = 0$ for $n < 0$), and $\tau(n)$ is the number of divisors of n (with $\tau(n) = 0$ for $n \leq 0$).

5. **Problem 11795** (Proposed by M. Merca, 121(7), 2014). Let p be the partition counting function on the set $\mathbb{Z}^+$ of positive integers, and let g be the function on $\mathbb{N}$ given by $g(n) = \frac{1}{2}\lceil n/2 \rceil \lceil (3n + 1)/2 \rceil$. Let $A(n)$ be the set of nonnegative integer triples (i, j, k) such that $g(i) + j + k = n$. Prove for $n \geq 1$ that

$$p(n) = \frac{1}{n} \sum_{(i,j,k) \in A(n)} (-1)^{\lceil i/2 \rceil - 1} g(i) p(j) p(k).$$

Hint. Rewrite the Euler's pentagonal number theorem as $\prod_{n=1}^{\infty}(1 - q^n) = \sum_{n=0}^{\infty} (-1)^{\lceil n/2 \rceil} q^{g(n)}$.

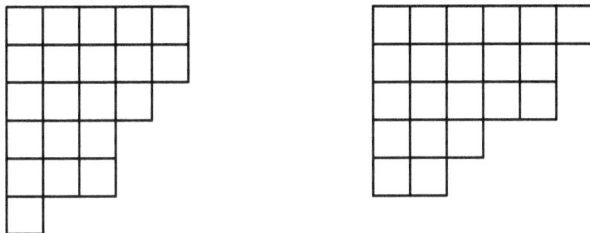

Figure 4.1: Partition of 21 with 6 parts and largest part 5 (left), and with 5 parts and largest part 6 (right)

4.3 Another equality for integer partitions

Problem 11908 (Proposed G. E. Andrews and E. Deutsch, 123(5), 2016). Let n and k be nonnegative integers. Show that the number of partitions of n having k even parts is the same as the number of partitions of n in which the largest repeated part is k (defined to be 0 if the parts are all distinct). For example, 7 has three partitions with two even parts $(4 + 2 + 1 = 3 + 2 + 2 = 2 + 2 + 1 + 1 + 1)$ and also three partitions in which the largest repeated part is 2 $(3 + 2 + 2 = 2 + 2 + 2 + 1 = 2 + 2 + 1 + 1 + 1)$.

Discussion.
We will show a more general statement by using generating functions: Given a positive integer d then the number of partitions of n having exactly k parts divisible by d is equal to the number of partitions of n in which k is the largest part that occurs at least d times.

Notice when $d = 1$, the equality is well known: The number of partitions of n having k parts is the same as the number of partitions of n in which the largest part is k. Indeed, if we interchange the rows and columns of the Young diagram we immediately have a bijective correspondence between the two kinds of partitions.

Solution.
Let $a(n, k)$ be the number of partitions of n having k parts divisible by d. Then

$$F(x, y) = \sum_{n,k \in \mathbb{N}} a(n, k) y^k x^n = P(x) \cdot \prod_{j=1}^{\infty} \frac{1 - x^{dj}}{1 - yx^{dj}}.$$

Furthermore, let $b(n, k)$ be the number of partitions of n in which the largest repeated part that occurs at least d times is k. Then

$$G(x, y) = \sum_{n,k} b(n,k) y^k x^n$$

$$= \sum_{k=0}^{\infty} y^k \left(\prod_{j=1}^{k} \frac{1}{1-x^j} \cdot x^{dk} \cdot \prod_{j=k+1}^{\infty} (1 + x^j + \cdots + x^{(d-1)j}) \right)$$

$$= \sum_{k=0}^{\infty} y^k \left(\prod_{j=1}^{k} \frac{1}{1-x^j} \cdot x^{dk} \cdot \prod_{j=k+1}^{\infty} \frac{1-x^{dj}}{1-x^j} \right)$$

$$= P(x) \cdot \prod_{j=1}^{\infty} (1 + x^{dj}) \cdot \sum_{k=0}^{\infty} \frac{y^k x^{dk}}{\prod_{j=1}^{k} (1 + x^{dj})}.$$

Now it suffices to show that $f(x, y) = g(x, y)$, that is

$$\sum_{k=0}^{\infty} \frac{y^k x^{dk}}{\prod_{j=1}^{k} (1 - x^{dj})} = \prod_{j=1}^{\infty} \frac{1}{1 - yx^{dj}}.$$

This holds because, after letting $z = x^d$,

$$\sum_{k=0}^{\infty} \frac{y^k z^k}{\prod_{j=1}^{k} (1 - z^j)} = \sum_{k=0}^{\infty} y^k \sum_{n=0}^{\infty} p(n,k) z^n$$

$$= \sum_{n=0}^{\infty} z^n \sum_{k=0}^{\infty} p(n,k) y^k = \prod_{j=1}^{\infty} \frac{1}{1 - yz^j},$$

where $p(n,k)$ denotes the number of partitions of n having k parts or, equivalently, the number of partitions into parts of which the largest part is k. This completes the proof. □

Remark. A quite similar approach can be used to prove a result due to J. W. L. Glaisher and published in 1883: The number of partitions of n in which no part is repeated d or more times is the same as the number of partitions of n into parts not divisible by d. Indeed the generating function of the first kind of partitions is

$$F(x) = \prod_{j=1}^{\infty} (1 + x^j + x^{2j} + \cdots + x^{(d-1)j}),$$

whereas the generating function of the second kind of partitions is

$$G(x) = \prod_{j=1}^{\infty} \frac{1 - x^{dj}}{1 - x^j}.$$

It is evident that $F(x) = G(x)$. Notice that for $d = 2$ we obtain an old statement which dates back to Euler: The number of partitions of n into distinct parts is the same as the number of partitions of n into odd parts. For more variations on this theme see [64].

Additional problems for practice.

1. (Due to S. Ramanujan) Prove that the number of partitions of n with unique smallest part and largest part at most twice the smallest part is equal to the number of partitions of n in which the largest part is odd and the smallest part is larger than half the largest part.

   Hint: Show that

   $$\sum_{n=0}^{\infty} \frac{q^n}{(1-q^{n+1})(1-q^{n+2})\cdots(1-q^{2n})}$$

   $$= 1 + \sum_{n=0}^{\infty} \frac{q^{2n+1}}{(1-q^{n+1})(1-q^{n+2})\cdots(1-q^{2n+1})}.$$

2. **Problem 10627** (Proposed by G. E. Andrews, 104(10), 1997). The Rogers-Ramanujan (RR) partitions of an integer are those that have no repetitions and no consecutive integers as parts. The RR$'$ partitions are those RR partitions that have no 1s.

   (a) For $n > 1$ prove that at least half of the RR partitions of n are RR$'$ partitions.

   (b) Let $Q(n)$ denote the number of RR$'$ partitions of n into at least two parts whose two largest parts differ by at most 2 more than the number of parts. For example, $Q(12) = 3$ because, of the nine RR partitions of 12, six are RR$'$ partitions, and of these only three ($8 + 4$, $7 + 5$, and $6 + 4 + 2$) meet the stated condition. For $n > 1$, prove that $Q(n)$ equals the difference between twice the number of RR$'$ partitions of n and the number of RR partitions of n.

3. **Problem 10629** (Proposed by F. Schmidt, 104(10), 1997). Let $p(n)$ denote the number of partitions of the integer n, and let $f(n)$ denote the number of partitions $\lambda_1 + \lambda_2 + \lambda_3 + \cdots$ satisfying $\lambda_1 > \lambda_2 > \lambda_3 > \cdots$ and $n = \lambda_1 + \lambda_3 + \lambda_5 + \cdots$. Prove that $p(n) = f(n)$ for every positive integer n. For example, $p(5)$ counts the 7 partitions 5, $4 + 1$, $3 + 2$, $3 + 1 + 1$, $2 + 2 + 1$, $2 + 1 + 1 + 1$, and $1 + 1 + 1 + 1 + 1$, and $f(5)$ the 7 partitions 5, $5 + 1$, $5 + 2$, $5 + 3$, $5 + 4$, $4 + 3 + 1$, and $4 + 2 + 1$.

4. **MM Problem 2124** (Proposed by M. Merca, 94(3), 2021). For a positive integer n, prove that

   $$\sum_{\substack{\lambda_1+\lambda_2+\cdots+\lambda_k=n \\ \lambda_1 \geq \lambda_2 \geq \cdots \geq \lambda_k > 0}} (-1)^{n-\lambda_1} \frac{\binom{\lambda_1}{\lambda_2}\binom{\lambda_2}{\lambda_3}\cdots\binom{\lambda_k}{0}}{1^{\lambda_1} 2^{\lambda_2} \cdots k^{\lambda_k}} = \frac{1}{n!},$$

   where the sum runs over all the partitions of n.

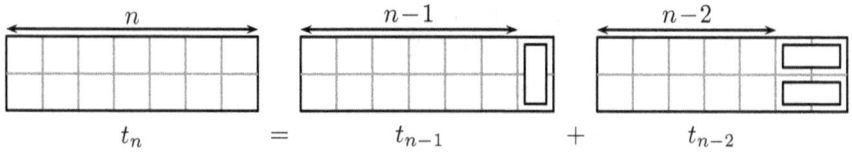

Figure 4.2: Recurrence $t_n = t_{n-1} + t_{n-2}$

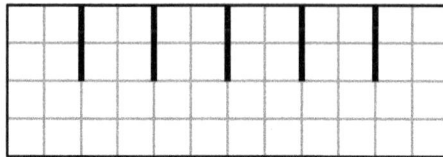

Figure 4.3: The domain for $n = 6$

4.4   Tiling a board with cuts

Problem 11241 (Proposed R. Tauraso, 113(7), 2006). Find a closed formula
for

$$\sum_{k=0}^{n} 2^{n-k} \sum_{x \in S[k,n]} \prod_{i=1}^{k+1} F_{1+2x_i} \, ,$$

where F_n denotes the nth Fibonacci number (that is, $F_0 = 0$, $F_1 = 1$, and
$F_j = F_{j-1} + F_{j-2}$ for $j \geq 2$) and $S[k,n]$ is the set of all $(k+1)$-tuples of
nonnegative integers that sum to $n - k$.

Discussion.
To find the closed formula, we give a combinatorial interpretation of the double sum as the number of domino tilings of a $4 \times 2n$ rectangle with some
restrictions. Recall that the number of domino tilings t_n of a $2 \times n$ rectangle is given by the Fibonacci number F_{n+1}. Indeed, it is easy to see that
$t_1 = 1 = F_2$, $t_2 = 2 = F_3$, and that the following recurrence holds for $n \geq 3$,
$t_n = t_{n-1} + t_{n-2}$. See Figure 4.2.

Solution.
We show that the given formula computes the total number a_n of domino
tilings of a $4 \times 2n$ board formed by joining along their vertical sides n copies
of a 4×2 rectangle that we call *tooth*.
The upper part of a single tooth can be tiled in three ways as shown in
Figure 4.5.
Let k be the number of teeth which are tiled as in the third case. These
k teeth induce a partition of the domain: there are the upper parts of the
remaining $n - k$ teeth and $k + 1$ horizontal strips.

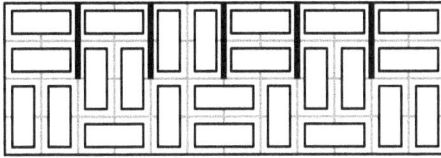

Figure 4.4: A tiling of the domain for $n = 6$

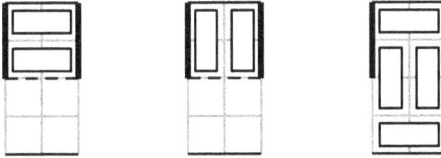

Figure 4.5: Tilings of a single tooth

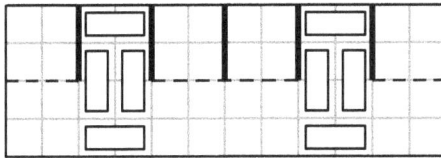

Figure 4.6: Partial tiling for $n = 6$ and $k = 2$

If we denote by $2x_i$ the size of the gap between the $(i-1)$th tooth and the ith tooth then

$$\sum_{i=1}^{k+1} 2x_i + 2k = 2n$$

that is

$$\sum_{i=1}^{k+1} x_i = n - k.$$

Moreover, any upper part of the $n - k$ teeth can be tiled in 2 ways and the $2 \times 2x_i$ horizontal strip can be tiled in F_{1+2x_i} ways. Therefore, the total number of these tilings is

$$a_n = \sum_{k=0}^{n} 2^{n-k} \sum_{x \in S[k,n]} \prod_{i=1}^{k+1} F_{1+2x_i}$$

where the internal sum runs over the set $S[k,n]$ of all nonnegative integer

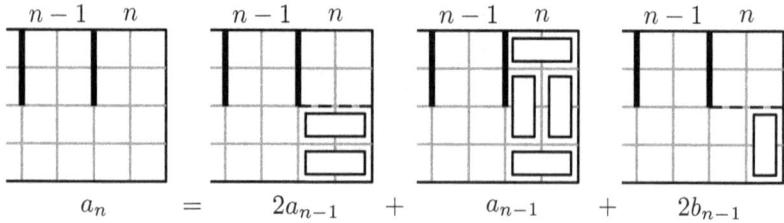

Figure 4.7: Recurrence for a_n

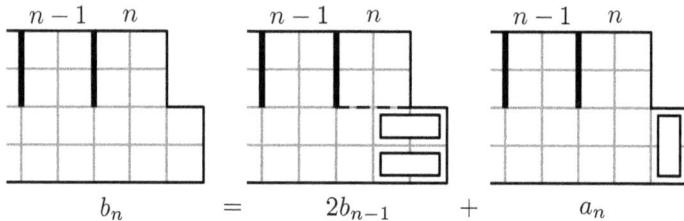

Figure 4.8: Recurrence for b_n

solutions of the equation

$$x_1 + \cdots + x_{k+1} = n - k.$$

Now, using the combinatorial interpretation, we determine two recurrences. Hence for $n \geq 2$,

$$\begin{cases} 2b_{n-1} = a_n - 3a_{n-1} \\ a_n = b_n - 2b_{n-1} \end{cases} \tag{4.3}$$

which implies

$$2a_n = 2b_n - 4b_{n-1} = (a_{n+1} - 3a_n) - 2(a_n - 3a_{n-1})$$

and so

$$a_{n+1} = 7a_n - 6a_{n-1}.$$

Since $a(0) = 1$ and $a(1) = 5$, solving this recurrence yields the required closed formula

$$a_n = \frac{4 \cdot 6^n + 1}{5}.$$

$\square$

Remark. It is worth noting the following curious coincidence. The formula $a_n = (4 \cdot 6^n + 1)/5$ gives also the number of tilings of a rectangle $3 \times 2n$ by

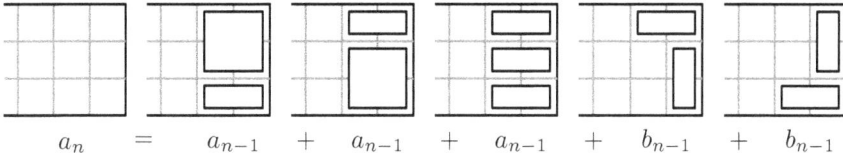

$$a_n \quad = \quad a_{n-1} \quad + \quad a_{n-1} \quad + \quad a_{n-1} \quad + \quad b_{n-1} \quad + \quad b_{n-1}$$

Figure 4.9: Recurrence for a_n

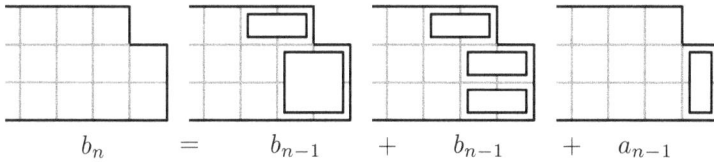

$$b_n \quad = \quad b_{n-1} \quad + \quad b_{n-1} \quad + \quad a_{n-1}$$

Figure 4.10: Recurrence for b_n

using dominoes and 2×2 squares. In fact, in this case again we find that $a(0) = 1$, $a(1) = 5$ and the same recurrences (4.3) hold for $n \geq 2$.

Additional problems for practice.

1. **MM Problem 1719** (Proposed by G.R.A.20 Problem Solving Group, 78(2), 2005). From an $(n + 4) \times (n + 4)$ checkerboard of unit squares, the central $n \times n$ square is removed to leave a square frame of width 2. In how many ways can the frame be tiled with 1×2 dominos? (Two different tilings rotation of the frame are considered different.)

2. **MM Problem 1868** (Proposed by D. E. Knuth, 85(2), 2012). Let $n \geq 2$ be an integer. Remove the central $(n - 2)^2$ squares from an $(n + 2) \times (n + 2)$ array of squares. In how many ways can the remaining squares be covered with $4n$ dominoes? (See [96])

3. **Problem E2508** (Proposed by S. Penner, 81(10), 1974). Let four squares, exactly two of each color be removed from a $2n \times 2n$ checkerboard. Show that if at least one of the deleted black squares and at least one of the deleted white squares are interior squares, then the remaining portion of the board can be tiled with 2×1 dominoes.

4. **Problem 11138** (Proposed by R. Tauraso, 112(3), 2005). For $n > 2$ consider the region obtained by removing from the square $[0, 2n] \times [0, 2n]$ the four $(n-2) \times (n-2)$ squares centered at the points $(n \pm n/2, n \pm n/2)$. Find the number of domino tilings of this region, and show that it is a perfect square.

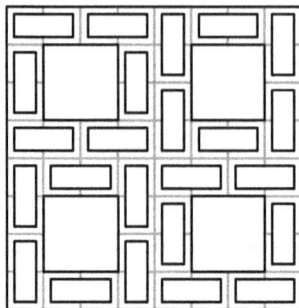

Figure 4.11: Problem 11138, a domino tiling of the region for $n = 4$

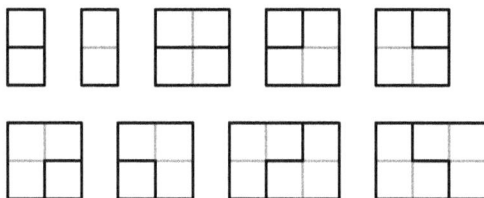

Figure 4.12: Unbreakable tilings without staggered horizontal dominoes

4.5 About the tilings of a $2 \times n$ strip with squares, dominoes, and trominoes

Problem 11996 (Proposed R. Tauraso, 124(7), 2017). Consider all the tilings of a 2-by-n rectangle comprised of tiles that are either a unit square, a domino, or a right tromino. Let f_n be the fraction of tiles among all such tilings that are unit squares. For example, $f_2 = 4/7$, because 16 out of the 28 tiles in the 11 tilings of a 2-by-2 rectangle are squares. What is $\lim_{n \to +\infty} f_n$?

Discussion.
A tiling of the $(2 \times n)$-board is called *unbreakable*, if it cannot be split into a tiling of a $(2 \times k)$-board and a tiling of a $(2 \times (n-k))$-board for some $0 < k < n$. They are of two types depending on whether they have a staggered horizontal dominoes.

If an unbreakable tiling consists of a non-empty sequence of staggered horizontal dominoes then at the ends we may find a unit square or a tromino.
Now we determine the generating function of such unbreakable tilings $U_{(s,d,t)}(x)$ where the powers of s, d, and t in a coefficient of x^n count the number of unit squares, dominoes, and right trominoes respectively in a tiling of a $2 \times n$ board. Looking at Figure 4.12, we obtain

$$(s^2 + d)x + (d^2 + 4st)x^2 + 2t^2 x^3. \tag{4.4}$$

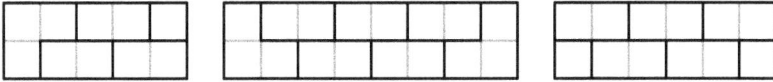

Figure 4.13: Some unbreakable tilings with staggered horizontal dominoes

As regards the tilings pictured in Figure 4.13, we get

$$(s+tx)\left(2\sum_{k=2}^{\infty} d^{k-1}x^k\right)(s+tx). \tag{4.5}$$

Then $U_{(s,d,t)}(x)$ is the sum of (4.4) and (4.5).

Moreover, since any valid tiling of the strip is obtained by appending an unbreakable tiling to a shorter tiling, it follows that the generating function of the number of tilings $T_{(s,d,t)}(x)$ satisfies the equality

$$T_{(s,d,t)}(x) = 1 + U_{(s,d,t)}(x) \cdot T_{(s,d,t)}(x). \tag{4.6}$$

After determining $T_{(s,d,t)}(x)$, it is straightforward to find the number of squares s_n and the total number of pieces p_n in all the tilings of the $(2 \times n)$-board. This enables us to evaluate the desired limit

$$\lim_{n\to\infty} \frac{s_n}{p_n}.$$

Solution.
From the discussion above we have

$$U_{(s,d,t)}(x) = (s^2+d)x + (d^2+4st)x^2 + 2t^2x^3 + (s+tx)\left(2\sum_{k=2}^{\infty} d^{k-1}x^k\right)(s+tx)$$

$$= (s^2+d)x + (d^2+4st+2s^2d)x^2 + \frac{2(sd+t)^2x^3}{1-dx} \tag{4.7}$$

and

$$T_{(s,d,t)}(x) = \frac{1}{1-U_{(s,d,t)}(x)}$$

$$= \frac{(1-dx)}{1-(s^2+2d)x-(4st+s^2d)x^2-(2t^2-d^3)x^3}. \tag{4.8}$$

Thus the number of unit squares s_n in all the tilings of the $(2 \times n)$-board is

$$s_n = [x^n]\frac{d}{ds}\left(T_{(s,1,1)}(x)\right)\Big|_{s=1} = [x^n]\frac{x(2+4x-6x^2)}{(1+x)^2(1-4x-x^2)^2}$$

$$= \frac{1}{2}(nF_{3n+1}+F_{3n-1}+(n-1)(-1)^n) \sim \frac{nF_{3n+1}}{2},$$

where F_n denotes the nth Fibonacci number.

On the other hand, the total number of pieces p_n is

$$p_n = [x^n] \frac{d}{dp} \left(T_{(p,p,p)}(x) \right) \Big|_{p=1} = [x^n] \frac{x(3 + 10x - 5x^2)}{(1+x)^2(1 - 4x - x^2)^2}$$

$$= \frac{1}{20} \left((17n + 10)F_{3n+1} + (4n - 12)F_{3n} + (15n - 10)(-1)^n \right)$$

$$\sim \frac{n(17F_{3n+1} + 4F_{3n})}{20}.$$

In summary, the required limit becomes

$$\lim_{n \to +\infty} f_n = \lim_{n \to +\infty} \frac{s_n}{p_n} = \lim_{n \to +\infty} \frac{10}{17 + 4F_{3n}/F_{3n+1}}$$

$$= \frac{10}{17 + 4/\phi} = \frac{30 - 4\sqrt{5}}{41} \approx 0.51355$$

where $\phi = (\sqrt{5} + 1)/2$ is the *golden ratio*. □

Remark. From (4.7) and (4.8), we have that

$$U_{1,1,1}(x) = \sum_{n=0}^{\infty} u_n x^n = 2x + 7x^2 + \frac{8x^3}{1 - x}$$

and

$$T_{1,1,1}(x) = \sum_{n=0}^{\infty} t_n x^n = \frac{1 - x}{(1+x)(1 - 4x - x^2)}$$

where u_n and t_n are the number of unbreakable tilings and the total number of tilings of the $2 \times n$ strip, respectively.

The identity (4.6) also provides a combinatorial interpretation of the following recurrence: for $n \geq 2$,

$$t_n - u_n = \sum_{k=1}^{n-1} t_k u_{n-k}.$$

Indeed, the left hand side counts the number of tilings with at least a vertical break line. The right hand side counts the same thing in different way: the index k indicates the location of the rightmost break, t_k is the number of ways to fill the first k columns and u_{n-k} is the number of ways to fill the remain part.

Since $t_n = (F_{3n+2} + (-1)^n)/2$, after a few steps, the above recurrence boils down to a nontrivial identity for the Fibonacci numbers

$$F_{3n+1} = 1 + 2 \sum_{k=0}^{n-1} F_{3k+2}.$$

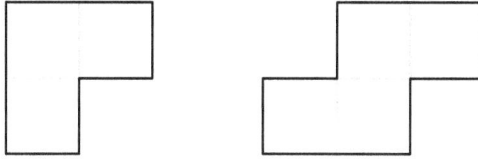

Figure 4.14: Two tiles

Additional problems for practice.

1. **Problem 11187** (Proposed by L. Zhou, 112(10), 2005). Find a closed formula for the number of ways to tile a 4 by n rectangle with 1 by 2 dominoes.

2. **Problem 10877** (Proposed by E. Deutsch, 108(5), 2001). An L-tile is a 2-by-2 square with the upper right 1-by-1 subsquare removed; no rotations are allowed. Let an be the number of tilings of a 4-by-n rectangle using tiles that are either 1-by-1 squares or L-tiles. Find a closed form for the generating function $1 + a_1x + a_2x^2 + a_3x^3 + \cdots$.

3. **Problem 10641** (Proposed by J. R. Grigg, 105(2), 1998). Determine all pairs m, n of positive integers such that an m-by-n rectangle can be tiled with congruent pieces formed by removing a 1-by-1 square from one corner of a 2-by-2 square.

4. **Putnam 2016-A4.** Consider a $(2m-1) \times (2n-1)$ rectangular region, where m and n are integers such that $m, n \geq 4$. This region is to be tiled using tiles of the two types shown in Figure 4.14. The tiles may be rotated and reflected, as long as their sides are parallel to the sides of the rectangular region. They must all fit within the region, and they must cover it completely without overlapping. What is the minimum number of tiles required to tile the region?

4.6   Fault-free tilings of 3-by-n strip

Problem 12141 (Proposed R. Tauraso, 126(9), 2019). Consider tilings of a rectangle with tiles each of which is 1-by-k for some positive integer k. A fault line through a tiling is a horizontal or vertical line through the tiling that cuts through no tile. A tiling is fault-free if it has no fault lines.
Let f_n be the number of fault-free tilings of a 3-by-n rectangle.
Find (a) $\lim\limits_{n\to\infty} \sqrt[n]{f_n}$, and (b) $\lim\limits_{n\to\infty} \sqrt[n]{f_{n+1}f_{n-1} - f_n^2}$.

Discussion.
We start by finding the generating function of the number of tilings which have no vertical fault line. We consider the sets of configurations shown in Figure 4.16.

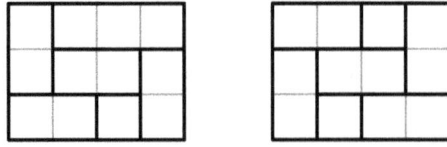

Figure 4.15: Two of the 22 fault-free tilings of the 3-by-4 rectangle

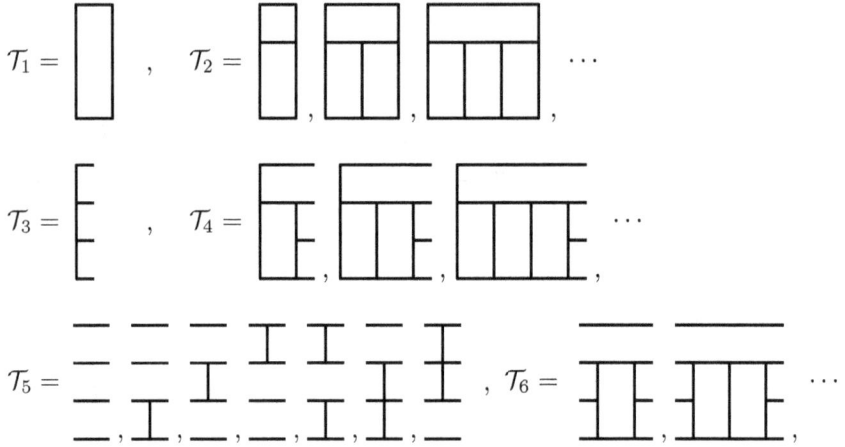

Figure 4.16: Sets of configurations

For $n \geq 1$, the combinatorial class of the $(3 \times n)$-tilings which are vertically fault-free, is given by

$$\mathcal{T}_1 + \mathcal{T}_2 + \overline{\mathcal{T}_2} + (\mathcal{T}_3 + \mathcal{T}_4 + \overline{\mathcal{T}_4}) \times \mathsf{Seq}\left(\mathcal{T}_5 \cup \mathcal{T}_6 \cup \overline{\mathcal{T}_6}\right) \times \left(\mathcal{T}_3^{|} + \mathcal{T}_4^{|} + \overline{\mathcal{T}_4}^{|}\right) \quad (4.9)$$

for any set of configurations $\mathcal{S}$, $\overline{\mathcal{S}}$ and $\mathcal{S}^{|}$ are the sets where each configuration of $\mathcal{S}$ is reflected across the horizontal middle axis or the vertical middle axis, respectively and the *sequence* class of $\mathcal{S}$ is defined as

$$\mathsf{Seq}\left(\mathcal{S}\right) = I + \mathcal{S} + \mathcal{S} \times \mathcal{S} + \mathcal{S} \times \mathcal{S} \times \mathcal{S} + \dots$$

(see [40, p. 102]).
From (4.9), it follows that the generating function which enumerates the tilings with no vertical fault line is given by

$$G(x) = x + \frac{2x}{1-x} + \frac{x\left(1 + \frac{2x}{1-x}\right)^2}{1 - 7x - \frac{2x^2}{1-x}}.$$

Notice that, for $n \geq 2$, a vertically fault-free tiling has no horizontal fault line if and only if it contains at least two vertical dominoes (a vertical domino is a tile 2×1), one touching the bottom border and the other touching the top border. Based on $G(x)$, we will be able to find the generating function of the number of fault-free tilings and use it to determine both required limits.

Solution.
For $n \geq 2$, the combinatorial subclass of (4.9) with no vertical domino touching the top line is

$$\mathcal{T}_2 + (\mathcal{T}_3 + \mathcal{T}_4) \times \mathsf{Seq}\,(\mathcal{T}_5 \cup \mathcal{T}_6) \times \left(\mathcal{T}_3^| + \mathcal{T}_4^|\right)$$

and therefore the corresponding generating function is

$$G_1(x) = \frac{x}{1-x} + \frac{x\left(1 + \frac{x}{1-x}\right)^2}{1 - 7x - \frac{x^2}{1-x}}.$$

By symmetry, G_1 is also equal the generating function of the number of vertically fault-free tilings with no vertical domino touching the bottom line. Furthermore, for $n \geq 2$, the combinatorial subclass of (4.9) with no vertical domino at all is

$$\mathcal{T}_3 \times \mathsf{Seq}(\mathcal{T}_5) \times \mathcal{T}_3^|$$

with the generating function

$$G_0(x) = \frac{x}{1 - 7x}.$$

Hence, by removing the whole subclass of configurations with no horizontal fault line from (4.9), we obtain the generating function of the number of fault-free tilings,

$$F(x) = \sum_{n=1}^{\infty} f_n x^n = G(x) - (2G_1(x) - G_0(x))$$

$$= x + \frac{2x^3(1 - 6x)^2}{(1 - 7x)(1 - 8x + 6x^2)(1 - 8x + 5x^2)}$$

$$= x + 2x^3 + 22x^4 + 204x^5 + 1804x^6 + 15538x^7 + \ldots.$$

Notice that the poles of the rational functions F are real and simple and

$$\frac{1}{4 + \sqrt{11}} < \frac{1}{4 + \sqrt{10}} < \frac{1}{7} < \frac{1}{4 - \sqrt{10}} < \frac{1}{4 - \sqrt{11}}.$$

Thus

$$f_n = A\,(4 + \sqrt{11})^n + B\,(4 + \sqrt{10})^n + O(7^n)$$

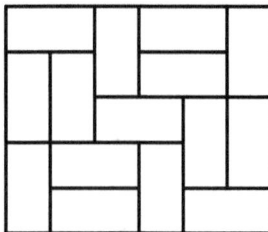

Figure 4.17: A fault-free domino tiling of a 5-by-6 rectangle

for some real numbers A and B. It can be verified that A and B are different from zero, and therefore we conclude that

$$\lim_{n\to\infty} \sqrt[n]{f_n} = 4 + \sqrt{11}$$

and

$$\lim_{n\to\infty} \sqrt[n]{f_{n+1}f_{n-1} - f_n^2} = \lim_{n\to\infty} \sqrt[n]{AB(\sqrt{11} - \sqrt{10})^2 \left((4 + \sqrt{11})(4 + \sqrt{10}) \right)^{n-1}}$$
$$= (4 + \sqrt{11})(4 + \sqrt{10}).$$

$\square$

Remark. It is known that there are no fault-free tiling of a 3-by-n rectangle where the tiles are only dominoes. More generally, R. L. Graham proved in [46] that necessary and sufficient conditions for the existence of a fault-free tiling of a m-by-n rectangle made all of dominoes are: mn is divisible by 2, $m \geq 5$, $n \geq 5$, and $(m, n) \neq (6, 6)$.
On the other hand, by [4], the generating function of the number of fault-free tilings t_n of a 3-by-n made by unit squares and dominoes is

$$\sum_{n=1}^{\infty} t_n = \frac{2x^3}{(1 + x - x^2)(1 - 2x - x^2)} = 2x^3 + 2x^4 + 10x^5 + 16x^6 + 52x^7 + 104x^8 + \ldots$$

see the sequence $A334396$ in the OEIS (https://oeis.org/A334396)).
It is easy to verify that in this case

$$\lim_{n\to\infty} \sqrt[n]{t_n} = 1 + \sqrt{2}.$$

Additional problems for practice.

1. **CMJ Problem 1216** (Proposed by O. Alabi, 53(1), 2022). For an integer $n \geq 3$, find a closed form expression for the number of ways to tile an $n \times n$ square with 1×1 squares and $(n-1) \times 1$ rectangles (each of which may be placed horizontally or vertically).

2. **Problem 12005** (Proposed by D. E. Knuth, 124(8), 2017). A tight m-by-n paving is a decomposition of an m-by-n rectangle into $m + n - 1$ rectangular tiles with integer sides such that each of the $m-1$ horizontal lines and $n-1$ vertical lines within the rectangle is part of the boundary of at least one tile. Let $a_{m,n}$ denote the number of tight m-by-n pavings.

   (a) Determine $a_{3,n}$ as a function of n.

   (b) Show for $m \geq 3$ that $\lim_{n\to\infty} a_{m,n}/m^n$ exists, and compute its value.

3. **Problem 11929** (Proposed by D. E. Knuth, 123(8), 2016). Let a_n be the number of ways in which a rectangular box that contains $6n$ square tiles in three rows of length $2n$ can be split into two connected pieces of size $3n$ without cutting any tiles. Thus $a_1 = 3$, $a_2 = 19$, and $a_3 = 85$. Taking $a_0 = 1$, find a closed form for the generating function $A(z) = \sum_{n=0}^{\infty} a_n z^n$. What is the asymptotic nature of a_n as $n \to \infty$?

4. **Problem 10883** (Proposed by N. MacKinnon, 108(6), 2001). (a) Let $R(n)$ be the set of all rectangles whose side lengths are natural numbers and whose area is at most n. Find an integer $n > 1$ such that the members of $R(n)$, each used exactly once, tile a square.

   (b) Let $C(n)$ be the set of all cuboids (rectangular parallelepipeds) whose side lengths are natural numbers and whose volume is at most n. Show that there is no integer $n > 1$ such that the members of $C(n)$, each used exactly once, tile a cube.

4.7 A sum involving a p-root of unity

Problem 11721 (Proposed R. Tauraso, 120(7), 2013). Let p be a prime greater than 3, and let q be a complex number other than 1 such that $q^p = 1$. Evaluate

$$\sum_{k=1}^{p-1} \frac{(1 - q^k)^5}{(1 - q^{2k})^3(1 - q^{3k})^2}.$$

Discussion.
Let a and b be nonnegative integers such that $\gcd(a, p) = 1$. Then for any integer j,

$$\sum_{k=0}^{p-1} q^{k(aj+b)} = \begin{cases} p & \text{if } p \mid (aj + b) \\ 0 & \text{otherwise} \end{cases}. \tag{4.10}$$

Notice that p divides $aj + b$ if and only if $j = j_0 + np$ for some integer n and $j_0 \in \{0, 1, \ldots, p-1\}$ such that $j_0 \equiv -b/a \pmod{p}$. Hence, by (4.10),

$$\sum_{k=1}^{p-1} \frac{q^{bk}}{(1 - q^{ak}z)^d} = \sum_{k=1}^{p-1} q^{bk} \sum_{j=0}^{\infty} \binom{j+d-1}{d-1} (q^{ak}z)^j$$

$$= \sum_{j=0}^{\infty} \binom{j+d-1}{d-1} z^j \sum_{k=1}^{p-1} q^{k(aj+b)}$$

$$= p \sum_{n=0}^{\infty} \binom{j_0 + np + d - 1}{d-1} z^{j_0+np} - \frac{1}{(1-z)^d}$$

$$= \frac{pz^{j_0}}{(1-z^p)^d} \sum_{n=0}^{d-1} c_n z^{np} - \frac{1}{(1-z)^d} \qquad (4.11)$$

where

$$c_n = [z^{np}](1-z^p)^d \sum_{k=0}^{\infty} \binom{j_0 + kp + d - 1}{d-1} z^{kp}$$

$$= \sum_{k=0}^{n} (-1)^{n-k} \binom{d}{n-k} \binom{j_0 + kp + d - 1}{d-1}.$$

Now the required problem turns into finding the limit of (4.11) as z goes to 1. From there, the partial fraction decomposition can simply be applied to the given summand and lead us to the required result.

Solution.
From (4.11), letting $z = 1 + w \to 1$, we obtain

$$\sum_{k=1}^{p-1} \frac{q^{bk}}{(1 - q^{ak})^d} = \lim_{z \to 1} \left(\frac{pz^{j_0}}{(1-z^p)^d} \sum_{n=0}^{d-1} c_n z^{np} - \frac{1}{(1-z)^d} \right)$$

$$= \lim_{w \to 0} \frac{p \sum_{n=0}^{d-1} c_n (1+w)^{j_0+np} - \left(\frac{1 - (1+w)^p}{-w} \right)^d}{(1 - (1+w)^p)^d}$$

$$= \lim_{w \to 0} \frac{p \sum_{n=0}^{d-1} c_n \binom{j_0+np}{d} w^d + o(w^d) - (-1)^d \sum_{n=0}^{d} (-1)^n \binom{d}{n} \binom{np}{2d} w^d + o(w^d)}{(-pw + o(w))^d}$$

$$= \frac{1}{p^d} \left((-1)^d p \sum_{n=0}^{d-1} c_n \binom{j_0 + np}{d} - \sum_{n=0}^{d} (-1)^n \binom{d}{n} \binom{np}{2d} \right) \qquad (4.12)$$

where we have applied

$$\sum_{n=0}^{d} (-1)^n \binom{d}{n} \binom{np}{k} = [w^k] \sum_{n=0}^{d} (-1)^n \binom{d}{n} (1+w)^{np} = [w^k] (1 - (1+w)^p)^d.$$

By plugging $a = 2$ and $d = 3$ into (4.12), we find that

$$\sum_{k=1}^{p-1} \frac{q^{bk}}{(1-q^{2k})^3} = \begin{cases} -(p-1)(p-3)/8 & \text{if } b = 0, \\ (p-1)(p+1)/24 & \text{if } b = 1, 4, \\ -(p-1)(p+1)/24 & \text{if } b = 2, 5, \\ 0 & \text{if } b = 3. \end{cases}$$

Moreover, for $a = 3$ and $d = 2$, (4.12) yields

$$\sum_{k=1}^{p-1} \frac{q^{bk}}{(1-q^{3k})^2} = \begin{cases} -(p-1)(p-5)/12 & \text{if } b = 0, \\ (p+5)(p-1)/36 & \text{if } b = 1, 5 \text{ and } p \equiv 1 \pmod 3, \\ (p-5)(p+1)/36 & \text{if } b = 1, 5 \text{ and } p \equiv -1 \pmod 3, \\ (p-1)^2/36 & \text{if } b = 2, 4 \text{ and } p \equiv 1 \pmod 3, \\ (p+1)^2/36 & \text{if } b = 2, 4 \text{ and } p \equiv -1 \pmod 3, \\ -(p-1)(p+1)/12 & \text{if } b = 3. \end{cases}$$

Finally, we perform the partial fraction decomposition and then apply these results above to yield

$$\sum_{k=1}^{p-1} \frac{(1-q^k)^5}{(1-q^{2k})^3(1-q^{3k})^2} = \sum_{k=1}^{p-1} \frac{4 - 8q^k + q^{2k} + 5q^{3k} - q^{4k} - q^{5k}}{(1-q^{2k})^3}$$

$$+ \sum_{k=1}^{p-1} \frac{-3 + 3q^k + 2q^{3k} - q^{4k} - q^{5k}}{(1-q^{3k})^2}$$

$$= \begin{cases} -\dfrac{(55p-1)(p-1)}{72} & \text{if } p \equiv 1 \pmod 3, \\ -\dfrac{(11p-1)(5p-1)}{72} & \text{if } p \equiv -1 \pmod 3. \end{cases}$$

Notice that the final result is always a negative integer. □

Remark. Two similar results, though more difficult, are left to the reader with references.

1. For any positive integer n,

$$\sum_{k=1}^{\lfloor \frac{n}{3} \rfloor} \frac{(-1)^k q^{\frac{k(3k-1)}{2}}}{1-q^{3k-1}} + \sum_{k=1}^{\lfloor \frac{n-1}{3} \rfloor} \frac{(-1)^k q^{\frac{k(3k+5)}{2}}}{1-q^{3k}}$$

$$= \begin{cases} -\dfrac{n-1}{6} & \text{if } n \equiv 1 \pmod 3, \\ 0 & \text{if } n \equiv -1 \pmod 3, \\ \dfrac{1}{3} + \dfrac{n+1}{6} q^{\frac{2n}{3}} & \text{if } n \equiv 0 \pmod 3. \end{cases}$$

where q is a primitive nth-root of unity. For a proof consult [68, (7)] and [8, (1.1)].

2. For any positive integer n,

$$\sum_{k=0}^{n-1} \begin{bmatrix} 2k \\ k \end{bmatrix}_q q^k = \begin{cases} q^{3r(r+1)/2} & \text{if } n \equiv 1 \pmod 3, \\ -q^{3r(r+1)/2} & \text{if } n \equiv -1 \pmod 3, \\ 0 & \text{if } n \equiv 0 \pmod 3. \end{cases}$$

where $r = \lfloor 2n/3 \rfloor$ and q is a primitive nth-root of unity. A proof is given in [97, Corollary 4.3]

Additional problems for practice.

1. Show that for any odd prime p,

$$\sum_{k=1}^{p-1} \csc^2 \left(\frac{\pi k}{p} \right) = p^2 - 1.$$

2. **Problem 12443** (Proposed by N. Osipov, 131(2), 2024). Let n be an odd positive integer. Evaluate

$$\sum_{k=1}^{n-1} \sin \left(\frac{k^2 \pi}{n} \right) \cot \left(\frac{k\pi}{n} \right).$$

3. **Problem 10937** (Proposed by David G. Wagner, 109(4), 2002). For $n \in \mathbb{N}$, let $[n] = \{1, \ldots, n\}$. For $A \subseteq [n]$, let $B = [n] - A$. Let $P(x) = \sum_{j \in A} x^j + \sum_{j \in B} x^{-j}$. Let $q = \exp(2\pi i/(2n + 1))$. Prove that

$$\sum_{k=0}^{2n} P(q^k)^3 = \frac{n(n+1)(2n+1)}{2}.$$

4.8   Sparse binary representation of an integer

Problem 11782 (Proposed I. Gessel, 121(6), 2014). A *signed binary representation* of an integer m is a finite list $a_0, a_1, \ldots$ of elements of $\{-1, 0, 1\}$ such that $\sum a_k 2^k = m$. A signed binary representation is *sparse* if no two consecutive entries in the list are nonzero.
(a) Prove that every integer has a unique sparse representation.
(b) Prove that for all $m \in \mathbb{Z}$, every non-sparse signed binary representation of m has at least as many nonzero terms as the sparse representation.

Discussion. As an example, we list the sparse representations of the numbers from 0 to 23. The digit -1 is given as $\bar{1}$.

m	sparse rep.	m	sparse rep.	m	sparse rep.
0	0	8	1000	16	10000
1	1	9	1001	17	10001
2	10	10	1010	18	10010
3	$10\bar{1}$	11	$10\bar{1}0\bar{1}$	19	$10\bar{1}0\bar{1}$
4	100	12	$10\bar{1}00$	20	10100
5	101	13	$10\bar{1}01$	21	10101
6	$10\bar{1}0$	14	$100\bar{1}0$	22	$10\bar{1}0\bar{1}0$
7	$100\bar{1}$	15	$1000\bar{1}$	23	$10\bar{1}00\bar{1}$

For example, to obtain the sparse representation of 23, we write down its ordinary binary representation and we make the following moves:

$$10111 \quad \longrightarrow \quad 1100\bar{1} \quad \longrightarrow \quad 10\bar{1}00\bar{1}$$

Here, at the first step we replaced 0111 with $100\bar{1}$ and then we replaced 011 with $10\bar{1}$. Notice that the sparse representation of -23 is the complement of the one of 23, i.e. $\bar{1}01001$.

Drawing from these examples, we will formulate a general conversion algorithm and use it to address questions (a) and (b).

Solution.

(a) We first show that any integer $m \in \mathbb{Z}$ has a sparse signed binary representation.

Consider the ordinary binary representation of $|m|$. Starting from the right end, we replace any group of $d \geq 2$ consecutive 1s and the 0 immediately to its left, with the equivalent string made of 1, $d-1$ consecutive 0s and a $\bar{1}$:

$$0\underbrace{1\ldots1}_{d} \quad \longrightarrow \quad 1\underbrace{0\ldots0}_{d-1}\bar{1}.$$

This procedure continues until all nonzero terms are separated by zeros. Finally, if $m < 0$ then we change sign to all digits. At the end, we get a sparse representation of m.

As regards uniqueness, if m has two sparse representations then, starting from the left end, we compare the corresponding digits and eliminate the digits which are equal. Thus the two representations are reduced to one of the following cases:

i) $2^r + x = -2^r + y$ where $|x| \leq \lfloor 2^r/3 \rfloor$ and $|y| \leq \lfloor 2^r/3 \rfloor$, which yields a contradiction

$$2^{r+1} = 2^r + 2^r = y - x \leq |x| + |y| \leq 2\lfloor 2^r/3 \rfloor < 2^r.$$

ii) $2^r + x = \pm 2^s + y$ where $r > s$ and $|x| \leq \lfloor 2^r/3 \rfloor$ and $|y| \leq \lfloor 2^s/3 \rfloor$, which yields a contradiction

$$2^r = \pm 2^s + y - x \leq 2^s + |x| + |y| \leq 2^{r-1} + \lfloor 2^r/3 \rfloor + \lfloor 2^{r-1}/3 \rfloor = 2^r - 1 < 2^r.$$

(b) Every signed binary representation can be reduced to the unique sparse representation by starting from the right end and by using these replacement rules:

$$\text{i) for } d \geq 2, \qquad 0\underbrace{1\ldots1}_{d} \quad \longrightarrow \quad 1\underbrace{0\ldots0}_{d-1}\bar{1},$$

$$\text{ii) for } d \geq 2, \qquad 0\underbrace{\bar{1}\ldots\bar{1}}_{d} \quad \longrightarrow \quad \bar{1}\underbrace{0\ldots0}_{d-1}1,$$

$$\text{iii) for } d \geq 1, \qquad \bar{1}\underbrace{1\ldots1}_{d} \quad \longrightarrow \quad \underbrace{0\ldots0}_{d}\bar{1},$$

$$\text{iv) for } d \geq 1, \qquad 1\underbrace{\bar{1}\ldots\bar{1}}_{d} \quad \longrightarrow \quad \underbrace{0\ldots0}_{d}1.$$

Notice that after any replacement the number of nonzero terms is not increase. Thus, in a finite number of steps, we obtain a sparse representation with no more nonzero terms than the original one. □

Remark. This *minimal* binary representation which uses the fewest nonzero digits and its main properties can be found in G. W. Reitwiesner's book [82, Section 8].

Additional problems for practice.

1. **Problem 10564** (Proposed by A. Fraenkel, 104(1), 1997). The *Nim-sum* of two positive integers with binary expansions $\sum_{i\geq0} a_i 2^i$ and $\sum_{i\geq0} b_i 2^i$ is the number with binary expansion $\sum_{i\geq0} c_i 2^i$, where a_i, b_i, c_i are in $\{0,1\}$ and $c_i \equiv a_i + b_i \pmod 2$. Let n be a positive integer and let j be a nonnegative integer. How many of the 2^n subsets of the set $\{1, 2, \ldots, n\}$ have the property that their elements have Nim-sum equal to j?

2. **Problem 10865** (Proposed by D. Beckwith, 108(4), 2001). The *binary sum* of a family of sets is the set whose elements appear in an odd number of members of the family. Given a finite set S with $|S| = n \geq 2$, let P be the set of nonempty proper subsets of S. For $A \subseteq S$, let $d_n(A)$ be the number of nonempty subsets of P whose binary sum is A. Show that $d_n(S) - d_n(\emptyset) = 1$.

3. **Problem 11336** (Proposed by D. E. Knuth, 115(1), 2008). A *near-deBruijn cycle of order d* is a cyclic sequence of $2^d - 1$ zeros and ones in which all $2^d - 1$ substrings of length d are distinct. For all $d > 0$, construct a near-deBruijn cycle of order $d + 1$ such that the front and back substrings of length $2^d - 1$ are both near-deBruijn cycles of order d. (Thus, for example, 1100010 is near-deBruijn of order 3, while 110 and 010 are both near-deBruijn of order 2.)

4. **Problem 12377** (Proposed by Li Zhou, 130(3), 2023). An integer is a *one-drop* number if its decimal digits $d_1 \cdots d_n$ satisfy

$$1 \leq d_1 \leq \cdots \leq d_i > d_{i+1} \leq \cdots \leq d_n$$

for some i. For $n \geq 2$, how many n-digit one-drop numbers are there?

5. **Putnam 2002-A6.** Fix an integer $b \geq 2$. Let $f(1) = 1, f(2) = 2$, and for each $n \geq 3$, define $f(n) = nf(d)$, where d is the number of base-b digits of n. For which values of b does $\sum_{n=1}^{\infty} 1/f(n)$ converge?

6. **Putnam 2021-B1.** For a positive integer n, define $d(n)$ to be the sum of the digits of n when written in binary (for example, $d(13) = 1+1+0+1 = 3$). Let

$$S = \sum_{k=1}^{2020} (-1)^{d(k)} k^3.$$

Determine S modulo 2020.

7. (Due to Erdös) Let $1 \leq a_1 < a_2 \cdots < a_n$ be a set of integers for which all the sums $\sum_{i=1}^{n} \epsilon_i a_i$ with $\epsilon_i = 0$ or 1 are distinct. Show that

$$\sum_{i=1}^{n} \frac{1}{a_i} \leq 2 - \frac{1}{2^{n-1}}$$

and equality holds only if $a_i = 2^{i-1}$ for $i = 1, 2, \ldots, n$.
Hint. Use $\prod_{i=1}^{n}(1 + x^{a_i}) < \sum_{k=0}^{\infty} x^k = 1/(1-x)$.

4.9   Subsets with equal sums of powers

Problem 12085 (Proposed by J. DeVincentis, S. Wagon, and M. Elgersma, 126(1), 2019). For which positive integers n can $\{1, \ldots, n\}$ be partitioned into two sets A and B of the same size so that

$$\sum_{k \in A} k = \sum_{k \in B} k, \quad \sum_{k \in A} k^2 = \sum_{k \in B} k^2, \quad \text{and} \quad \sum_{k \in A} k^3 = \sum_{k \in B} k^3 ?$$

Discussion. Let n be a positive integer such that the required partition exists. We make some observations on n. Since $|A| = |B|$ and $|A| + |B| = n$, then n has to be even. Therefore, $n + 1$ is odd and

$$\sum_{k \in A} k = \frac{1}{2} \sum_{k=1}^{n} k = \frac{n(n+1)}{4} \in \mathbb{N}.$$

This implies that 4 divides n. Let $n = 4N$. Then

$$\sum_{k \in A} k^3 = \frac{1}{2} \sum_{k=1}^{n} k^3 = \frac{n^2(n+1)^2}{8} = 2N^2(4N+1)^2 \equiv 0 \pmod{2}.$$

Moreover, since $k^3 \equiv k \pmod 2$, we have

$$\sum_{k \in A} k = \frac{n(n+1)}{4} = N(4N+1) \equiv N \pmod 2$$

which implies that $N \equiv 0 \pmod 2$. So we conclude that n has to be a multiple of 8.

In the following, we will show that, with the exception of $n = 8$, the condition of n to be a multiple of 8 is also sufficient.

Solution.

We show that such partition exists if and only if n is a multiple of 8 with $n \geq 16$.

A short brute force search reveals that there is no such partition for $n = 8$. It remains to show that a partition exists when n is multiple of 8 and $n \geq 16$.

For any integer $x \geq 0$, we have a partition of $\{1 + x, \ldots, 8 + x\}$,

$$A_8(x) := \{1 + x, 4 + x, 6 + x, 7 + x\}$$

and

$$B_8(x) := \{2 + x, 3 + x, 5 + x, 8 + x\}$$

and a partition of $\{1 + x, \ldots, 12 + x\}$,

$$A_{12}(x) := \{1 + x, 3 + x, 7 + x, 8 + x, 9 + x, 11 + x\}$$

and

$$B_{12}(x) := \{2 + x, 4 + x, 5 + x, 6 + x, 10 + x, 12 + x\}$$

such that $|A_8(x)| = |B_8(x)|$, $|A_{12}(x)| = |B_{12}(x)|$ and for $j \in \{1, 2\}$,

$$\sum_{k \in A_8(x)} k^j = \sum_{k \in B_8(x)} k^j \quad \text{and} \quad \sum_{k \in A_{12}(x)} k^j = \sum_{k \in B_{12}(x)} k^j.$$

Therefore, if m is positive integer of the form $8a + 12b = 4(2a + 3b)$ with $a, b \geq 0$, i.e., m is a multiple of 4 with $m \geq 8$, then we have a partition of $\{1, \ldots, m\}$,

$$A' := \bigcup_{j=0}^{a-1} A_8(8j) \cup \bigcup_{j=0}^{b-1} A_{12}(8a + 12j)$$

and

$$B' := \bigcup_{j=0}^{a-1} B_8(8j) \cup \bigcup_{j=0}^{b-1} B_{12}(8a + 12j)$$

such that $|A'| = |B'|$ and

$$\sum_{k \in A'} k = \sum_{k \in B'} k, \quad \text{and} \quad \sum_{k \in A'} k^2 = \sum_{k \in B'} k^2.$$

Finally we define a partition of $\{1, \ldots, n\}$ with $n = 2m$,

$$A := A' \cup (m + B') \quad \text{and} \quad B := B' \cup (m + A').$$

Then it is straightforward to show that

$$|A| = |B|, \quad \sum_{k \in A} k = \sum_{k \in B} k, \quad \sum_{k \in A} k^2 = \sum_{k \in B} k^2.$$

Moreover,

$$\sum_{k \in A} k^3 = \sum_{k \in A'} k^3 + \sum_{k \in B'} (m + k)^3$$

$$= \sum_{k \in A'} k^3 + m^3 |B'| + 3m^2 \sum_{k \in B'} k + 3m \sum_{k \in B'} k^2 + \sum_{k \in B'} k^3$$

$$= \sum_{k \in B'} k^3 + m^3 |A'| + 3m^2 \sum_{k \in A'} k + 3m \sum_{k \in A'} k^2 + \sum_{k \in A'} k^3 = \sum_{k \in B} k^3.$$

This completes the proof. For example, for $n = 16$, we have the partitions:

$$A = \{1, 4, 6, 7, 10, 11, 13, 16\} \quad \text{and} \quad B = \{2, 3, 5, 8, 9, 12, 14, 15\}.$$

$\square$

Remark. The problem recalls the Prouhet-Tarry-Escott problem: given positive integers n and k with $n > k$, find two disjoint multisets (repetitions are allowed) A and B of n integers such that for any j from 1 to k

$$\sum_{a \in A} a^j = \sum_{b \in B} b^j \quad \text{for } j = 0, 1, 2, \ldots, k. \tag{4.13}$$

The Prouhet-Tarry-Escott problem has a vast literature (for a survey see, for example, Chapter 11 in [24]) and a long history. It is remarkable that (4.13) can be reformulated as a question about polynomials, namely $(z - 1)^{k+1}$ divides

$$\sum_{a \in A} z^a - \sum_{b \in B} z^b.$$

The ubiquitous Thue-Morse sequence

$$(t_n)_{n \geq 0} = 0110100110010110100101100110\ldots$$

defined recursively by $t_0 = 0$, $t_{2n} = t_n$, $t_{2n+1} = 1 - t_n$ for $n \geq 0$, has the following nice property proved by Eugene Prouhet in 1851: if

$$A = \{j \in \{0, 1, 2, \ldots, 2^m - 1\} : t_j = 0\}$$

and

$$B = \{j \in \{0, 1, 2, \ldots, 2^m - 1\} : t_j = 1\}.$$

then (4.13) holds for $k = m - 1$ (see [5]).

Additional problems for practice.

1. **Problem 10284** (Proposed by L.-S. Hahn, 100(2), 1993). For each positive integer l, show that there exists a positive integer n and a partition of $\{1, \ldots, n\}$ as a disjoint union of two sets A and B, such that for $1 \leq i \leq l$,

$$\sum_{a \in A} a^i = \sum_{b \in B} b^i.$$

2. **Problem 6574** (Proposed by M. Laub, 95(6), 1988). Put $M_r = \sum_{j=1}^{r} j^2$ for $r = 1, 2, \ldots$. If $n \in \mathbb{N}$ (the set of positive integers), define $m(n)$ to be the unique positive integer r such that $M_r < n < M_{r+1}$.

   (a) Let S_1, be the set $n \in \mathbb{N}$ such that n is expressible as a sum of $m(n)$ distinct squares of positive integers. Prove that S_1, has asymptotic density one but that $\mathbb{N} \setminus S_1$, is infinite.

   (b) Let S_2 be the set of $n \in \mathbb{N}$ such that n^2 is expressible as a sum of $m(n^2)$ distinct squares of positive integers. Prove that S_2 has asymptotic density one. Is $\mathbb{N} \setminus S_2$ infinite?

3. **Problem 11002** (Proposed by Y.-F. S. Pétermann, 110(3), 2003). Pooh Bear has $2N + 1$ honey pots. No matter which one of them he sets aside, he can split the remaining $2N$ pots into two sets of the same total weight, each consisting of N pots. Must all $2N+1$ pots weigh the same?

4. **MM Problem 1768** (Proposed by G.R.A.20 Problem Solving Group, 80(2), 2007). For which positive integers n can the set $1, 2, \ldots, 2n$ be partitioned into n two element subsets so that the sum of the two numbers in each subset is a perfect square?

4.10 Equality of two sums of reciprocals

Problem 12227 (Proposed G. Galperin and Y. J. Ionin, 128(1), 2021). Prove that for any integer n with $n \geq 3$ there exist infinitely many pairs (A, B) such that A is a set of n consecutive positive integers, B is a set of fewer than n positive integers, A and B are disjoint, and

$$\sum_{k \in A} \frac{1}{k} = \sum_{k \in B} \frac{1}{k}.$$

Discussion.
As a warm-up, let's solve the case $n = 4$. For all $N \geq 2$, let

$$A_N = \{4N - 2, 4N - 1, 4N, 4N + 1\}$$

and

$$B_N = \{N, 2N(4N-2), 2N(4N-1)(4N+1)\}.$$

Then $|A_N| = 4 > 3 = |B_n|$ and $A_N \cap B_N = \emptyset$. Moreover

$$\frac{1}{2N(4N-2)} = \frac{1}{4N-2} - \frac{1}{4N}$$

and

$$\frac{1}{2N(4N-1)(4N+1)} = \frac{1}{4N-1} + \frac{1}{4N+1} - \frac{1}{2N}$$

and it follows

$$\sum_{k \in B_N} \frac{1}{k} = \frac{1}{N} + \frac{1}{4N-2} - \frac{1}{4N} + \frac{1}{4N-1} + \frac{1}{4N+1} - \frac{1}{2N} = \sum_{k \in A_N} \frac{1}{k}.$$

Solution.
We prove by distinguishing the following three cases.

1) $n = 3m$ with $m \geq 1$.
For all $N \geq 2$, letting

$$A_N = \{k : k \in [3N-1, 3(N+m) - 2]\}$$

and

$$B_N = \{k : k \in [N, N+m-1]\} \cup \left\{ \frac{3k(3k-1)(3k+1)}{2} : k \in [N, N+m-1] \right\},$$

we have $n = 3m = |A_N| > 2m = |B_n|$, $A_N \cap B_N = \emptyset$, and $\sum_{k \in A_N} \frac{1}{k} = \sum_{k \in B_N} \frac{1}{k}$. Since

$$\frac{2}{3k(3k-1)(3k+1)} = \frac{1}{3k-1} + \frac{1}{3k} + \frac{1}{3k+1} - \frac{1}{k},$$

we get

$$\sum_{k \in B_N} \frac{1}{k} = \sum_{k=N}^{N+m-1} \left(\frac{1}{3k-1} + \frac{1}{3k} + \frac{1}{3k+1} \right) = \sum_{k=3N-1}^{3(N+m-1)+1} \frac{1}{k} = \sum_{k \in A_N} \frac{1}{k}.$$

Note that $k(3k-1)(3k+1)/2$ is integer because any triple of consecutive integers, at least one is even.

2) $n = 3m+1$. Since $m = 1$ is set in Discussion above, we assume that $m \geq 2$.
For all $N \geq 2$, letting

$$A_N = \{k : k \in [3N-1, 3(N+m) - 1]\}$$

and

$$B_N = \{k : k \in [N, N + m - 1]\} \cup \left\{ \frac{3k(3k-1)(3k+1)}{2} : k \in [N, N + m - 1] \right\}$$
$$\cup \{3(N + m), 3(N + m)(3(N + m) - 1)\},$$

we find that $n = 3m + 1 = |A_N| > 2m + 2 = |B_n|$, $A_N \cap B_N = \emptyset$, and $\sum_{k \in A_N} \frac{1}{k} = \sum_{k \in B_N} \frac{1}{k}$.

3) $n = 3m + 2$ with $m \geq 1$.

For all $N \geq 2$ such that $N + m$ is even, letting

$$A_N = \{k : k \in [3N - 1, 3(N + m)]\}$$

and

$$B_N = \{k : k \in [N, N + m - 1]\} \cup \left\{ \frac{3k(3k-1)(3k+1)}{2} : k \in [N, N + m - 1] \right\}$$
$$\cup \left\{ \frac{3(N + m)}{2}, 3(N + m)(3(N + m) - 1) \right\},$$

we obtain that $n = 3m + 2 = |A_N| > 2m + 2 = |B_n|$, $A_N \cap B_N = \emptyset$, and $\sum_{k \in A_N} \frac{1}{k} = \sum_{k \in B_N} \frac{1}{k}$. $\qquad\square$

Remark. The Erdős-Straus conjecture is an open statement in number theory which involve unit fractions: for every integer $n > 1$, there are positive integers a, b and c such that

$$\frac{4}{n} = \frac{1}{a} + \frac{1}{b} + \frac{1}{c}. \tag{4.14}$$

In 2014, S. E. Salez checked the conjecture up to $n = 10^{17}$. It is easy to see that it suffices to consider the case when n is a prime number. Moreover, for $n > 2$, we may require that a, b and c are distinct because

$$\frac{1}{2k} + \frac{1}{2k} = \frac{1}{k} + \frac{1}{k(k+1)} \quad \text{and} \quad \frac{1}{2k-1} + \frac{1}{2k-1} = \frac{1}{k} + \frac{1}{k(2k-1)}.$$

Another identity

$$\frac{4}{3k-1} = \frac{1}{k} + \frac{1}{3k-1} + \frac{1}{k(3k-1)}$$

easily shows that the conjecture holds when 3 divides $n + 1$.

Notice that the problem becomes significantly more approachable if we change a single sign on the right-hand side of (4.14). In fact, for every $n > 2$, there exist positive integers a, b and c such that

$$\frac{4}{n} = \frac{1}{a} + \frac{1}{b} - \frac{1}{c}.$$

For a proof see [59].

Additional problems for practice.

1. **Problem 10992** (Proposed by D. Lubell, 110(2), 2003). Let N and n be positive integers, and let

$$S = \{(w_1, \ldots, w_n) \in \mathbb{N}^n : 0 < w_1 + \cdots + w_n \le N\}$$

and

$$T = \{(w_1, \ldots, w_n) \in \mathbb{N}^n : w_1, \ldots, w_n \text{ are distinct and bounded by } N\}.$$

Show that

$$\sum_S \frac{1}{w_1 \cdots w_n} = \sum_T \frac{1}{w_1 \cdots w_n}.$$

2. **Problem 10294** (Proposed by D. A. Holton, 100(3), 1993). Given a positive integer n, let $[n] = \{1, \ldots, n\}$. For a positive integer k, say that n is k-good if $[n]$ can be partitioned into k sets each with the same sum. Show that n is k-good if k divides $\binom{n}{2}$ and n is sufficiently large.

3. **Problem 12055** (Proposed by D. E. Knuth, 125(7), 2018). Let $(a_i)_{i \ge 1}$ be a sequence of nonnegative integers with $a_1 \ge a_2 \ge \cdots$ and with finite sum. For a positive integer j, let b_j be the number of indices i such that $a_i \ge j$. Prove that the multisets $\{a_1 + 1, a_2 + 2, \ldots\}$ and $\{b_1 + 1, b_2 + 2, \ldots\}$ are equal.

4. **Problem 12233** (Proposed by C. R. Pranesachar, 128(2), 2021). Let n and k be positive integers with $1 \le k < (n+1)/2$. For $1 \le r \le n$, let $h(r)$ be the number of k-element subsets of $\{1, \ldots, n\}$ that do not contain consecutive elements but that do contain r. Prove

   (a) $h(r) = h(r+1)$ when $r \in \{k, \ldots, n-k\}$.

   (b) $h(k-1) = h(k) \pm 1$.

   (c) $h(r) > h(r+2)$ when $r \in \{1, \ldots, k-2\}$ and r is odd.

   (d) $h(r) < h(r+2)$ when $r \in \{1, \ldots, k-2\}$ and r is even.

5. **Putnam 2018-B1.** Let $\mathcal{P}$ be the set of vectors defined by

$$\mathcal{P} = \left\{ \begin{pmatrix} a \\ b \end{pmatrix} : 0 \le a \le 2, 0 \le b \le 100, \text{ and } a, b \in \mathbb{Z} \right\}.$$

Find all $\mathbf{v} \in \mathcal{P}$ such that the set $\mathcal{P} \setminus \{\mathbf{v}\}$ obtained by omitting vector $\mathbf{v}$ from $\mathcal{P}$ can be partitioned into two sets of equal size and equal sum.

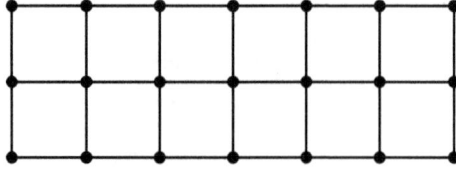

Figure 4.18: A 3×7 grid

4.11 A monochromatic pentagons with given area

Problem 12251 (Proposed R. Tauraso, 128(5), 2021). Each point in the plane is colored either red or blue. Show that for any positive real number S, there is a proper convex pentagon of area S all five of whose vertices have the same color. (By a proper convex pentagon we mean a convex pentagon whose internal angles are less than π.)

Discussion.

We consider a 3×7 grid where each unit square has side $l = \frac{\sqrt{S}}{6}$.

By the pigeonhole principle, out of the 7 vertices in the first row, at least four have to be of the same color, say, blue. Now we look just at the corresponding 4 columns. If the second or the third row has at least 2 blue vertices, we have a rectangle with all blue vertices. Otherwise, the second and the third row have at least 3 red vertices each and we find a rectangle with all red vertices. Therefore, after this first step, we have a rectangle of sides x and y where $x \in \{l, 2l, 3l, 4l, 5l, 6l\}$ and $y \in \{l, 2l\}$, of area $xy \leq 6l \cdot 2l = \frac{S}{3}$ whose 4 vertices are of the same color. Without loss of generality, we may assume that their color is blue.

Solution.

We draw two open horizontal segments of length x at distance $h = \frac{2(S-xy)}{x} > 0$ from the two horizontal sides of the blue rectangle. If there is at least a blue point along the top open segment or along the bottom open segment, then this point together with the 4 vertices of the blue rectangle form a proper convex pentagon. The vertices of the pentagon are all blue and its area is

$$xy + \frac{xh}{2} = xy + \frac{x}{2} \cdot \frac{2(S - xy)}{x} = S.$$

Now we assume that the points of the two open segments are all red. Notice that the area of the rectangle $x \times (y + 2h)$ is

$$x(y + 2h) = xy + 4(S - xy) = 4S - 3xy \geq 4S - S = 3S > 2S. \qquad (4.15)$$

We distinguish two cases.

(1) If there is at least a red point in the open rectangle $x \times (y + 2h)$, then, by (4.15), we can choose two pairs of points along each of the two red horizontal

segments in such a way that, together with the red point inside the open rectangle, those 5 points form a red pentagon whose area is S.

(2) On the other hand, if all the points in the open rectangle $x \times (y + 2h)$ are blue, then, again by (4.15), we are able to find 5 points inside it such that the area of the resulting blue pentagon is equal to S. $\qquad\square$

Remark. More generally, any sufficiently large grid where each point is colored with $c \geq 2$ colors contains a monochromatic rectangle. We show that a $(c + 1) \times \left(c\binom{c+1}{2} + 1\right)$ grid works.

Indeed, each column has $c + 1$ points, which means that there are a pair of points of the same color. The number of ways to choose a pair of points of the same color along a column is $c \cdot \binom{c+1}{2}$. Having $c\binom{c+1}{2} + 1$ columns implies that, by the pigeonhole principle, there are two columns which have the a pair of points in the same position of the same color, making monochromatic rectangle. For an extensive discussion on monochromatic rectangles in a grid we refer to [16].

Actually, the above statement is a particular case of Gallai's theorem which states that if the points in the plane are colored with finitely many colors, then for every finite subset of the plane there exists a monochromatic homothetic copy of that set. So it happens that if the colored grid is much bigger than before, then we can also find a monochromatic square in it.

This kind of results is typical of an intriguing branch of Combinatorics called *Euclidean Ramsey Theory*. For a survey on this subject we recommend [61, Section 8] and A. Soifer's book [87].

Additional problems for practice.

1. **Problem E3378** (Proposed by M. Bòna, 97(3), 1990). Suppose the points of $\mathbb{Z}^2$ (the set of points in the plane with integer coordinates) are colored with a finite number of colors. For every $n > 3$, prove that there exists a convex n-gon with vertices and centroid in $\mathbb{Z}^2$ such that all $n + 1$ points have the same color.

2. **Putnam 1988-A4.** (a) If every point of the plane is painted one of three colors, do there necessarily exist two points of the same color exactly one inch apart?

   (b) What if "three" is replaced by "nine"?

3. **Problem 12449** (Proposed by V. Jungić, 131(3), 2024). Let n be a positive integer with $n \geq 2$. The squares of an (n^2+n-1)-by-(n^2+n-1) grid are colored with up to n colors. Prove that there exist two rows and two columns whose four squares of intersection have the same color.

4. **Putnam 2004-A5.** An $m \times n$ checkerboard is colored randomly: each square is independently assigned red or black with probability $1/2$. We say that two squares, p and q, are in the same connected monochromatic region if there is a sequence of squares, all of the same color, starting

at p and ending at q, in which successive sequences in the sequence share a common side. Show that the expected number of connected monochromatic regions is greater than $mn/8$.

4.12 Coloring a graph

Problem 12296 (Proposed D. A. Kalarkop, R. Rangarajan, and D. B. West, 129(1), 2022). For $t \leq n/2$, let $H(n,t)$ be the graph obtained from the complete graph on n vertices by deleting t pairwise disjoint edges. Determine the number of ways to assign each vertex of $H(n,t)$ a color from a set of k available colors so that vertices forming an edge receive distinct colors.

Discussion.
We have to find the so-called *chromatic polynomial* $P_G(x)$ of the graph $G = H(n,t)$ with $x = k$, where $P_G(x)$ counts the number of ways we can color the vertices of a graph G with x of colors in such a way that no two adjacent vertices have the same color. For instance, for the complete graph on n vertices K_n, we have

$$P_{K_n}(x) = x(x-1)\ldots(x-n+1).$$

Moreover, a fundamental property of $P_G(x)$ is the deletion-contraction recurrence: If G is a simple graph and e is one of its edges, then

$$P_{G-e}(x) = P_G(x) + P_{G/e}(x) \tag{4.16}$$

where $G - e$ is the graph obtained from G by deleting the edge e and G/e is the graph obtained from G by contracting the edge e. Now we have everything we need to determine $P_{H(n,t)}(x)$.

Solution.
We will show, by double induction with respect to (n,t) for $n \geq 1$ and $0 \leq t \leq n/2$, that

$$P_{H(n,t)}(k) = \sum_{j=0}^{t} \binom{t}{j} P_{H(n-j,0)}(k) = \sum_{j=0}^{t} \binom{t}{j} k(k-1)\ldots(k-(n-j)+1)$$

where we define $H(m,0)$ as the complete graph K_m.
The claim is trivially true for all $n \geq 1$ and $t = 0$.

For the inductive step, we assume $t > 1$. By (4.16), it follows that

$$P_{H(n,t)}(k) = P_{H(n,t-1)}(k) + P_{H(n-1,t-1)}(k)$$

$$= \sum_{j=0}^{t-1} \binom{t-1}{j} P_{H(n-j,0)}(k) + \sum_{j=0}^{t-1} \binom{t-1}{j} P_{H(n-1-j,0)}(k)$$

$$= \sum_{j=0}^{t-1} \binom{t-1}{j} P_{H(n-j,0)}(k) + \sum_{j=1}^{t} \binom{t-1}{j-1} P_{H(n-j,0)}(k)$$

$$= P_{H(n,0)}(k) + \sum_{j=1}^{t-1} \left(\binom{t-1}{j} + \binom{t-1}{j-1} \right) P_{H(n-j,0)}(k) + P_{H(n,t)}(k)$$

$$= \sum_{j=0}^{t} \binom{t}{j} P_{H(n-j,0)}(k)$$

and the proof is complete. $\square$

Remark. The deletion-contraction recurrence (4.16) enables to find the chromatic polynomial of several other graphs. For example, if T is any tree with n vertices then by choosing at each step a pendant edge and applying (4.16), by induction, we find that

$$P_{T_n}(x) = x(x-1)^{n-1}.$$

Noticing that if we remove an edge in a cycle on n vertices C_n we obtain a path P_n (which is a particular tree). Then, again by (4.16), for $n \geq 2$, we have

$$P_{C_n}(x) = P_{P_n}(x) - P_{C_{n-1}}(x) = P_{P_n}(x) - (P_{P_{n-1}}(x) - P_{C_{n-2}}(x))$$

$$= \cdots = \sum_{j=3}^{n} (-1)^{n-j} P_{P_j}(x) + (-1)^n P_{C_2}(x)$$

$$= \sum_{j=3}^{n} (-1)^{n-j} x(x-1)^{j-1} + (-1)^n x(x-1)$$

$$= (x-1)^n + (-1)^n (x-1).$$

For more problems on graph coloring consult [69, Chapter 9] and [98, Chapter 10].

Additional problems for practice.

1. A *ladder graph* L_n is a graph with $2n$ vertices and $3n-2$ edges which is obtained as the Cartesian product of the path graphs P_2 and P_n. Show that the chromatic polynomial of L_n is

$$P_{L_n}(x) = x(x-1)(x^2 - 3x + 3)^{n-1}.$$

2. **Problem E3409** (Proposed by I. Tomescu, 97(10), 1990). Suppose G is a connected k-chromatic graph which is neither a complete graph nor a cycle on m vertices with $m \equiv 3 \pmod 6$. Prove that in any k-coloring of G there exist two vertices of the same color having a common neighbor.

3. **MM Problem 1953** (Proposed by R. Tauraso, 87(4), 2014). Given a graph $G = (V, E)$, a perfect matching M of G is a subset of the set of edges E such that every vertex $v \in V$ lies on exactly one edge in M. Prove that for each positive integer n there is a planar connected graph G whose total number of perfect matchings is equal to n.

4. **Problem 11086** (Proposed by S. C. Locke, 111(5), 2004). Let G be a finite loopless graph in which no vertex is on more than N odd cycles. Prove that the (vertex) chromatic number of G is $O(\sqrt{N})$. (A *loopless graph* is one in which no edge joins a vertex to itself.)

5. **Problem 11898** (Proposed by R. P. Stanley, 123(3), 2016). Let n and k be integers, with $n \geq k \geq 2$. Let G be a graph with n vertices whose components are cycles of length greater than k. Let $f_k(G)$ be the number of k-element independent sets of vertices of G. Show that $f_k(G)$ depends only on k and n. (A set of vertices is independent if no two of them are adjacent.)

6. **Problem 12039** (Proposed by S. Silwal, 125(5), 2018). Let G be a graph with an even number of vertices. Show that there are two vertices in G with an even number of common neighbors.

4.13 Arranging coins along a line

Problem 12316 (Proposed H. A. ShahAli and M. Shahali, 129(4), 2022). For each i in $\{1, 2, \ldots, C\}$, we have $2i$ coins with color i. Place these $C(C+1)$ coins in a line. A move consists of the transposition of two adjacent coins. Let m be the minimum number of moves required to reach a configuration where all coins of the same color are together in a run of consecutive coins. Show that

$$\frac{(C-1)C(C+1)(3C+2)}{12}.$$

is the maximum value of m over all initial configurations.

Discussion.
Fix any permutation P of the colors $1, \ldots, C$. Given any arrangement, the ordered pair of positions $\langle i, j \rangle$ is an *inversion* if it satisfies the following properties: 1) $i < j$, 2) the color c_i of the coin at position i differs from the color c_j of the coin at position j, and 3) c_i appears after c_j in P. Since one adjacent transposition reduces the number of inversions by at most one, the number of

inversions is a lower bound for the number of adjacent transpositions needed to arrange the coins into groups of colors in the order given by P. For example, the arrangement

$$1, 2, 3, 2, 3, 3, 1, 3, 2, 3, 3, 2$$

where an integer i represents a coin of color i, has 16 inversion with respect to the order $1, 2, 3$ and 26 with respect to the order $3, 1, 2$. After checking all the 6 permutations of the colors, we find that the minimal number of inversion is 16. For $C = 3$ we have to show that this minimal number of inversions is less than $(2 \cdot 3 \cdot 4 \cdot 11)/12 = 22$. We notice that the upper bound 22 is attained for the arrangement

$$1, 2, 2, 3, 3, 3, 3, 3, 3, 2, 2, 1. \tag{4.17}$$

Actually, the number of inversion is always 22 regardless of the order of the colors.

Solution.
Let $f(C)$ be the minimum number of moves needed to reach the required configuration. If $C > 1$ and the $2C$ coins of color C are placed along the line as shown below,

$$\underbrace{*, \ldots, *}_{n_0}, C, \underbrace{*, \ldots, *}_{n_1}, C, \underbrace{*, \ldots, *}_{n_2}, C, *, \ldots, *, C, \underbrace{*, \ldots, *}_{n_{2C-1}}, C, \underbrace{*, \ldots, *}_{n_{2C}}$$

where $*$ represents a coin of a color different from C, then we need

$$n_0 + (n_0 + n_1) + \cdots + (n_0 + n_1 + n_2 + \cdots + n_{2C-1}) = \sum_{k=0}^{2C-1}(2C - k)n_k$$

adjacent transpositions in order to move the $2C$ coins to the far left side. Similarly, in order to move the $2C$ coins to the far right side, we need

$$n_{2C} + (n_{2C} + n_{2C-1}) + \cdots + (n_{2C} + n_{2C-1} + n_{2C-2} + \cdots + n_1) = \sum_{k=1}^{2C} kn_k$$

adjacent transpositions.
Since the smallest of those two numbers is less or equal to their arithmetic mean, which is

$$\frac{1}{2}\left(\sum_{k=0}^{2C-1}(2C - k)n_k + \sum_{k=1}^{2C} kn_k\right) = \frac{2C}{2}\sum_{k=0}^{2C} n_k$$

$$= C \cdot (C(C+1) - 2C) = C^2(C-1),$$

it follows that

$$f(C) \le f(C-1) + C^2(C-1).$$

Noticing that $f(1)$ is trivially zero, by induction, we find

$$f(C) = \sum_{i=2}^{C}(f(i) - f(i-1)) \leq \sum_{i=2}^{C} i^2(i-1) = \frac{(C-1)C(C+1)(3C+2)}{12}.$$

It remains to show that there is an arrangement of coins where the number of adjacent transpositions needed to group all the coins of the same color together is at least $(C-1)C(C+1)(3C+2)/12$.

We claim that such *worst case* is given by placing along the line, from left to right, 1 coin of color 1, 2 coins of color 2, ..., C coins of color C, and then mirroring the arrangement so far obtained (as we did in (4.17) for $C = 3$). We note that regardless of the choice of P, the number of inversions of our worst-case arrangement is always the same. Indeed, for any pair of coins of different colors at positions i, j with $i < j$, we have that $\langle i, j \rangle$ is an inversion if and only if the symmetric pair $\langle C(C+1) - j, C(C+1) - i \rangle$ is not an inversion. Hence the number of inversions for the worst case is half of the number of unordered pairs of coins of different colors, i.e.,

$$\frac{1}{2} \sum_{1 \leq i < j \leq C} (2i)(2j) = 2 \sum_{1 \leq i < j \leq C} ij = \left(\sum_{i=1}^{C} i\right)^2 - \sum_{i=1}^{C} i^2$$
$$= \frac{(C-1)C(C+1)(3C+2)}{12}$$

as expected. □

Remark. Incidentally the number

$$\frac{(C-1)C(C+1)(3C+2)}{12} = \sum_{1 \leq i \neq j \leq C} ij$$

can be interpreted as the number of nonsquare rectangles on a $C \times C$ square board (see the sequence $A052149$ in the OEIS (https://oeis.org/A052149)). Indeed if $1 \leq i \neq j \leq C$, then the product ij is the number of ways to choose the top-left corner of the rectangle $(C+1-i) \times (C+1-j)$ inside the square $C \times C$.

Additional problems for practice.

1. Ten coins of two colors are laid out along a strip of eleven squares as shown in Figure 4.19.

   We want to interchange the black and the gray coins, but I am only allowed to move coins into an adjacent empty square or to jump over one coin into an empty square. Can I make the interchange? If possible, what is the minimal number of moves? For a discussion of this solitaire game see [71, p. 52]. A partisan version of this game is discussed in [17, p. 14]

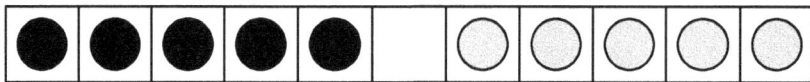

Figure 4.19: Swap the coins

2. **Problem 10197** (Proposed by U. Peled, 99(2), 1992).
Light bulbs $L_1, L_2, \ldots, L_n$ are controlled by switches $S_1, S_2, \ldots, S_n$. Switch S_i changes the on/off status of light L_i and possibly the status of some other lights. Assume that if S_i changes the status of light L_j, then S_j changes the status of light L_i. Initially all the lights are off. Prove that it is possible to operate the switches in such a way that all the lights are on.

3. **Problem 10390** (Proposed by O. Enchev, 101(6), 1994). A standard deck of 52 playing cards is arranged at random in 4 rows and 13 columns. Show that with finitely many transpositions of cards of the same value (e.g., 7♣ and 7♡, K♢ and K♠, and so on) all cards can be arranged in such a way that each column contains one club, one diamond, one heart, and one spade.

4. **Problem 10459** (Proposed by D. Beckwith, 102(6), 1995). A game is played with n disks ($n \geq 3$), each having a black face and a red face. Initially, then disks are arranged in a circle showing a random pattern of black and red faces. A move consists of taking away a black disk (i.e., one with its black face exposed) and inverting its neighbors (if any). The resulting gap is not closed up, so the remaining disks do not acquire new neighbors. The goal is to remove all the disks. For which initial patterns is this possible?

5. **Problem 10960** (Proposed by O. Eng, 109(7), 2002). Let n be an odd positive integer, and let $\{1, 2, \ldots, n\} \times \{1, 2, \ldots, n\}$ be a "game board" of n^2 positions called "squares". Let each of the $((n+1)/2)^2$ squares with both coordinates odd be colored black; the other squares are white. A *domino* is a set of two adjacent squares, that is, a pair of the form $\{(i, j), (i, j+1)\}$ or the form $\{(i, j), (i+1, j)\}$. Initially, the board is covered by $(n^2 - 1)/2$ disjoint dominoes, except for one black square on the boundary. A *move* translates a domino along its length by one square to cover the empty square, (uncovering in its wake another square two places away). Show that for every black square on the board there is a sequence of moves that uncovers it.

6. **Problem 11712** (Proposed by D. W. Cranston and D. B. West, 120(6), 2013). In the game of *Bulgarian solitaire*, n identical coins are distributed into two piles, and a move takes one coin from each existing pile to form a new pile. Beginning with a single pile of size n, how many

moves are needed to reach a position on a cycle (a position that will eventually repeat)? For example, $5 \to 41 \to 32 \to 221 \to 311 \to 32$, so the answer is 2 when $n = 5$.

7. **Problem 12218** (Proposed by R. Stong and S. Wagon, 127(10), 2020). For which positive integers n does there exist an ordering of all permutations of $\{1, \ldots, n\}$ so that their composition in that order is the identity?

4.14 Averaging the number of fixed points of permutations

Problem 12349 (Proposed R. Tauraso, 129(9), 2022). Let A_n be the set of permutations of $\{1, \ldots, n\}$ that have at least one fixed point. For $\pi \in A_n$ we write $\mathrm{Fix}(\pi)$ for $\{j : \pi(j) = j\}$. Evaluate

$$\sum_{\pi \in A_n} \left(\frac{\mathrm{sgn}(\pi)}{|\mathrm{Fix}(\pi)|} \sum_{j \in \mathrm{Fix}(\pi)} j \right).$$

Discussion.
We first rewrite the double sum in a suitable way

$$\sum_{\pi \in A_n} \frac{\mathrm{sgn}(\pi)}{|\mathrm{Fix}(\pi)|} \sum_{j \in \mathrm{Fix}(\pi)} j = \sum_{k=1}^{n} \frac{1}{k} \sum_{\substack{\pi \in A_n \\ |\mathrm{Fix}(\pi)|=k}} \mathrm{sgn}(\pi) \sum_{j \in \mathrm{Fix}(\pi)} j$$

$$= \sum_{k=1}^{n} \frac{1}{k} \cdot \left(\sum_{\substack{J \subseteq \{1,\ldots,n\} \\ |J|=k}} \sum_{j \in J} j \right) \cdot \sum_{\pi \in D_{n-k}} \mathrm{sgn}(\pi)$$

where D_{n-k} is the set of permutations of $\{1, \ldots, n-k\}$ with no fixed points. Then we calculate these two inner sums

$$\sum_{\substack{J \subseteq \{1,\ldots,n\} \\ |J|=k}} \sum_{j \in J} j \quad \text{and} \quad \sum_{\pi \in D_{n-k}} \mathrm{sgn}(\pi)$$

respectively. From which the expected result will follow.

Solution.
We begin with

$$\sum_{\substack{J \subseteq \{1,\ldots,n\} \\ |J|=k}} \sum_{j \in J} j = \sum_{j=1}^{n} j \sum_{\substack{J \subseteq \{1,\ldots,n\} \\ |J|=k, j \in J}} 1 = \frac{n(n+1)}{2} \binom{n-1}{k-1}. \tag{4.18}$$

Recall the Leibniz formula for the determinant of a $n \times n$ matrix A:

$$\det(A) = \sum_{\pi \in S_n} \operatorname{sgn}(\pi) \prod_{i=1}^{n} a_{i,\pi_i}.$$

It implies that

$$\sum_{\pi \in D_{n-k}} \operatorname{sgn}(\pi) = \det(M_{n-k}) = (-1)^{n-1-k}(n-1-k) \qquad (4.19)$$

where M_{n-k} is the $(n-k) \times (n-k)$ matrix with 0s along the main diagonal and 1s elsewhere. Hence, by (4.18) and (4.19), we obtain

$$\sum_{\pi \in A_n} \frac{\operatorname{sgn}(\pi)}{|\operatorname{Fix}(\pi)|} \sum_{j \in \operatorname{Fix}(\pi)} j = \sum_{k=1}^{n} \frac{1}{k} \cdot \frac{n(n+1)}{2} \binom{n-1}{k-1} \cdot (-1)^{n-1-k}(n-1-k)$$

$$= (-1)^n \frac{n^2-1}{2} \sum_{k=1}^{n} \binom{n}{k}(-1)^{k+1} - (-1)^n \frac{n(n+1)}{2} \sum_{k=1}^{n} \binom{n-1}{k-1}(-1)^{k-1}$$

$$= (-1)^n \frac{n^2-1}{2}(1 - (1-1)^n) - (-1)^n \frac{n(n+1)}{2}(1-1)^{n-1}$$

$$= \begin{cases} (-1)^n \dfrac{n^2-1}{2} & \text{if } n > 1, \\ 1 & \text{if } n = 1. \end{cases}$$

$\square$

Remark. The evaluation of the determinant in (4.19) can be extended in the following way.

Let A be a $n \times n$ matrix which has x along its main diagonal and $y \neq 0$ elsewhere:

$$A = y J_n + (x - y) I_n$$

where J_n is the matrix with all 1s and I_n is the identity matrix. The matrix J_n has rank 1 and therefore its eigenvalue 0 has multiplicity $n-1$ whereas the nonzero eigenvalue is $\operatorname{Tr}(J_n) = n$ with multiplicity 1. If v is an eigenvector of A such that $Av = \lambda v$ then v is also an eigenvector of $y J_n$ because

$$y J_n v = (\lambda - x + y)v.$$

This equality indicates that the eigenvalues of A are $x - y$ with multiplicity $n - 1$ and $x + (n-1)y$ with multiplicity 1. Hence we conclude that

$$\det(A) = (x + (n-1)y)(x - y)^{n-1}.$$

In particular, letting $x = 0$ and $y = 1$ gives (4.19).

Additional problems for practice.

1. **Putnam 2005-B6.** Let S_n denote the set of all permutations of the numbers $\{1, 2, \ldots, n\}$. For $\pi \in S_n$, let $|\text{Fix}(\pi)|$ denote the number of fixed points of π. Show that

$$\sum_{\pi \in S_n} \frac{\text{sgn}(\pi)}{|\text{Fix}(\pi)| + 1} = (-1)^{n+1} \frac{n}{n+1}.$$

2. **Putnam 2006-A4.** Let $S = \{1, 2, \ldots, n\}$ for some integer $n > 1$. Say a permutation π of S has a *local maximum* at $k \in S$ if

   (i) $\pi(k) > \pi(k+1)$ for $k = 1$;
   (ii) $\pi(k-1) < \pi(k)$ and $\pi(k) > \pi(k+1)$ for $1 < k < n$;
   (iii) $\pi(k-1) < \pi(k)$ for $k = n$.

   (For example, if $n = 5$ and π takes values at $1, 2, 3, 4, 5$ of $2, 1, 4, 5, 3$, then π has a local maximum of 2 at $k = 1$, and a local maximum of 5 at $k = 4$.) What is the average number of local maxima of a permutation of S, averaging over all permutations of S?

3. **Problem E3371** (Proposed by A. M. Garsia and B. A. Sethurama, 97(2), 1990). Let S_n denote the symmetric group on the n symbols $1, 2, \ldots n$. Given σ in S_n, let

   $$M(\sigma) = \{i : 1 \le i \le n, i > \sigma(i)\}.$$

   Prove that

   $$\sum_{\pi \in S_n} \frac{1}{|M(\sigma)|} \sum_{i \in M(\sigma)} (i + \sigma(i)) = (n+1)!.$$

4. **Problem 12432** (Proposed by E. Vigren, 130(10), 2023). Suppose that k and n are integers with $n \ge 2$ and $1 \le k \le n$. What is the average value of $\sum_{i=1}^{\pi(k)} \pi(i)$ over all permutations π of $\{1, \ldots, n\}$?

5. **Problem 12219** (Proposed by B. Isaacson, 127(10), 2020). Let k and m be positive integers with $k < m$. Let $c(m, k)$ be the number of permutations of $\{1, \ldots, m\}$ consisting of k cycles. The numbers $c(m, k)$ are known as unsigned Stirling numbers of the first kind. Prove

   $$\sum_{j=k}^{m} \frac{(-2)^j \binom{m}{j} c(j, k)}{(j-1)!} = 0$$

   whenever m and k have opposite parity.

6. **Problem 12430** (Proposed by A. Dzhumadil'daev, 130(10), 2023). Let $P(n)$ denote the set of all partitions of $\{1, \ldots, n\}$. For $A = \{A_1, \ldots, A_k\} \in P(n)$, let $f(A) = \prod_{i=1}^{k} |A_i|$ and $g(A) = \prod_{i=1}^{k} \sum_{m \in A_i} m$.

(a) Prove $\sum_{A\in P(n)} f(A) = \sum_{k=1}^{n} k^{n-k}\binom{n}{k}$.

(b) Prove $\sum_{A\in P(n)} g(A) = \sum_{k=1}^{n} (n+1-k)^{k-1}c(n+1,k)$, where $c(n,k)$ is the unsigned Stirling number of the first kind, the number of permutations of $\{1,\ldots,n\}$ with exactly k cycles.

4.15 A one-sided inverse involving Motzkin numbers

Problem 12356 (Proposed I. Gessel, 129(10), 2022). Let $A(z) = z^3 - z^2$ and

$$B(z) = 1 + \sum_{n=0}^{\infty} \frac{(-1)^n}{n+1}\binom{3n+1}{n}z^{n+1}.$$

Prove that B is a one-sided inverse to A in the sense that $A(B(z)) = z$. Also, prove $B(A(z)) = 1 - z^2 M(-z)$, where

$$M(z) = \frac{1 - z - \sqrt{1 - 2z - 3z^2}}{2z^2}. \tag{4.20}$$

The coefficients of $M(z)$ are the Motzkin numbers $1, 1, 2, 4, 9, 21, \ldots$

Discussion.
Let $f(w, z) = A(w) - z = w^3 - w^2 - z$. Then $f(1,0) = 0$ and

$$f_w(1,0) = (3w^2 - 2w)\big|_{(1,0)} = 1 \neq 0.$$

Thus, by the implicit function theorem, there is a unique analytic function $B(z)$ such that

$$f(B(z), z) = A(B(z)) - z = 0$$

in a neighborhood of $z = 0$ with $B(0) = 1$:

$$B(z) = 1 + \sum_{n=0}^{\infty} b_{n+1}z^{n+1}$$

where the coefficients of the one-sided inverse to A have to be determined.

Solution.
By the Lagrange inversion theorem,

$$
\begin{aligned}
b_{n+1} &= \frac{1}{(n+1)!}\lim_{w\to 1}\frac{d^n}{dw^n}\left(\left(\frac{w-1}{A(w)-A(1)}\right)^{n+1}\right)\\
&= \frac{1}{(n+1)!}\lim_{w\to 1}\frac{d^n}{dw^n}\left(w^{-(2n+2)}\right)\\
&= \frac{(-1)^n(2n+2)(2n+3)\cdots(2n+2+n-1)}{(n+1)!} = \frac{(-1)^n}{n+1}\binom{3n+1}{n}.
\end{aligned}
$$

We now verify that $B(A(z)) = 1 - z^2 M(-z)$, that is

$$B(z^3 - z^2) = F(z) \tag{4.21}$$

where

$$F(z) = 1 - z^2 M(-z) = \frac{1 - z + \sqrt{1 + 2z - 3z^2}}{2}.$$

It is straightforward to check that $F^3(z) - F^2(z) = z^3 - z^2$ in a neighborhood of $z = 0$.

Moreover, from $B^3(z) - B^2(z) = A(B(z)) = z$ we find

$$B^3(z^3 - z^2) - B^2(z^3 - z^2) = z^3 - z^2 = F^3(z) - F^2(z)$$

which implies

$$B^3(z^3 - z^2) - F^3(z) = B^2(z^3 - z^2) - F^2(z),$$

and it follows that

$$(B(z^3 - z^2) - F(z)) \cdot (B^2(z^3 - z^2) + B(z^3 - z^2) F(z) + F^2(z) - B(z^3 - z^2) - F(z)) = 0.$$

Since $B(0) = F(0) = 1$ then the second factor at $z = 0$ is equal to

$$B^2(0) + B(0)F(0) + F^2(0) - B(0) - F(0) = 3 - 2 = 1 \neq 0$$

and, by continuity, it differs from zero in a neighborhood of $z = 0$. Therefore, in such neighborhood, the first factor must be identically zero. This shows (4.21) and completes the proof. $\qquad\square$

Remark. Motzkin numbers $(M_n)_{n\geq 0} = 1, 1, 2, 4, 9, 21, \ldots$ appear as the sequence A001006 in the OEIS (https://oeis.org/A001006)) and they have many combinatorial properties (for a survey see [36]). For example M_n is the number of ways of drawing any number of nonintersecting chords joining n labeled points along a circle.

Notice that given n points on the circle then we have two possibilities depending on whether the point with label n is an endpoint of a chord or not. In the first case, the remaining $n - 1$ points can be connected by chords in M_{n-1} ways. In the second case, the point with label n is connected by a chord to another point by with label $(k + 1)$ such that $0 \leq k \leq n - 2$. This chord divides the circle into two parts, whose points can be connected in M_k and M_{n-2-k} respectively. This argument leads us to the conclusion that the following recurrence relation holds for $n \geq 2$,

$$M_n = M_{n-1} + \sum_{k=0}^{n-2} M_k M_{n-2-k}$$

with $M_0 = M_1 = 1$. The above recurrence implies that the generating function $M(z) = \sum_{n\geq 0} M_n z^n$ satisfies the quadratic equation

$$M(z) = 1 + zM(z) + z^2 M(z)^2$$

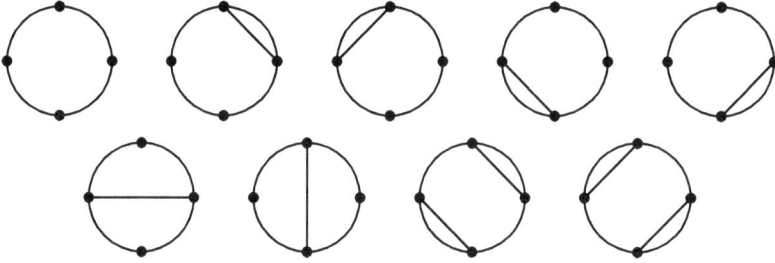

Figure 4.20: The 9 ways of drawing any number of nonintersecting chords for $n = 4$

and by solving it we easily find (4.20).

Furthermore, Motzkin numbers are related to the Catalan numbers $C_n = \frac{1}{n+1}\binom{2n}{n}$ through the formula,

$$M_n = \sum_{k=0}^{\lfloor n/2 \rfloor} \binom{n}{2k} C_k$$

which has a nice combinatorial proof due to the fact that C_{2k} is the number of ways to connect $2k$ points on a circle to form k disjoint chords. The reader is invited to complete the details.

Additional problems for practice.

1. Let $G(x) = \sum_{n=0}^{\infty} a_n x^n$. For any positive integer m, show that

$$\sum_{n=0}^{\infty} a_{\lfloor n/m \rfloor} x^n = (1 + x + \cdots + x^m) G(x^m).$$

2. Let $F(x) = \sum_{n=0}^{\infty} \binom{3n}{n} x^n$. Show that $F(x)$ satisfies the functional equation

$$(27x - 4)F^3(x) + 3F(x) + 1 = 0.$$

3. Let $F(x) = \sum_{n=1}^{\infty} \sigma(n) x^n$ where $\sigma(n)$ is the sum of the divisors of n. Prove that

$$F(x) = \sum_{n=1}^{\infty} \frac{nx^n}{1 - x^n} = \sum_{n=1}^{\infty} \frac{x^n}{(1 - x^n)^2}.$$

4. **Problem 10750** (Proposed by L. Smiley, 106(7), 1999). For a positive integer m, express

$$\sum_{n=1}^{\infty} \frac{nx^n}{\gcd(m, n)}$$

as a rational function of x.

5. **Problem 11151** (Proposed by D. E. Knuth, 112(4), 2005). Suppose n people are sitting at a circular table. Let $e_{m,n}$ denote the number of ways to partition them into m affinity groups with no two members of a group seated next to each other. For example, $e_{3,4} = 2$, $e_{3,5} = 5$, and $e_{3,6} = 10$. For $m \geq 2$, find the generating function $\sum_{n=0}^{\infty} e_{m,n} z^n$.

6. **Problem 11610** (Proposed by R. P. Stanley, 118(10), 2011). Let $f(n)$ be the number of binary words $a_1 \cdots a_n$ of length n that have the same number of pairs $a_i a_{i+1}$ equal to 00 as 01. Show that

$$\sum_{n \geq 0} f(n) x^n = \frac{1}{2} \left(\frac{1}{1-x} + \frac{1+2x}{\sqrt{(1-x)(1-2x)(1+x+2x^2)}} \right).$$

7. **Problem 11757** (Proposed by I. Gessel, 121(2), 2014). Let $[x^a y^b] f(x, y)$ denote the coefficient of $x^a y^b$ in the Taylor series expansion of f. Show that

$$[x^n y^n] \frac{1}{(1-3x)(1-y-3x+3x^2)} = 9^n.$$

8. **Problem 12489** (Proposed by A. Burstein, L. Shapiro, J. Jang and M. Song, 131(9), 2024). Let $B(z) = 1/\sqrt{1-4z}$. The Maclaurin series of $\sqrt{2B(z) - 1}$ is

$$1 + 2z + 4z^2 + 12z^3 + 38z^4 + \cdots.$$

Do the coefficients of this series form a strictly increasing sequence of positive integers?

9. **Problem 12513** (Proposed by R. P. Stanley, 132(2), 2025). Let $F(x) = \sum_{n \geq 0} a_{2n+1} x^{2n+1}$ where the coefficients $a_1, a_3, \ldots$ are indeterminates over the complex numbers. Let $e^{F(x)} = \sum_{m \geq 0} b_m x^m$. Find an expression for b_{2n} as a polynomial in $b_1, b_3, b_5, \ldots, b_{2n-1}$, giving an explicit formula for the coefficients. For example, $b_6 = (1/16)b_1^6 - (1/2)b_1^3 b_3 + (1/2)b_3^2 + b_1 b_5$.

Chapter 5

Number Theory

This chapter presents 16 *Monthly* problems, which concentrate on the central area of number theory. We will come across divisibility, multiplication functions, congruence, quadratic forms, quadratic residues and sums of squares. During the journey, we will also revisit many important results such as Euler's criterion, Gauss quadratic reciprocity law, Legendre formula, Lucas' theorem and Wolstenholme's theorem. Most of problems introduce a variety of ideas not treated in regular textbooks. Some problems are even related to the Riemann hypothesis and $3n + 1$ conjecture, affirming once again the vitality of the *Monthly* problems as a catalyst for research activity.

5.1 Divisibility of a central binomial sum

Problem 11292 (Proposed by D. Callan, 114(5), 2007). Show that if p is a prime and $p \geq 5$ then p^2 divides

$$\sum_{k=1}^{p^2-1} \binom{2k}{k}.$$

Discussion.
Since the nineteenth century, the problems involving binomial coefficients modulo a prime power have been studied by many famous mathematicians including Gauss, Kummer, Legendre and Lucas. They discovered many beautiful and surprising theorems. For example, Lucas discovered an elegant method in 1878 easily to find the value of $\binom{n}{k}$ (mod p): Let

$$n = n_r p^r + n_{r-1} p^{r-1} + \cdots + n_1 p + n_0 \quad \text{and} \quad k = k_r p^r + k_{r-1} p^{r-1} + \cdots + k_1 p + k_0$$

be the base p-expansions of n and k, respectively. Then

$$\binom{n}{k} \equiv \prod_{j=0}^{r} \binom{n_j}{k_j} \pmod{p}. \tag{5.1}$$

For completeness, here we present a proof of Lucas' theorem based on the generating function for $\binom{n}{k}$. Note that $(1+x)^{p^j} \equiv 1 + x^{p^j} \pmod{p}$. Using the base p-expansions of n and k, we have

$$\sum_{k=0}^{n} \binom{n}{k} x^k = (1+x)^n = \prod_{j=0}^{r} \left(1 + x\right)^{p^j}\Big)^{n_j}$$

$$\equiv \prod_{j=0}^{r} (1 + x^{p^j})^{n_j} = \prod_{j=0}^{r} \left(\sum_{k_j=0}^{n_j} \binom{n_j}{k_j} x^{k_j p^j} \right)$$

$$= \sum_{k=0}^{n} \left(\prod_{j=0}^{r} \binom{n_j}{k_j} \right) x^k \pmod{p}$$

and the formula (5.1) follows.

For this proposed problem, we first consider the weaker congruence modulo p. By Lucas' theorem,

$$\sum_{k=0}^{p^2-1} \binom{2k}{k} = \sum_{j=0}^{p-1} \sum_{k=0}^{p-1} \binom{2(pj+k)}{pj+k}$$

$$\equiv \sum_{j=0}^{p-1} \sum_{k=0}^{p-1} \binom{2j}{j} \binom{2k}{k} = \left(\sum_{k=0}^{p-1} \binom{2k}{k} \right)^2 \pmod{p}.$$

Moreover, p divides $\binom{2k}{k}$ for $\frac{p-1}{2} < k \le p-1$, whereas for $0 \le k \le \frac{p-1}{2}$,

$$\binom{2k}{k} = \frac{2^k (2k-1)!!}{k!} \equiv \frac{(p-1)(p-3)\dots(p-(2k-1))}{2^k k!}$$

$$\equiv \binom{\frac{p-1}{2}}{k} (-4)^k \pmod{p}$$

which implies that

$$\sum_{k=0}^{p-1} \binom{2k}{k} \equiv \sum_{k=0}^{\frac{p-1}{2}} \binom{\frac{p-1}{2}}{k} (-4)^k \equiv (1-4)^{\frac{p-1}{2}} = (-3)^{\frac{p-1}{2}} \pmod{p}.$$

Hence

$$\sum_{k=1}^{p^2-1} \binom{2k}{k} \equiv \left((-3)^{\frac{p-1}{2}} \right)^2 - 1 = 0 \pmod{p}.$$

We are now ready to prove the stronger congruence modulo p^2.

Solution.

We start by proving the following identity (see also [78]): For any integer $n \geq 1$, we have

$$\sum_{k=0}^{n-1} \binom{2k}{k} = \sum_{k=1}^{n} r_{n-k} \binom{2n}{k} \tag{5.2}$$

where $r_{3k} = 0$, $r_{3k+1} = 1$, and $r_{3k+2} = -1$ for all $k \geq 0$. Indeed,

$$\sum_{k=0}^{n-1} \binom{2k}{k} = [x^n] \sum_{k=0}^{n-1} x^{n-k} (1+x)^{2k} = [x^n] \sum_{j=1}^{\infty} x^j (1+x)^{2(n-j)}$$

$$= [x^n] \sum_{j=1}^{\infty} \left(\frac{x}{(1+x)^2} \right)^j \cdot (1+x)^{2n}$$

$$= [x^n] \frac{x}{1+x+x^2} \cdot (1+x)^{2n} = \sum_{k=1}^{n} r_{n-k} \binom{2n}{k}.$$

Letting $n = p^2$ in (5.2), we find that

$$\sum_{k=0}^{p^2-1} \binom{2k}{k} = \sum_{k=0}^{p^2-1} r_{p^2-k} \binom{2p^2}{k}$$

By Lucas' theorem, $\binom{2p^2}{k} \equiv 0 \pmod{p^2}$ when p does not divide k, and, if $k = pj$ with $0 < j < p$ then

$$\binom{2p^2}{pj} = \frac{2p^2}{pj} \binom{2p^2 - 1}{pj - 1} = \frac{2p}{j} \left(\frac{p^2 + p(p-1) + p - 1}{0p^2 + p(j-1) + p - 1} \right)$$

$$\equiv \frac{2p}{j} \binom{p-1}{j-1} = 2 \binom{p}{j} \pmod{p^2}.$$

Moreover, $r_k = \frac{\omega^k - \omega^{-k}}{i\sqrt{3}}$ with $\omega = \frac{1}{2}(-1 + i\sqrt{3})$. Hence, since $p^2 \equiv 1 \pmod 3$, it follows that

$$\sum_{k=1}^{p^2-1} \binom{2k}{k} = -1 + \sum_{k=0}^{p^2-1} r_{p^2-k} \binom{2p^2}{k}$$

$$\equiv r_{p^2} - 1 + 2 \sum_{j=1}^{p-1} r_{p^2-pj} \binom{p}{j} = r_1 - 1 + 2 \sum_{j=1}^{p-1} r_{pj} \binom{p}{p-j}$$

$$\equiv \frac{2}{i\sqrt{3}} \left(A_p - \overline{A_p} \right) \equiv 0 \pmod{p^2}$$

where $p \equiv \pm 1 \pmod 3$, $1 + \omega + \omega^{-1} = 0$, and

$$A_p := \sum_{j=1}^{p-1} \omega^{pj} \binom{p}{j} = \sum_{j=1}^{p-1} \omega^{\pm j} \binom{p}{j}$$

$$= (1 + \omega^{\pm 1})^p - \omega^{\pm p} - 1 = -\omega^{\mp p} - \omega^{\pm p} - 1 = 0.$$

$\square$

Remark. A more general congruence has been proved in [93]: If p is a prime and a is a positive integer then the following congruence modulo p^2 holds,

$$\sum_{k=1}^{p^a - 1} \binom{2k}{k} \equiv \begin{cases} -1 & \text{if } p = 3, \\ -2 & \text{if } a \text{ is odd and } 3 \mid (p-2), \\ 0 & \text{otherwise.} \end{cases}$$

As regards Lucas' theorem, we warmly recommend A. Granville's elegant paper [48] in which he gives a generalization to arbitrary prime powers: If p is a prime which does not divide $\binom{n}{k}$ then for any positive integer r,

$$\binom{n}{k} \equiv \binom{\lfloor n/p \rfloor}{\lfloor k/p \rfloor} \cdot \binom{n_0}{k_0} \Big/ \binom{\lfloor n_0/p \rfloor}{\lfloor k_0/p \rfloor} \pmod{p^r}$$

where n_0 and k_0 are the least non-negative residue of n and k modulo p^r.

In [49], the same author discusses connections of the properties of the binomial coefficients modulo prime powers with cellular automata, Fermat's last theorem and the prime recognition problem.

Additional problems for practice.

1. Let $p \geq 5$ be a prime number. Show that

$$\sum_{k=1}^{p-1} \frac{\binom{2k}{k}}{k} \equiv 0 \pmod{p^2}.$$

See [73, (41)] for a more general result.

2. Let p be an odd prime. Show that

$$\sum_{k=0}^{p-1} \frac{\binom{2k}{k}}{2^k} \equiv (-1)^{(p-1)/2} \pmod{p^2}.$$

3. **Problem E3457** (Proposed by H. Wilf, 98(8), 1991). Find all positive integers k such that the sequence

$$\left\{ \binom{2n}{n} \right\}_{n \geq 0}$$

is periodic modulo k from some point onward.

4. **Problem 11165** (Proposed by Y. More, 112(6), 2005). Let C_k be the kth Catalan number, $\frac{1}{k+1}\binom{2k}{k}$. Prove that, for each positive integer n,

$$\sum_{1}^{n} C_k \equiv 1 \pmod{3}$$

if and only if the base 3 expansion of $n + 1$ contains the digit 2. Find similar characterization for the other two cases, in which the sum is congruent to 0 or 2 modulo 3.

5. **Putnam 1996-A5**. If p is a prime number greater than 3 and $k = \lfloor 2p/3 \rfloor$, prove that the sum

$$\binom{p}{1} + \binom{p}{2} + \cdots + \binom{p}{k}$$

is divisible by p^2.

6. **MM Problem 1392** (Proposed by G. E. Andrews, 65(1), 1992). Prove that for any prime p in the interval $(n, 4n/3]$, p divides $\sum_{k=0}^{n} \binom{n}{k}^4$.

5.2 A congruence implied by Wolstenholme's theorem

Problem 11382 (Proposed by R. Tauraso, 115(7), 2008). For $k \geq 1$, let H_k be the kth harmonic number, $H_k = \sum_{j=1}^{k} 1/j$. Show that if p is prime and $p > 5$, then

$$\sum_{k=1}^{p-1} \frac{H_k^2}{k} \equiv \sum_{k=1}^{p-1} \frac{H_k}{k^2} \pmod{p^2}.$$

Discussion.
Our main tool will be Wolstenholme's theorem about congruences for harmonic numbers. It states that if d is a positive integer and p is prime such that $p \geq d + 3$, then

$$H_{p-1}(d) := \sum_{k=1}^{p-1} \frac{1}{k^d} \equiv \begin{cases} 0 \pmod{p^2}, & \text{if } d \text{ is odd,} \\ 0 \pmod{p}, & \text{if } d \text{ is even.} \end{cases} \tag{5.3}$$

where $H_{p-1}(1) = H_{p-1}$. Recall that two rationals are congruent modulo q means that their difference can be expressed as a reduced fraction of the form qa/b, with b relatively prime to a and q. For example, for $p = 11$, we have $H_{10} = 7381/2520$, $\gcd(11, 2520) = 1$ and $7381 = 11^2 \cdot 61$. Actually, in 1862,

J. Wolstenholme proved (5.3) just for $d = 1$ and $d = 2$. For $d > 2$ and for results about multiple harmonic sums we refer to [107, 81].

Now we are ready to show that the proposed problem is a simple consequence of (5.3).

Solution.

We first note that

$$H_{k-1}^3 = \left(H_k - \frac{1}{k}\right)^3 = H_k^3 - 3\frac{H_k^2}{k} + 3\frac{H_k}{k^2} - \frac{1}{k^3}.$$

Therefore,

$$\sum_{k=1}^{p-1}\frac{H_k^2}{k} - \sum_{k=1}^{p-1}\frac{H_k}{k^2} = \frac{1}{3}\left(\sum_{k=1}^{p-1}(H_k^3 - H_{k-1}^3) - \sum_{k=1}^{p-1}\frac{1}{k^3}\right) = \frac{1}{3}\left(H_{p-1}^3 - H_{p-1}(3)\right).$$

Moreover, by (5.3), for $p > 5$, we have

$$H_{p-1}(1) = H_{p-1} \equiv 0 \pmod{p^2} \quad \text{and} \quad H_{p-1}(3) \equiv 0 \pmod{p^2}.$$

This concludes that

$$\sum_{k=1}^{p-1}\frac{H_k^2}{k} - \sum_{k=1}^{p-1}\frac{H_k}{k^2} = \frac{1}{3}\left(H_{p-1}^3 - H_{p-1}(3)\right) \equiv 0 \pmod{p^2}.$$

□

Remark. For $d = 2$, the congruence (5.3) a simple consequence of the fact that the inversion mapping in the multiplicative group $\mathbb{Z}_p^*$ induces a natural bijection between $\{1, 2, \ldots, (p-1)\}$ and $\{1^{-1}, 2^{-1}, \ldots, (p-1)^{-1}\}$. Hence, by rearranging the terms in the sum, we find for $p \geq 5$,

$$H_{p-1}(2) = \sum_{k=1}^{p-1}(k^{-1})^2 \equiv \sum_{k=1}^{p-1}k^2 = \frac{p(p-1)(2p-1)}{6} \equiv 0 \pmod{p}.$$

Then, for $d = 1$, it suffices to observe that

$$2H_{p-1}(1) = \sum_{k=1}^{p-1}\left(\frac{1}{k} + \frac{1}{p-k}\right) = \sum_{k=1}^{p-1}\frac{p}{k(p-k)} \equiv -pH_{p-1}(2) \equiv 0 \pmod{p^2}.$$

Notice that for any odd prime p, the harmonic sums H_{p-1} and $H_{p-1}(2)$ are strictly related to the binomial coefficient $\binom{2p-1}{p-1}$:

$$\binom{2p-1}{p-1} = \prod_{k=1}^{p-1}\left(1 - \frac{2p}{k}\right) \equiv 1 - 2p\sum_{0<k<p}\frac{1}{k} + 4p^2\sum_{0<j<k<p}\frac{1}{jk}$$

$$\equiv 1 - 2pH_{p-1} + 2p^2(H_{p-1}^2 - H_{p-1}(2)) \pmod{p^3}.$$

It follows from (5.3) that for a prime $p \geq 5$,

$$\binom{2p-1}{p-1} \equiv 1 \pmod{p^3}.$$

More generally, Glaisher established that for a prime $p \geq 5$ and an integer $m \geq 0$,

$$\binom{mp+p-1}{p-1} \equiv 1 \pmod{p^3}.$$

It is worth noticing that, in the last two decades, many papers have been contributed to our understanding of such congruences. In [42], Gessel studied the congruences modulo n and n^2 of

$$\sum_{\substack{1 \leq k \leq n-1 \\ \gcd(k,n)=1}} \frac{1}{k^d}.$$

The case $d = 1$ is also considered in [56, p. 130]. Other extensions that concerns the q-binomial coefficients can be found in G. E. Andrews's elegant paper [12]. For example, he gives a q-generalization of Glaisher which states that for any odd prime p,

$$\begin{bmatrix} mp+p-1 \\ p-1 \end{bmatrix}_q \equiv q^{mp(p-1)/2} \pmod{[p]_q^2}$$

where $[n]_q = 1+q+\cdots+q^{n-1}$. Moreover, an analog of (5.3) for the q-harmonic sums becomes

$$\sum_{k=1}^{p-1} \frac{1}{[k]_q} \equiv \frac{p-1}{2}(1-q) \pmod{[p]_q}.$$

The above q-congruences returns the original result as soon as we evaluate the limit for $q \to 1$.

Additional problems for practice.

1. Let p be a prime number. Show that

$$q^{\binom{k+1}{2}} \begin{bmatrix} p-1 \\ k \end{bmatrix}_q \equiv (-1)^k \pmod{[p]_q}.$$

2. Let p and q be odd primes. Prove that for any odd integer $d > 0$ there is an integer r such that

$$\sum_{n=1}^{p-1} \frac{[n \equiv r \pmod{q}]}{n^d} \equiv 0 \pmod{p}$$

where $[Q]$ is equal to 1 or 0 as the proposition Q is true or false.

3. **Putnam 1991-B4.** Suppose p is an odd prime. Prove that

$$\sum_{j=0}^{p} \binom{p}{j}\binom{p+j}{j} \equiv 2^p + 1 \pmod{p^2}.$$

4. **Putnam 1997-B3.** For each positive integer n, write $H_n = 1 + \frac{1}{2} + \cdots + \frac{1}{n}$ in the form p_n/q_n, where p_n and q_n are relative prime. Determine all n such that 5 does not divide q_n.

5. **Problem 10723** (Proposed by C. J. Hillar, 106(2), 1999). Let p be an odd prime. Prove that

$$\sum_{k=1}^{p-1} 2^k k^{p-2} \equiv \sum_{k=1}^{(p-1)/2} k^{p-2} \pmod{p}.$$

6. **Problem 11118** (Proposed by V. Dimitrov, 111(10), 2004). Let p be an odd prime, and let k be a positive integer. Prove that

$$\sum_{j=0}^{k} \binom{k(p-1)}{j(p-1)} \equiv 2 + p(1-k) \pmod{p^2}.$$

7. **Problem 11364** (Proposed by P. P. Dályay, 115(5), 2008). Let p be a prime greater than 3, and t the integer nearest $p/6$.

   (a) Show that if $p = 6t + 1$, then

   $$(p-1)! \left(\sum_{j=0}^{2t-1} \frac{(-1)^j}{3j+1} + \sum_{j=0}^{2t-1} \frac{(-1)^j}{3j+2} \right) \equiv 0 \pmod{p}.$$

   (b) Show that if $p = 6t - 1$, then

   $$(p-1)! \left(\sum_{j=0}^{2t-1} \frac{(-1)^j}{3j+1} + \sum_{j=0}^{2t-2} \frac{(-1)^j}{3j+2} \right) \equiv 0 \pmod{p}.$$

8. **Problem 11411** (Proposed by A. W. Jones, 116(2), 2009). For positive integers k and n, let

$$L_k(n) = \sum_{j=1}^{n-1} (-1)^j j^k.$$

   (a) Show that $L_1(n) \equiv L_5(n) \pmod{n}$ if and only if n is not a multiple of 4.

   (b) Given distinct, odd, positive integers i and j with $\{i,j\} \neq \{1,5\}$, show that the set of n such that $L_i(n) \equiv L_j(n) \pmod{n}$ is finite.

9. **Problem 12148** (Proposed by T. Beke, 126(10), 2019). Let p be a prime number, and let f be a symmetric polynomial in $p-1$ variables with integer coefficients. Suppose that f is homogeneous of degree d and that $p-1$ does not divide d. Prove that p divides $f(1, 2, \ldots, p-1)$.

10. **Open Problem**. A prime p is called a *Wolstenholme prime* if

$$\binom{2p-1}{p-1} \equiv 1 \pmod{p^4}.$$

So far the only known Wolstenholme primes are 16843 and 2124679, an entry $A088164$ in the OEIS. McIntosh and Roettger showed that the next Wolstenholme primes, if it exists, must be greater than 10^9. Similar to the collaborative project GIMPS to search for the Mersenne primes (https://www.mersenne.org), can we have an efficient algorithm in polynomial time to find the new Wolstenholme primes?

5.3   A congruence of a multiple sum involving Fibonacci numbers

Problem 11602 (Proposed by R. Tauraso, 118(9), 2011). Let p be a prime. Let F_n denote the nth Fibonacci number. Show that

$$\sum_{0<i<j<k<p} \frac{F_i}{ijk} \equiv 0 \pmod{p}.$$

Discussion.
As a warm-up, let's begin with showing the following easier congruence holds for any prime $p > 2$,

$$\sum_{0<i<p} \frac{F_k}{k} \equiv 0 \pmod{p}.$$

To this end, we apply the known identity

$$\sum_{k=0}^{n} (-1)^{n-k} \binom{n}{k} F_{n-k} = -F_n. \tag{5.4}$$

Indeed, by letting $n = p$ in (5.4), we have that

$$\sum_{k=1}^{p-1} (-1)^k \frac{p}{k} \cdot \binom{p-1}{k-1} F_{p-k} = F_p - F_p = 0.$$

Moreover, noting that $\binom{p-1}{k-1} \equiv (-1)^{k-1}$ modulo p, we find

$$-\sum_{k=1}^{p-1} \frac{F_{p-k}}{k} \equiv \sum_{k=1}^{p-1} \frac{F_k}{k-p} \equiv \sum_{k=1}^{p-1} \frac{F_k}{k} \equiv 0 \pmod{p}.$$

Solution.
Let $p > 3$ be prime (if $p = 2, 3$ the above sum is empty and it gives 0). For $k = 1, \dots, p - 1$, since

$$\binom{p}{k} = (-1)^{k-1} \frac{p}{k} \prod_{j=1}^{k-1} \left(1 - \frac{p}{j}\right)$$

$$\equiv (-1)^{k-1} \frac{p}{k} \left(1 - p H_{k-1}(1) + p^2 H_{k-1}(1,1)\right) \pmod{p^4}$$

where

$$H_{k-1}(1) = \sum_{0 < i < k} \frac{1}{i} \quad, \quad H_{k-1}(1,1) = \sum_{0 < i < j < k} \frac{1}{ij},$$

by letting $p = n$ in (5.4), we obtain

$$S_1 - p S_2 + p^2 S_3 \equiv 0 \pmod{p^3}, \tag{5.5}$$

where

$$S_1 = \sum_{0 < i < p} \frac{F_{p-i}}{i}, \quad S_2 = \sum_{0 < i < j < p} \frac{F_{p-j}}{ij}, \quad S_3 = \sum_{0 < i < j < k < p} \frac{F_{p-k}}{ijk}.$$

Next, we recall another known binomial identity involving the Fibonacci numbers, namely

$$\sum_{k=0}^{n} \left[\binom{n+k-1}{k} + (-1)^{n-k}\binom{2n}{k}\right] F_{n-k} = 0. \tag{5.6}$$

Along the same lines, we find that, for $k = 1, \dots, p - 1$,

$$\binom{p+k-1}{k} = \frac{p}{k} \prod_{j=1}^{k-1} \left(1 + \frac{p}{j}\right)$$

$$\equiv \frac{p}{k} \left(1 + p H_{k-1}(1) + p^2 H_{k-1}(1,1)\right) \pmod{p^4}$$

and

$$\binom{2p}{k} = (-1)^{k-1} 2pk \prod_{j=1}^{k-1} \left(1 - \frac{2p}{j}\right)$$

$$\equiv (-1)^{k-1} \frac{2p}{k} \left(1 - 2p H_{k-1}(1) + 4p^2 H_{k-1}(1,1)\right) \pmod{p^4}.$$

Letting $p = n$ in (5.6), we get

$$(S_1 + p S_2 + p^2 S_3) + (2S_1 - 4p S_2 + 8p^2 S_3) \equiv 0 \pmod{p^3}. \tag{5.7}$$

Thus, by subtracting 3 times the first congruence (5.5) from the second one (5.7), we obtain $6p^3 S_3 \equiv 0 \pmod{p^4}$, that is $S_3 \equiv 0 \pmod{p}$. Therefore,

$$0 \equiv S_3 = \sum_{0<i<j<k<p} \frac{F_{p-k}}{ijk} = \sum_{0<i<j<k<p} \frac{F_i}{(p-k)(p-j)(p-i)}$$

$$\equiv - \sum_{0<i<j<k<p} \frac{F_i}{ijk} \pmod{p}.$$

This proves the desired congruence. □

Remark. More similar congruences can be found in [72]. As regards the binomial identities (5.4) and (5.6), they can be verified by using Binet's formula $F_n = \frac{1}{\sqrt{5}}(\varphi^n - (-\frac{1}{\varphi})^n)$. Another possible method is through generating functions. Let

$$F(x) = \sum_{k \geq 0} F_k x^k = \frac{x}{1 - x - x^2},$$

then

$$\frac{1}{1-x} F\left(\frac{x}{x-1}\right) + F(x) = 0. \tag{5.8}$$

Since

$$(-1)^{n-k}\binom{n}{k} = [x^k] \frac{(-1)^{n-k}}{(1-x)^{n-k+1}} = [x^n] \frac{1}{1-x}\left(\frac{x}{x-1}\right)^{n-k},$$

by (5.8),

$$0 = [x^n]\left[\frac{1}{1-x} F\left(\frac{x}{x-1}\right) + F(x)\right] = \sum_{k=0}^{n} (-1)^{n-k}\binom{n}{k} F_{n-k} + F_n.$$

In a similar way, since

$$\binom{n+k-1}{k} = [x^k]\frac{1}{(1-x)^n} = [x^n]\frac{1}{(1-x)^n} \cdot x^{n-k}$$

and

$$(-1)^{n-k}\binom{2n}{k} = [x^k]\frac{(-1)^{n-k}}{(1-x)^{2n-k+1}} = [x^n]\frac{1}{(1-x)^{n+1}} \cdot \left(\frac{x}{x-1}\right)^{n-k},$$

by (5.8),

$$0 = [x^n]\frac{1}{(1-x)^n} \cdot \left[F(x) + \frac{1}{1-x} F\left(\frac{x}{x-1}\right)\right]$$

$$= \sum_{k=0}^{n}\left[\binom{n+k-1}{k} + (-1)^{n-k}\binom{2n}{k}\right] F_{n-k},$$

and the identity follows because, by (5.8), the expression between brackets is identically zero.

Additional problems for practice.

1. Let F_n be n-th Fibonacci number. If $p \neq 5$ is a prime, show that

$$F_{p-\left(\frac{5}{p}\right)} \equiv 0 \pmod{p}$$

   where $\left(\frac{\cdot}{p}\right)$ is the Legendre symbol.

2. Let L_n be the n-th Lucas number defined by $L_0 = 2, L_1 = 1, L_{n+1} = L_n + L_{n-1}$ for $n \geq 1$. Show that for any prime $p > 5$,

$$\sum_{k=1}^{p-1} \frac{L_k}{k^2} \equiv 0 \pmod{p}.$$

3. **Problem 10971** (Proposed by C. P. Rupert, 109(9), 2002). Prove that for every prime p, F_{p-1} is the first Fibonacci number divisible by p if and only if there is a primitive root r (mod p) satisfying $(r+1)(r+2) \equiv 1$ (mod p).

4. **Problem 10748** (Proposed by I. Borosh, D. A. Hensley, and J. Zinn, 106(7), 1999). Let p and q be prime numbers such that q divides $p-1$. Let r be a positive integer such that q does not divide r and $p > r^{q-1}$. Show that for any integers $a_1, a_2, \ldots, a_r$, if $\sum_{j=1}^{r} a_j^{(p-1)/q} \equiv 0 \pmod{p}$, then $\prod_{j=1}^{r} a_j \equiv 0 \pmod{p}$.

5. **Problem 11968** (Proposed by C. J. Hillar and R. Krone, 124(3), 2017). Let F_n be the nth Fibonacci number. For $n \geq 1$, prove that $F_{5n}/(5F_n)$ is an integer congruent to 1 modulo 10.

6. **Problem 12463** (Proposed by R. Tauraso, 131(8), 2024). For a positive integer n, let F_n be the nth Fibonacci number. Show that when p is prime,

$$\left(\sum_{k=0}^{p-1} \binom{2k}{k} F_k 2^{p-k-1} \right) \cdot \left(\sum_{k=0}^{p-1} \binom{2k}{k}^2 8^{p-k-1} \right)$$

   is divisible by p.

5.4   A congruence for a product of quadratic forms

Problem 12234 (Proposed by N. Osipov, 128(2), 2021). Let p be an odd prime, and let $Ax^2 + Bxy + Cy^2$ be a quadratic form with A, B, and C in $\mathbb{Z}$ as soon as such that $B^2 - 4AC$ is neither a multiple of p nor a perfect square modulo p. Prove that

$$\prod_{0 < x < y < p} (Ax^2 + Bxy + Cy^2)$$

is 1 modulo p if exactly one or all three of A, C, and $A + B + C$ are perfect squares modulo p and is -1 modulo p otherwise.

Discussion.
To get acquainted with this problem, it seems appropriate to review how to convert a quadratic congruence to a quadratic residue. Consider the congruence

$$ax^2 + bx + c \equiv 0 \pmod{p}$$

where p is an odd prime and $a \not\equiv 0 \pmod{p}$. Since $\gcd(4a, p) = 1$, the above congruence is equivalent to

$$4a(ax^2 + bx + c) \equiv 0 \pmod{p}.$$

In view of the identity

$$4a(ax^2 + bx + c) = (2ax + b)^2 - (b^2 - 4ac),$$

if let $y = 2ax + b$ and $d = b^2 - 4ac$, then the quadratic congruence may be expressed as

$$y^2 \equiv d \pmod{p}.$$

Thus, the problem of finding a solution to the quadratic congruence becomes that of finding a solution of quadratic residue. For the latter case, the following *Euler's criterion* provides a simple test to decide whether an integer a is a quadratic residue of a given odd prime p: Let $\gcd(a, p) = 1$. Then a is a quadratic residue of p if and only if $a^{(p-1)/2} \equiv 1 \pmod{p}$.

We now show how the same transform goes through in the required problem.

Solution.
We first note that A, C, and $A + B + C$ are not multiple of p, otherwise $D := B^2 - 4AC$ is equal respectively to B^2, B^2, $(A - C)^2$ modulo p which are perfect squares. Moreover $Ax^2 + Bxy + Cy^2$ is a multiple of p if and only if both x and y are multiple of p, otherwise again D is a perfect square modulo p.

Since $x^2 \equiv a \pmod{p}$ has $1 + \left(\frac{a}{p}\right)$ solutions in $\{0, 1, \ldots, p-1\}$, it follows that for any integer $0 < n < p$ the number of solutions in $\{0, 1, \ldots, p-1\}^2$ of the congruence $Ax^2 + Bxy + Cy^2 \equiv n \pmod{p}$, i.e. $(2Ax + By)^2 \equiv Dy^2 + 4An \pmod{p}$ is

$$\sum_{y=0}^{p-1} \left(1 + \left(\frac{Dy^2 + 4An}{p}\right)\right) = p + \left(\frac{D}{p}\right) \sum_{y=0}^{p-1} \left(\frac{y^2 + 4AnD^{-1}}{p}\right)$$

$$= p - \left(\frac{D}{p}\right) = p + 1$$

where we have applied the known fact that

$$\sum_{y=0}^{p-1}\left(\frac{y^2+a}{p}\right)=-1 \tag{5.9}$$

when p is an odd prime and a is not a multiple of p.

Therefore, the number of solutions of the same congruence with the restriction $0<x<y<p$ is

$$\frac{1}{2}\Big(p+1-\underbrace{\Big(1+\Big(\frac{A}{p}\Big)\Big)\Big(\frac{n}{p}\Big)\Big)}_{y=0}-\underbrace{\Big(1+\Big(\frac{C}{p}\Big)\Big)\Big(\frac{n}{p}\Big)\Big)}_{x=0}$$

$$-\underbrace{\Big(1+\Big(\frac{A+B+C}{p}\Big)\Big(\frac{n}{p}\Big)\Big)\Big)}_{x=y}=\begin{cases}\frac{p-2-M}{2} & \text{if }n\text{ is a perfect square modulo }p,\\[4pt]\frac{p-2+M}{2} & \text{otherwise,}\end{cases}$$

with $M:=\left(\frac{A}{p}\right)+\left(\frac{C}{p}\right)+\left(\frac{A+B+C}{p}\right)$ (which is an odd integer).

By Wilson's theorem $(p-1)!\equiv-1\pmod p$, we have

$$\prod_{0<n<p,\left(\frac{n}{p}\right)=1}n\equiv\prod_{k=1}^{\frac{p-1}{2}}k^2\equiv(-1)^{\frac{p-1}{2}}\prod_{k=1}^{\frac{p-1}{2}}k\prod_{k=1}^{\frac{p-1}{2}}(p-k)$$

$$=(-1)^{\frac{p-1}{2}}(p-1)!\equiv(-1)^{\frac{p+1}{2}}\pmod p.$$

In summary, we get

$$\prod_{0<x<y<p}(Ax^2+Bxy+Cy^2)\equiv\left(\prod_{0<n<p,\left(\frac{n}{p}\right)=1}n\right)^{\frac{p-2-M}{2}}\cdot\left(\prod_{0<n<p,\left(\frac{n}{p}\right)=-1}n\right)^{\frac{p-2+M}{2}}$$

$$=\left(\prod_{0<n<p,\left(\frac{n}{p}\right)=1}n\right)^{-M}\cdot((p-1)!)^{\frac{p-2+M}{2}}$$

$$\equiv(-1)^{-\frac{p+1}{2}M+\frac{p-2+M}{2}}=(-1)^{\frac{M-3}{2}}\pmod p.$$

This equals to 1 if exactly one or all three of A, C, and $A+B+C$ are perfect squares modulo p and to -1 otherwise, as desired. $\qquad\square$

Remark. For completeness we give a proof of the identity (5.9). By Euler's criterion, and the congruence

$$\sum_{y=0}^{p-1}y^k\equiv\begin{cases}0\pmod p & \text{if }0\le k<p-1,\\-1\pmod p & \text{if }k=p-1,\end{cases}$$

it follows that

$$\sum_{y=0}^{p-1} \left(\frac{y^2 + a}{p} \right) \equiv \sum_{y=0}^{p-1} (y^2 + a)^{\frac{p-1}{2}} \equiv \sum_{y=0}^{p-1} (y^{p-1} + P(y)) \equiv -1 \pmod{p}$$

where $P(y)$ is a polynomial in y with degree less than $p - 1$ and integer coefficients.

Since $\sum_{y=0}^{p-1} \left(\frac{y^2 + a}{p} \right)$ is an integer in $[-p, p]$, its value should be -1 or $p - 1$. It remains to show that sum cannot be equal to $p - 1$. In fact, in this case $p - 1$ terms have to be equal to 1 and exactly 1 term have to be equal to 0. If $\left(\frac{y^2 + a}{p} \right) = 0$ then also $\left(\frac{(-y)^2 + a}{p} \right) = 0$ which means that $y = -y$, that is $y = 0$.

But $\left(\frac{a}{p} \right) \neq 0$ because p does not divide p and leads to a contradiction.

Additional problems for practice.

1. Consider $ax^2 + bx + c \equiv 0 \pmod{n}$. Let $d = b^2 - 4ac$. If $\gcd(2ad, n) = 1$, then the equation has a solution if and only if d is a square modulo p^m for each power dividing n. In this case, if k is the number of distinct prime divisors of n, the equation has exactly 2^k incongruent solutions modulo n.

2. **Problem 10258** (Proposed by H. Liebeck and A. Osborne, 99(9), 1992). Let a, b and c be positive integers which are pairwise relatively prime. Prove that if the congruences

$$A^2 = -bc \pmod{a}, \quad B^2 = -ca \pmod{b}, \quad C^2 = -ab \pmod{c}$$

are solvable for A, B and C, then the equation

$$ax^2 + by^2 + cz^2 = abc$$

has a solution in integers x, y and z.

3. **Problem 12161** (Proposed by J. Hernández, 127(2), 2020). Let $N(C)$ be the number of pairs $(u, v) \in \mathbb{Z} \times \mathbb{Z}$ satisfying $u^2 + uv + v^2 = C$. Prove that 6 divides $N(C)$ for every positive integer C.

4. **Problem 12200** (Proposed by I. S. Evren, 127(7), 2020). Prove that for every positive integer m, there is a positive integer k such that k does not divide $m + x^2 + y^2$ for any positive integers x and y.

5. **Problem 12328** (Proposed by P. Koymans and J. Lagarias, 129(6), 2022). An integer binary quadratic form is a function $f : \mathbb{Z}^2 \to \mathbb{Z}$ defined by $f(m, n) = am^2 + bmn + cn^2$ for some $a, b, c \in \mathbb{Z}$. The value set $V(f)$ of such a form is defined to be $\{f(m, n) : (m, n) \in \mathbb{Z}^2\}$.

(a) Prove that if $f_1(m, n) = m^2 - mn - 3n^2$ and $f_2(m, n) = m^2 - 13n^2$, then $V(f_1) = V(f_2)$.

(b) Prove that if $f_1(m, n) = m^2 - mn - 4n^2$ and $f_2(m, n) = m^2 - 17n^2$, then $V(f_1) \supseteq V(f_2)$ but $V(f_1) \neq V(f_2)$.

6. **Problem 12350** (Proposed by N. MacKinnon, 129(9), 2022). What is the smallest positive integer k such that for any quadratic polynomial P with integer coefficients, one of the integers $P(1), \ldots, P(k)$ has a zero digit when written in base two?

5.5 Powers of a prime dividing a product of binomial coefficients

Problem 12041 (Proposed by R. P. Stanley, 125(5), 2018). For a positive integer c, let $\nu_p(c)$ denote the largest integer d such that p^d divides c. Let

$$G_m = \prod_{i=0}^{m} \prod_{j=0}^{m} \binom{i+j}{i}.$$

For $n \geq 1$, prove

$$\nu_p(G_{p^n - 1}) = \frac{1}{2} \left(\left(n - \frac{1}{p-1} \right) p^{2n} + \frac{p^n}{p-1} \right).$$

Discussion.
Notice that G_m is a fraction, for which its numerator and denominator can be expressed as products of factorials. Thus, it is natural to apply the following Legendre's formula: If N is a positive integer then, for any prime p,

$$\nu_p(N!) = \sum_{k \geq 1} \left\lfloor \frac{N}{p^k} \right\rfloor = \frac{N - s_p(N)}{p - 1} \tag{5.10}$$

where $s_p(N)$ is the sum of all the digits in the expansion of N in base p.

Solution.
Since

$$\nu_p(a \cdot b) = \nu_p(a) + \nu_p(b) \qquad \text{and} \qquad \nu_p(a/b) = \nu_p(a) - \nu_p(b),$$

we find that

$$\nu_p(G_m) = \sum_{i=0}^{m} \sum_{j=0}^{m} (\nu_p((i+j)!) - \nu_p(i!) - \nu_p(j!))$$

$$= \sum_{i=0}^{m} (i+1)\nu_p(i!) + \sum_{i=0}^{m-1} (m-i)\nu_p((i+m+1)!) - 2(m+1) \sum_{i=0}^{m} \nu_p(i!)$$

$$= \sum_{i=0}^{m} (m-i) \left(\nu_p((i+m+1)!) - \nu_p(i!) \right) - (m+1) \sum_{i=0}^{m} \nu_p(i!).$$

Thus, by Legendre's formula, for $m = p^n - 1$, we get

$$\sum_{i=0}^{p^n-1} (p^n - 1 - i)\left(\nu_p((i+p^n)!) - \nu_p(i!)\right)$$

$$= \sum_{k=1}^{n} \sum_{i=0}^{p^n-1} (p^n - 1 - i)\left(\left\lfloor \frac{i+p^n}{p^k} \right\rfloor - \left\lfloor \frac{i}{p^k} \right\rfloor\right)$$

$$= \sum_{k=1}^{n} \sum_{i=0}^{p^n-1} (p^n - 1 - i)p^{n-k} = \frac{p^n(p^n-1)^2}{2(p-1)}.$$

Moreover, since

$$\sum_{i=0}^{p^n-1} s_p(i) = \frac{1}{2}np^n(p-1),$$

it follows that

$$\sum_{i=0}^{p^n-1} \nu_p(i!) = \sum_{i=0}^{p^n-1} \frac{i - s_p(i)}{p-1} = \frac{p^n(p^n-1)}{2(p-1)} - \frac{np^n}{2}.$$

Hence

$$\nu_p(G_{p^n-1}) = \frac{p^n(p^n-1)^2}{2(p-1)} - p^n\left(\frac{p^n(p^n-1)}{2(p-1)} - \frac{np^n}{2}\right)$$

$$= \frac{1}{2}\left(\left(n - \frac{1}{p-1}\right)p^{2n} + \frac{p^n}{p-1}\right).$$

$\square$

Remark. Both forms of Legendre's formula (5.10) enable us to determine valuations of the natural number, binomial coefficients, Catalan numbers, and many other numbers involving factorials. We single out the following three useful results:

1. Let p be a prime number. Then for $n \in \mathbb{N}$,

$$n = (p-1)\nu_p(n!) + s_p(n).$$

2. For prime $p > 2$,

$$\nu_p((2n-1)!!) = \frac{n + s_p(n) - s_p(2n)}{p-1}.$$

3. Let $C_n = \frac{1}{n+1}\binom{2n}{n}$ be the nth Catalan number. Then

$$\nu_p\left(\binom{2n}{n}\right) = \frac{2s_p(n) - s_p(2n)}{p-1}$$

and

$$\nu_p(C_n) = \frac{s_p(n) + s_p(n+1) - s_p(2n) - 1}{p-1}.$$

Additional problems for practice.

1. Show that

   (i) the entire $(2^k - 1)$-th row of Pascal's triangle consists of odd numbers only.

   (ii) every entry in 2^k-th row of Pascal's triangle is even, except for 1s at the beginning and the end.

2. **Problem E1408** (Proposed by J. L. Selfridge, 67(3), 1960). Find the highest power of 2 which divides the numerator of

$$1 + \frac{1}{3} + \frac{1}{5} + \cdots + \frac{1}{2n-1}.$$

3. **Problem 6625** (Proposed by N. Strauss and J. Shallit, 97(3), 1990). If k is a positive integer, let $3^{\nu(k)}$ be the highest power of 3 dividing k. Put

$$r(n) = \sum_{i=0}^{n-1} \binom{2i}{i}$$

   for positive integers n. Prove that

   (i) $\nu(r(n)) \geq 2\nu(n)$,

   (ii) $\nu(r(n)) = \nu\left(\binom{2n}{n}\right) + 2\nu(n)$.

4. **Problem 11158** (Proposed by D. Beckwith, 112(5), 2005). Let n be a positive integer, and let p be a prime number. Prove that $p^p \mid n!$ implies that $p^{p+1} \mid n!$.

5. **Problem 11546** (Proposed by K. MacMillan and J. Sondow, 118(1), 2011). Let d, k, and q be positive integers, with k odd. Find the highest power of 2 that divides

$$S_q(2^d k) = \sum_{n=1}^{2^d k} n^q.$$

6. **Problem 11568** (Proposed by K. Foster, 118(4), 2011). For $n \geq 1$, let $f(n)$ be the least-significant nonzero decimal digit of $n!$. For $n \geq 2$, show that $f(625n) = f(n)$.

7. **Problem 11864** (Proposed by B. Farhi, 122(8), 2015). Let p be a prime number, and let $\{u_n\}_{n\geq0}$ be the sequence given by $u_n = n$ for $0 \leq n \leq p-1$ and by $u_n = pu_{n+1-p} + u_{n-p}$ for $n \geq p$. Prove that for each positive integer n, the greatest power of p dividing u_n is the same as the greatest power of p dividing n.

8. **Problem 12128** (Proposed by O. Kouba, 126(8), 2019). Let F_n be the nth Fibonacci number. Find, in terms of n, the number of trailing zeros in the decimal representation of F_n.

9. **Putnam 2022-A3**. Let p be a prime number greater than 5. Let $f(p)$ denote the number of infinite sequences $a_1, a_2, a_3, \ldots$ such that $a_n \in \{1, 2, \ldots, p-1\}$ and $a_n a_{n+2} \equiv 1 + a_{n+1} \pmod{p}$ for all $n \geq 1$. Prove that $f(p)$ is congruent to 0 or 2 (mod 5).

5.6 Divisibility of coefficients of powers of polynomials

Problem 12286 (Proposed by I. Gessel, 128(10), 2021). Let p be a prime number, and let m be a positive integer not divisible by p. Show that the coefficients of $(1+x+\cdots+x^{m-1})^{p-1}$ that are not divisible by p are alternately 1 and -1 modulo p. For example,

$$(1 + x + x^2 + x^3)^4 \equiv 1 - x + x^4 - x^6 + x^8 - x^{11} + x^{12} \pmod{5}.$$

Discussion.
For this problem, it will be useful to recall the so-called *Freshman's dream* theorem: for any prime p,

$$(x_1 + x_2)^p \equiv x_1^p + x_2^p \pmod{p}$$

due to the fact that p divides each binomial coefficient $\binom{p}{k}$ for $k = 1, 2, \ldots, n$. By induction, we easily find that, for any integer $n \geq 2$,

$$\left(\sum_{i=1}^n x_i\right)^p \equiv \sum_{i=1}^n x_i^p \pmod{p}.$$

Solution.
For $0 \leq n \leq (m-1)(p-1)$, the coefficient of x^n in $(1+x+\cdots+x^{m-1})^{p-1}$ is

$$[x^n]\left(\sum_{k=0}^{m-1} x^k\right)^{p-1} = [x^n]\left(\sum_{k=0}^{m-1} x^k\right)^p \cdot \frac{1-x}{1-x^m}$$

$$= [x^n]\left(\sum_{k=0}^{m-1} x^k\right)^p \cdot \frac{1-x^{mp}}{1-x^m} \cdot (1-x)$$

$$= [x^n]\left(\sum_{k=0}^{m-1} x^k\right)^p \cdot \sum_{j=0}^{p-1} x^{mj} \cdot (1-x)$$

$$\equiv [x^n]\sum_{k=0}^{m-1} x^{pk} \cdot \sum_{j=0}^{p-1} x^{mj} \cdot (1-x) \pmod{p}$$

$$\equiv [x^n]\sum_{k=0}^{m-1}\sum_{j=0}^{p-1} x^{pk+mj} \cdot (1-x) \pmod{p}.$$

We notice that the map $(k, j) \to pk + mj$ is injective in $[0, m - 1] \times [0, p - 1]$ because $\gcd(p, m) = 1$, hence the terms

$$\sum_{k=0}^{m-1} \sum_{j=0}^{p-1} x^{pk+mj}$$

are all distinct. Therefore the double sum can be written as the sum of several blocks of consecutive powers $x^a + x^{a+1} + \cdots + x^{a+b}$. By multiplying each block by $1 - x$ we obtain

$$(x^a + x^{a+1} + \cdots + x^{a+b})(1 - x) = x^a - x^{a+b+1}$$

and it follows that, modulo p, the coefficients of $\left(\sum_{k=0}^{m-1} x^k \right)^{p-1}$ are alternately 1 and -1. □

Remark. From the generating function point of view, the motivation of the above solution is as follows: let $f(x) = \sum_{n=1}^{\infty} a_n x^n$ be a formal power series with all $a_n \in \mathbb{Z}_p$. Since

$$\frac{f(x)}{1 - x} = \sum_{n=0}^{\infty} \left(\sum_{k=0}^{n} a_k \right) x^n,$$

to show that the nonzero coefficients of f alternate between 1 and -1, it suffices to show that each coefficient of $f(x)/(1 - x)$ is 0 or 1.

Additional problems for practice.

1. If p and q are relatively prime, show that the nonzero coefficients of

$$P(x, p, q) = \frac{(1 - x^{pq})(1 - x)}{(1 - x^p)(1 - x^q)}$$

alternate between 1 and -1. For example, $P(x, 3, 7) = 1 - x + x^3 - x^4 + x^6 - x^8 + x^9 - x^{11} + x^{12}$.

2. Prove that any polynomial of the form

$$P(x) = (x - a_0)^2 (x - a_1)^2 \cdots (x - a_n)^2 + 1$$

where $a_0, a_1, \ldots, a_n$ are all integers, cannot be factorized into two non-trivial polynomials, each with integer coefficients. For a proof see [95, p. 45].

3. **Putnam 1992-B2.** For nonnegative integers n and k, define $Q(n, k)$ to be the coefficient of x^k in the expansion of $(1 + x + x^2 + x^3)^n$. Prove that

$$Q(n, k) = \sum_{j=0}^{k} \binom{n}{j} \binom{n}{k - 2j},$$

where $\binom{a}{b} = 0$ for $0 \leq a < b$.

4. **Putnam 2008-B4**. Let p be a prime number. Let $h(x)$ be a polynomial with integer coefficients such that $h(0), h(1), \ldots, h(p^2 - 1)$ are distinct modulo p^2. Show that $h(0), h(1), \ldots, h(p^3 - 1)$ are distinct modulo p^3.

5. **Putnam 2021-A6**. Let $P(x)$ be a polynomial whose coefficients are all either 0 or 1. Suppose that $P(x)$ can be written as a product of two nonconstant polynomials with integer coefficients. Does it follow that $P(2)$ is a composite integer?

6. **Putnam 2022-B1**. Suppose that $P(x) = a_1 x + a_2 x^2 + \cdots + a_n x^n$ is a polynomial with integer coefficients, with a_1 odd. Suppose that $e^{P(x)} = b_0 + b_1 x + b_2 x^2 + \cdots$ for all x. Prove that b_k is nonzero for all $k \geq 0$.

7. **Problem 11577** (Proposed by P. P. Dályay, 118(5), 2011). Let n be a positive even integer and let p be prime. Show that the polynomial f given by $f(x) = p + \sum_{k=1}^{n} x^k$ is irreducible over $\mathbb{Q}$.

8. **Problem 12392** (Proposed by C. J. Hillar and L. Levine, 130(5), 2023). For which positive integers a is there a nonzero polynomial $P(x, y) \in \mathbb{C}[x, y]$ such that
$$P\left(\binom{2n}{n}, \binom{an}{n}\right) = 0$$
for all nonnegative integers n?

5.7 A Diophantine equation with powers

Problem 12019 (Proposed by N. Safaei, 125(1), 2018). Find all positive integers n such that $(2^n - 1)(5^n - 1)$ is a perfect square.

Discussion.
It is easy to see that $(2^1 - 1)(5^1 - 1) = 4$ is a perfect square, but
$$(2^2 - 1)(5^2 - 1) = 72 \quad \text{and} \quad (2^3 - 1)(5^3 - 1) = 868$$
are not. Indeed we will prove that $(2^n - 1)(5^n - 1)$ is not a perfect square for all $n \geq 2$. Recall that q is quadratic residue modulo n if there exists an integer x such that $x^2 \equiv q \pmod{n}$. To show this impossibility, we naturally reduce the problem to congruences. Note that if a number is not a perfect square modulo some positive integer then it can't be a perfect square at all.

Solution.
We distinguish two cases according to the parity of $n \geq 2$. We start with the case in which n is odd.

If $n = 4j + 1$ with $j \geq 1$ then
$$(2^n - 1)(5^n - 1) = (2 \cdot 16^j - 1)(5 \cdot 625^j - 1) \equiv (-1)(5 - 1) \equiv 12 \pmod{16}$$
which is not a quadratic residue modulo 16.

If $n = 4j + 3$ with $j \geq 0$ then

$$(2^n - 1)(5^n - 1) = (8 \cdot 16^j - 1)(5^n - 1) \equiv (3 - 1)(-1) \equiv 3 \pmod 5$$

which is not a quadratic residue modulo 5.

Now we assume that n is even, i.e., $n = 2j$ with $j \geq 1$.

The number $(2^n - 1)(5^n - 1)$ is a perfect square if and only if $2^n - 1 = du^2$ and $5^n - 1 = dv^2$ with u, v are positive integers and d is a positive square-free integer (an integer which is divisible by no perfect square other than 1).

If $d = 1$ then $2^n - 1 = u^2$ and $5^n - 1 = v^2$ imply

$$1 = 2^n - u^2 = (2^j - u)(2^j + u) > 2 \quad \text{and} \quad 1 = 5^n - v^2 = (5^j - v)(5^j + v) > 5$$

which are both impossible.

4) If $d > 1$ then $2^n - 1 = du^2$ and $5^n - 1 = dv^2$ then it follows that d, u are odd, and v is even. Note that $(x, y) = (2^j, u)$ and $(x, y) = (5^j, v)$ are solutions of the Pell's equation (d is not a perfect square),

$$x^2 - dy^2 = 1.$$

Let (x_1, y_1) be the fundamental solution, then any other solution (x_k, y_k) is such that

$$x_k + y_k \sqrt{d} = (x_1 + y_1 \sqrt{d})^k$$

for some positive integer k. We have that

$$x_{2k} + y_{2k} \sqrt{d} = (x_k + y_k \sqrt{d})^2 = x_k^2 + dy_k^2 + 2x_k y_k \sqrt{d} = (2x_k^2 - 1) + 2x_k y_k \sqrt{d}$$

which implies that x_{2k} is odd and y_{2k} is even. Moreover

$$\begin{aligned} x_{2k+1} + y_{2k+1} \sqrt{d} &= (x_{2k} + y_{2k} \sqrt{d})(x_1 + y_1 \sqrt{d}) \\ &= (x_{2k} x_1 + dy_{2k} y_1) + (x_{2k} y_1 + y_{2k} x_1) \sqrt{d} \end{aligned}$$

which implies that x_{2k+1} has the same parity of x_1, and y_{2k+1} has the same parity of y_1.

Hence $(2^j, u) = (x_i, y_i)$ for some odd integer i and $(5^j, v) = (x_l, y_l)$ for some even integer $l = 2k$. Finally

$$5^j = x_{2k} = 2x_k^2 - 1$$

which implies that $x_k^2 \equiv 3 \pmod 5$, against the fact that 3 is not a quadratic residue modulo 5. $\square$

Remark. It is worth noting that $(2^n - 1)(5^n - 1)/4$ is the sum of divisors of 10^{n-1}. The sequence of integers which satisfy this property is the sequence A006532 in the OEIS (https://oeis.org/A006532). The first few terms are

$$1, 3, 22, 66, 70, 81, 94, 115, 119, 170, 210, 214, 217, 265, 282, 310, 322, \ldots$$

For related results see [19]. In [94] it is shown that for all positive integers n the number $(2^n - 1)(3^n - 1)$ is never a perfect square.

Additional problems for practice.

1. Prove that $4mn - m - n$ is not a perfect square for any $m, n \in \mathbb{N}$.

2. **Problem 10510** (Proposed by E. Proth, 103(3), 1996). Show that

$$\frac{x(x+1)}{2} + y^2 = z^3$$

has infinitely many integer solutions with $x > 0$.

3. **Problem 11023** (Proposed by W. W. Chao, 110(6), 2003). Find all pairs (x, y) of integers such that

$$x^2 + 3xy + 4006(x + y) + 2003^2 = 0.$$

4. **Problem 11125** (Proposed by S. Amrahov, 112(1), 2005). Find all solutions in integers m and n to

$$1997^{1001}(m^2 - 1) - 2m + 5 = 3\binom{2003^{2004}}{n}.$$

5. **Problem 11355** (Proposed by J. C. Lagarias, 115(4), 2008). Determine for which integers a the Diophantine equation

$$\frac{1}{x} + \frac{1}{y} + \frac{1}{z} = \frac{a}{xyz}$$

has infinitely many integer solutions (x, y, z) such that $\gcd(a, xyz) = 1$.

6. **Problem 11827** (Proposed by G. Stoica, 122(3), 2015). Show that there are infinitely many rational triples (a, b, c) such that $a + b + c = abc = 6$.

7. **Problem 11992** (Proposed by N. Safaei, 124(8), 2017). Prove that, for every positive integer n, there is a positive integer m such that $3^m + 5^m - 1$ is divisible by 7^n.

8. **Problem 12239** (Proposed by D. Altizio, 128(3), 2021). Determine all positive integers r such that there exist at least two pairs of positive integers (m, n) satisfying the equation $2^m = n! + r$.

9. **Problem 12346** (Proposed by N. Q. Minh, 129(8), 2022). Prove that there are infinitely many integers A such that, for every nonzero integer x and distinct positive odd integers m and n, the integer $x^m + Ax^n$ is not a perfect square.

5.8   Multiples with small digits

Problem 12034 (Proposed by G. Galperin and Y. J. Ionin, 125(4), 2018). Let N be any natural number that is not a multiple of 10. Prove that there is a multiple of N each of whose digits in base 10 is 1, 2, 3, 4, or 5.

Discussion.
We begin by examining what happens when N is coprime with 10. In particular, we consider the n-*repunit*

$$R_n := \sum_{k=0}^{n-1} 10^k = \overbrace{11\ldots1}^{n \text{ digits } 1}.$$

By the pigeonhole principle there are two distinct positive integers $n_1 < n_2$ such that

$$R_{n_2} \equiv R_{n_1} \pmod{N}.$$

This implies N divides $R_{n_2} - R_{n_1}$. Since

$$R_{n_2} - R_{n_1} = 10^{n_1} R_{n_2 - n_1},$$

and $\gcd(N, 10) = 1$, it follows that N divides $R_{n_2 - n_1}$. In the following, we show how to refine this approach to include the cases where N is divisible by 5 or 2 for proceeding with the proposed problem.

Solution.
We first show three claims.

i) 2^d has a multiple $m_2(d)$ of d digits containing only digits 1 or 2.
Let $m_2(1) = 2$ and

$$m_2(d+1) = \begin{cases} 2 \cdot 10^d + m_2(d) & \text{if } m_2(d)/2^d \equiv 0 \pmod 2, \\ 1 \cdot 10^d + m_2(d) & \text{if } m_2(d)/2^d \equiv 1 \pmod 2. \end{cases}$$

Hence if $m_2(d)/2^d \equiv r \pmod 2$ then

$$m_2(d+1) = (2-r)10^d + m_2(d) = (2-r)10^d + (2k+r)2^d$$

$$= 2^{d+1}\left((5^d + k) - \frac{(5^d - 1)}{2}r\right)$$

and, since 2 divides $5^d - 1$, it follows that 2^{d+1} divides $m_2(d+1)$.

ii) 5^d has a multiple $m_5(d)$ of d digits containing only digits 1, 2, 3, 4, or 5.
Let $m_5(1) = 5$ and

$$m_5(d+1) = \begin{cases} 5 \cdot 10^d + m_5(d) & \text{if } 3^d\, m_5(d)/5^d \equiv 0 \pmod 5, \\ 4 \cdot 10^d + m_5(d) & \text{if } 3^d\, m_5(d)/5^d \equiv 1 \pmod 5, \\ 3 \cdot 10^d + m_5(d) & \text{if } 3^d\, m_5(d)/5^d \equiv 2 \pmod 5, \\ 2 \cdot 10^d + m_5(d) & \text{if } 3^d\, m_5(d)/5^d \equiv 3 \pmod 5, \\ 1 \cdot 10^d + m_5(d) & \text{if } 3^d\, m_5(d)/5^d \equiv 4 \pmod 5. \end{cases}$$

Hence if $3^d m_5(d)/5^d \equiv r \pmod 5$ then

$$3^d m_5(d+1) = 3^d(5-r)10^d + 3^d m_5(d) = (5-r)30^d + (5k+r)5^d$$

$$= 5^{d+1}\left((6^d + k) - \frac{(6^d - 1)}{5}r\right)$$

and, since 5 divides $6^d - 1$, it follows that 5^{d+1} divides $m_5(d+1)$.

iii) Let a be a positive integer such that $\gcd(a, 10) = 1$. Then for any $d \geq 1$ there is a positive integer n such that

$$R_n(d) := \sum_{k=0}^{n-1} 10^{kd} = 1\underbrace{00\ldots01}_{d}\ldots\overbrace{\underbrace{00\ldots01}_{d}\underbrace{00\ldots01}_{d}}^{n \text{ digits } 1}$$

is a multiple a.

Since the sequence $(R_n(d))_n$ is infinite, there are two distinct positive integers $n_1 < n_2$ such that

$$R_{n_2}(d) \equiv R_{n_1}(d) \pmod a$$

which implies that a divides

$$R_{n_2}(d) - R_{n_1}(d) = 10^{n_1 d} R_{n_2 - n_1}(d).$$

Since $\gcd(a, 10) = 1$ it follows that a divides $R_{n_2 - n_1}(d)$.

Finally, let N be a positive integer which is not a multiple of 10. Then N can be written as $2^d \cdot a$ or $5^d \cdot a$ with $d \geq 0$ and $\gcd(a, 10) = 1$. If $d = 0$ then, by iii), $R_n(1)$ is a multiple of N which contains only the digit 1. If $d \geq 1$ and $N = 2^d \cdot a$ then, by i) and ii), $R_n(d)m_2(d)$ is a multiple of N which contains only the digits 1 or 2. If $d \geq 1$ and $N = 5^d \cdot a$ then, by i) and iii), $R_n(d)m_5(d)$ is a multiple of N which contains only the digits 1, 2, 3, 4, or 5. $\square$

Remark. Along the same lines of the proof above, we can show a more general statement: For every natural number N that is not a multiple of b has a multiple whose base b expansion has entries in $\{1, 2, \ldots, d\}$ with $d = b/p$, where p is the smallest prime divisor of b. For this proposed problem, we have $b = 10, p = 2, d = 5$.

A closely related problem is on digit sum of multiples of an integer. For positive integer n, let $s_b(n)$ be the sum of digits in base b expansion of n. If m is a nontrivial multiple of n (that is, $m/n \neq b^l$), define

$$\mathcal{N}_b = \{n : s_b(n) = s_b(kn) \text{ for some integer } k \neq b^l\}.$$

Clearly, no power of b belongs to $\mathcal{N}_b$. But $\mathcal{N}_b$ contains the rest of all positive integers.

For $n \in \mathcal{N}_b$, let

$$A_b(n) = \min\{kn : k \neq b^l, s_b(n) = s_b(kn)\}.$$

If $b = 10$, we have $s_{10}(2) = 2 = s_{10}(110), s_{10}(3) = 3 = s_{10}(12), s_{10}(4) = 4 = s_{10}(112)$. In general,

$$\{\mathcal{A}_{10}(n)\}_{n \geq 1} = \{0, 110, 12, 112, 140, 24, 133, 152, 18, 0, 1001, 300, 2002, \ldots\},$$

which is the entry $A087303$ (https://oeis.org/A087303) in the OEIS.

Let $n = b^m + 1$. Since $b^{3m} + 1 = (b^m + 1)(b^{2m} - b^m + 1)$, we obtain $\mathcal{A}_b(n) = b^{3k} + 1$, which indicates $\mathcal{A}_b(n) \sim n^3$ for infinitely many n. On the other hand, for prime p, let $n = 1 + b + \cdots + b^{p-1}$ and $N = 1 + b^2 + \cdots b^{p-1} + b^{p+1}$. Then

$$N = n + b^{p+1} - b = n(1 + b(b-1)), \quad s_b(n) = s_b(N) = p,$$

and so $\mathcal{A}_b(n) \sim cn$ for infinitely many n. Consequently, determining an optimal bound for $\mathcal{A}_n(n)$ seems pretty challenge.

Additional problems for practice.

1. **Putnam 1987-A6.** For each positive integer n, let $a(n)$ be the number of zeros in the base 3 representation of n. For which positive real numbers x does the series

$$\sum_{n=1}^{\infty} \frac{x^{a(n)}}{n^3}$$

converge? *Hint*: The positive integer n has exactly $k + 1$ digits in base 3 if and only if $3^k \leq n < 3^{k+1}$.

2. **Putnam 1989-A1.** How many primes among the positive integers, written as usual in base 10, are such that their digits are alternating 1's and 0's, beginning and ending with 1.

3. **Putnam 1998-B5.** Let N be the positive integer with 1998 decimal digits, all of them 1; that is,

$$N = 1111 \cdots 11.$$

Find the thousandth digit after the decimal point of $\sqrt{N}$.

4. **Problem 10439** (Proposed by C. Vanden Eynde, 102(3), 1995). The rational number $1/9$ is an example of a number c in $[0, 1]$ such that the decimal representation of neither c nor $\sqrt{c}$ contains the digit 0. Find an irrational number with the same property.

5. **Problem 10758** (Proposed by M. Sapir, 106(8), 1999). Prove that the sum of the (decimal) digits of 9^n cannot equal 9 when $n > 2$.

6. **Problem 12174** (Proposed by G. Galperin and Y. J. Ionin, 127(3), 2020). (a) Let n be a positive integer, and suppose that the three leading digits of the decimal expansion of 4^n are the same as the three leading digits of 5^n. Find all possibilities for these three leading digits.

   (b) Prove that for any positive integer k there exists a positive integer n such that the k leading digits of the decimal expansion of 4^n are the same as the k leading digits of 5^n.

7. **Problem 12368** (Proposed by C. Chiser, 130(1), 2023). According to problem A3 in the 1970 Putnam Competition, no perfect square can have a decimal representation ending in 4444. There are, however, perfect squares with a decimal representation ending in 444. For $n \geq 4$, how many perfect squares k have a decimal representation that consists of n digits ending in 444?

5.9 Another property of the taxicab number 1729

Problem 12048 (Proposed by J. C. Lagarias, 125(6), 2018). Call an integer a special Carmichael number if it can be written as, $(6k+1)(12k+1)(18k+1)$ for some integer k, with each of $6k+1$, $12k+1$, and $18k+1$ being prime. Call an integer a taxicab number if it can be written as the sum of two positive integer cubes in two different ways. Show that 1729 is the only positive integer that is both a special Carmichael number and a taxicab number.

Discussion.
Observe that if $n = x^3 + y^3$, then $d = x + y$ is one of its divisors which admits some specific bounds. On the other hand, if n is the product of three distinct primes then n has exactly 8 explicit divisors. To conclude the proof, it suffices to test if these divisors fall within the specific bounds.

Solution.
Let n be a positive integer which can be written as the sum of the cubes of two positive integers x and y:

$$n = x^3 + y^3 = (x + y)(x^2 - xy + y^2).$$

Then $d = x + y$ is a positive divisor of n such that

$$n = x^3 + y^3 = x^3 + (d - x)^3 = d(3x^2 - 3dx + d^2),$$

and rearranging the terms we find the quadratic equation

$$x^2 - dx + \frac{d^3 - n}{3d} = 0.$$

Since this equation has two positive integer solutions x and y then $d^3 - n > 0$ and

$$3d\Delta = 3d\left((-d)^2 - \frac{4(d^3 - n)}{3d}\right) = 3d^3 - 4(d^3 - n) = 4n - d^3 \geq 0.$$

Therefore the divisor d satisfies the following bounds

$$n < d^3 \leq 4n. \tag{5.11}$$

Moreover, we know that n is the product of three primes $p < q < r$ with $p = 6k + 1 \geq 7$, $q = 2p - 1$ and $r = 3p - 2$. Now, for each divisor $d \in \{1, p, q, r, pq, pr, qr, pqr\}$, we check whether the necessary condition (5.11) is satisfied.

1) If $d \in \{1, p\}$ then $n = pqr > p^3 \geq d^3$ and (5.11) does not hold.

2) If $d = q$ then (5.11) is satisfied if and only if $pr < q^2 \leq 4pr$ that is

$$p(3p - 2) < (2p - 1)^2 \leq 4p(3p - 2)$$

which are fulfilled for $p \geq 2$.

3) If $d = r$ then (5.11) is satisfied if and only if $pq < r^2 \leq 4pq$ that is

$$4pq - r^2 = 4p(2p - 1) - (3p - 2)^2 = -p^2 + 8p - 4 = 12 - (p - 4)^2 \geq 0$$

which holds only when $2 \leq p = 6k + 1 \leq 7$, that is $p = 7$.

4) If $d \in \{pq, pr, qr, pqr\}$ then $d^3 \geq p^3 q^3 > 4pqr = 4n$ and (5.11) does not hold because for $p \geq 2$,

$$p^2 q^2 - 4r = p^2(2p - 1)^2 - 4(3p - 2) = 4p^4 - 4p^3 + p^2 - 12p + 8$$
$$= 4p^3(p - 1) + (p - 6)^2 - 28 \geq 32 - 28 > 0.$$

Hence, in order to have at least two different ways to write n as the sum of two positive integer cubes, it is necessary that both 2) and 3) hold, which happens just for $p = 7$. Notice that for $p = 7$ we have that $q = 2p - 1 = 13$, $r = 3p - 2 = 19$ and $d = q = 13$ or $d = r = 19$:

$$n = 7 \cdot 13 \cdot 19 = 1729 = 1^3 + 12^3 = 9^3 + 10^3.$$

$\square$

Remark. The proof of (5.11) may be shortened using the following elementary inequality:
$$x^2 - xy + y^2 < (x + y)^2 \leq 4(x^2 - xy + y^2).$$

Multiplying by $x + y$ immediately yields

$$n = x^3 + y^3 < (x + y)^3 = d^3 \leq 4(x^3 + y^3) = 4n.$$

A Carmichael number is a composite number n such that

$$a^{n-1} \equiv 1 \pmod{n}$$

for any integer $a \in (1, n)$ which is coprime with n. The number $(6k+1)(12k+1)(18k+1)$ with each of the factors $6k+1$, $12k+1$, and $18k+1$ being prime is a Carmichael number because the least common multiple of $6k$, $12k$, and $18k$ is $36k$ and

$$n - 1 = (6k + 1)(12k + 1)(18k + 1) = 36k \cdot (36k^2 + 11k + 1)$$

and therefore $a^{n-1} \equiv 1$ modulo each of the primes $6k+1$, $12k+1$, and $18k+1$.

The number 1729 was made famous because of a story where G. H. Hardy mentioned to S. Ramanujan that the taxi he took to visit him had the dull number 1729, which Ramanujan immediately declared that on the contrary it was a very interesting number because it is the smallest number being expressible as the sum of two cubes in two different ways : $1729 = 1^3 + 12^3 = 9^3 + 10^3$. We refer to the article [85] for more details on taxicab numbers.

Additional problems for practice.

1. (Characterizing the sum of two cubes, due to K. Broughan). Let $n \in \mathbb{N}$. Show that the equation $n = x^3 + y^3$ has positive integer solutions if and only if there exists a divisor d of n with $n \le d^3 \le 4n$ such that for some $l \in \mathbb{N}, d^2 - n/d = 3l$ and $d^2 - 4l$ is a perfect square.

2. **MM Problem 829** (Proposed by J. D. Baum, 45(2), 1972). It is well known that a positive integer can be written as sum of consecutive integers if and only if it is not a power of two. If a positive integer is so expressible its representation is not necessarily unique. For example,

$$15 = 7 + 8 = 4 + 5 + 6 = 1 + 2 + 3 + 4 + 5.$$

For integers of what form are their expressions as sums of positive consecutive integers unique?

3. **Problem 10338** (Proposed by C. Vanden Eynde, 100(9), 1993). Given an integer $n > 1$, determine the set of integers which can be written as a sum of two integers relatively prime to n.

4. **Problem 10454** (Proposed by H. Tamvakis, 102(5), 1995). We say that a natural number n is *amenable* if there exist integers $a_1, a_2, \ldots, a_n$ such that

$$a_1 + a_2 + \cdots + a_n = a_1 a_2 \cdots a_n = n.$$

Find all amenable numbers.

5. **Problem 10711** (Proposed by F. Luca, 106(2), 1999). A natural number is *perfect* if it is the sum of its proper divisors. Prove that two consecutive numbers cannot both be perfect.

6. **Problem 10844** (Proposed by N. MacKinnon, 108(1), 2001). The Fibonacci numbers are

$$1, 2, 3, 5, 8, 13, 21, 34, 55, 89, 144, 233, 377, 610, 987, \ldots,$$

and the twin primes are

$$3, 5, 7, 11, 13, 17, 19, 29, 31, 41, 43, 59, 61, 71, 73, 101, \ldots.$$

Which numbers are members of both sequences?

7. **Problem 12167** (Proposed by N. MacKinnon, 127(3), 2020). Let S be the set of positive integers expressible as the sum of two nonzero Fibonacci numbers. Show that there are infinitely many six-term arithmetic progressions of numbers in S but only finitely many such seven-term arithmetic progressions.

8. **Problem 12258** (Proposed by J. C. Lagarias, 128(6), 2021). Let S be the set of positive integers n such that $n!$ is not the sum of three squares. Show that S has bounded gaps, i.e., there is a positive constant C such that for every positive integer n, there is an element of S between n and $n + C$.

5.10   An application of prime density

Problem 11781 (Proposed by R. Tauraso, 121(5), 2014). For $n \geq 2$, call a positive integer n-*smooth* if none of its prime factors is larger than n. Let S_n be the set of all n-smooth positive integers. Let C be a finite, nonempty set of the nonnegative integers, and let a and d be positive integers. Let M be the set of all positive integers of the form $m = \sum_{k=1}^{d} c_k s_k$ where $c_k \in C$ and $s_k \in S_n$ for $k = 1, \ldots, d$. Prove that there are infinitely many primes p such that $p^a \notin M$.

Discussion.
Let p_k be the k-th prime number. By the *fundamental theorem of arithmetic*, then
$$S_{p_k} = \{p_1^{e_1} p_2^{e_2} \cdots p_k^{e_k} : e_1, e_2, \ldots, e_k \geq 0\}.$$
The following table gives the first few p-smooth numbers.

S_n	OEIS	n-smooth numbers
S_2	A000079	1, 2, 4, 8, 16, 32, 64, 128, 256, 512, 1024, 2048, $\ldots$
S_3	A003586	1, 2, 3, 4, 6, 8, 9, 12, 16, 18, 24, 27, 32, 36, 48, $\ldots$
S_5	A051037	1, 2, 3, 4, 5, 6, 8, 9, 10, 12, 15,16, 18, 20, 24, 25,$\ldots$
S_7	A002473	1, 2, 3, 4, 5, 6, 7, 8, 9, 10, 12, 14, 15, 16, 18, 20, $\ldots$

Clearly, 2-smooth numbers are simply all the powers of 2. As regards 3-smooth numbers, we would like to mention the intriguing paper intriguing [20].

The statement of the problem would be trivially false without the finiteness assumption of d. In fact since S_n contains all the powers of n, if $C = \{0, 1, \ldots, n-1\}$ then the sums $\sum_k c_k s_k$ generate all the nonnegative integers. If M is finite then the property is straightforwardly true. On the other hand, if M is an infinite sequence of positive integers $m_1 < m_2 < m_3 < \cdots$, then it suffices to show that, for all $\alpha > 0$,
$$\sum_{k=1}^{\infty} \frac{1}{m_k^{\alpha}} < +\infty.$$

In fact, if $p_k^a \in M$ for any $k \geq k_0$, where p_k is the k-th prime number, then, for $\alpha = 1/a$, we get a contradiction due to the divergence of the series of reciprocals of the primes:

$$+\infty = \sum_{k=k_0}^{\infty} \frac{1}{p_k} = \sum_{k=k_0}^{\infty} \frac{1}{(p_k^a)^\alpha} \leq \sum_{k=1}^{\infty} \frac{1}{m_k^\alpha} < +\infty.$$

Solution.

For $x > 1$, let us consider the cardinality of the set $M \cap [1, x]$, that is

$$f(x) = |M \cap [1, x]|.$$

Then $f(x)$ is a non-decreasing function such that

$$f(x) \leq (|C| + 1)^d |S_n \cap [1, x]|^d \leq (|C| + 1)^d \prod_{i=1}^{\pi(n)} \left(1 + \frac{\ln(x)}{\ln(p_i)}\right)^d$$

where $\pi(n)$ is the number of primes less than or equal to n. Hence, for all $\alpha > 0$,

$$\lim_{x \to +\infty} \frac{f(x)/x^{1+\alpha}}{1/x^{1+\alpha/2}} = \lim_{x \to +\infty} \frac{f(x)}{x^{\alpha/2}} = 0$$

and therefore, since $\sum_{j=1}^{\infty} 1/x^{1+\alpha/2} < +\infty$,

$$+\infty > \sum_{j=1}^{\infty} \frac{f(j)}{j^{1+\alpha}} = \sum_{k=1}^{\infty} \sum_{j=m_k}^{m_{k+1}-1} \frac{f(j)}{j^{1+\alpha}} = \sum_{k=1}^{\infty} k \sum_{j=m_k}^{m_{k+1}-1} \frac{1}{j^{1+\alpha}}$$

$$\geq \sum_{k=1}^{\infty} k \int_{m_k}^{m_{k+1}} \frac{dx}{x^{1+\alpha}} = \frac{1}{\alpha} \sum_{k=1}^{\infty} k \left(\frac{1}{m_k^\alpha} - \frac{1}{m_{k+1}^\alpha}\right)$$

$$= \frac{1}{\alpha} \left(\sum_{k=1}^{\infty} \frac{1}{m_k^\alpha} - \lim_{k \to +\infty} \frac{k}{m_{k+1}^\alpha}\right) = \frac{1}{\alpha} \sum_{k=1}^{\infty} \frac{1}{m_k^\alpha}$$

because

$$\lim_{k \to +\infty} \frac{k}{m_{k+1}^\alpha} = \lim_{k \to +\infty} \frac{k}{k+1} \cdot \frac{f(m_{k+1})}{m_{k+1}^\alpha} = \lim_{x \to +\infty} \frac{f(x)}{x^\alpha} = 0.$$

$\square$

Remark. To design efficient number theoretic algorithms, it has turned out to be important to have accurate estimates for the number of smooth numbers in various sequences. Let $\Psi(x, n)$ be the number of n-smooth integers less than or equal to x. Similar to the prime number theorem $\pi(x) \sim x/\ln(x)$, it is known that

$$\Psi(x, n) \sim \frac{1}{\pi(n)!} \prod_{p \leq n} \frac{\ln(x)}{\ln(p)}.$$

For more details, Granville's survey paper [50] is a "must-read".

The divergence of the series of reciprocals of the primes, that was a key point of our solution, is also a consequence of the prime number theorem. An alternative short proof in Erdős style is the following (see also [2, pp. 5-6]). Assume that the series is convergent. Given $k_0 \geq 1$, then each positive integer in the interval $[1, n]$ falls into one of these cases: i) it is a multiple of a square j^2 with $j \geq 2$, or ii) it is a multiple of a prime greater than p_{k_0}, or iii) it is a squarefree product of distinct primes all less than or equal to p_{k_0}. Recalling that the number of multiples of d less than n are $\lfloor n/d \rfloor$, we find that

$$n \leq \sum_{j \geq 2} \left\lfloor \frac{n}{j^2} \right\rfloor + \sum_{k > k_0} \left\lfloor \frac{n}{p_k} \right\rfloor + 2^{k_0} \leq \sum_{j \geq 2} \frac{n}{j^2} + \sum_{k > k_0} \frac{n}{p_k} + 2^{k_0},$$

that is

$$1 - \sum_{j \geq 2} \frac{1}{j^2} - \sum_{k > k_0} \frac{1}{p_k} \leq \frac{2^{k_0}}{n}. \tag{5.12}$$

It suffices to choose k_0 large enough so that the tail of the series $\sum_{k > k_0}^{\infty} \frac{1}{p_k}$ is less than the positive number $1 - \sum_{j \geq 2} \frac{1}{j^2}$ and then the inequality (5.12) yields a contradiction. Indeed, the left-hand side is positive which does not depend on n, whereas the right-hand side goes to zero as $n \to \infty$.

Additional problems for practice.

1. Let $\Psi(x, n)$ be the number of n-smooth integers less than or equal to x. Show that

$$\Psi(x, \ln^\alpha(x)) = x^{1 - 1/\alpha + 0(1)} \quad \text{for any } \alpha > 1.$$

2. (Due to P. Erdős). Let $(a_k)_{k \geq 1}$ be the sequence of 3-smooth numbers, that is the positive integers of the form $2^\alpha 3^\beta$, where α and β are non-negative integers. Prove that every positive integer is expressible in the form

$$a_{k_1} + a_{k_2} + \cdots + a_{k_n},$$

where no summand is a multiple of any other.

3. **Problem 10847** (Proposed by R. Chapman, K. Eriksson and R. P. Stanley, 108(1), 2001). Given that

$$\sum_{k=1}^{\infty} (-1)^{k-1} \frac{q^{\binom{k+1}{2}}}{1 - q^k} = \sum_{k=1}^{\infty} a_n q^n,$$

show that a_n is the number of divisors of n in the interval $[\sqrt{n/2}, \sqrt{2n})$.

4. **Problem 11109** (Proposed by S. W. Golomb, 111(8), 2004). Prove that for every odd prime and for every positive integer k there are infinitely many primes q such that

$$q^{p-1} \equiv 1 \pmod{p^k}.$$

5. **Problem 11831** (Proposed by R. Ozols, 122(4), 2015). Prove that for $\varepsilon > 0$ there exists an integer n such that the greatest prime divisor of $n^2 + 1$ is less than εn.

6. **Problem 12435** (Proposed by R. Tauraso, 131(1), 2024). For a positive integer n, let $d(n)$ be the number of positive divisors of n, let $\phi(n)$ be Euler's totient function (the number of integers in $\{1, \ldots, n\}$ that are relatively prime to n), and let $q(n) = d(\phi(n))/d(n)$. Find $\inf_n q(n)$ and $\sup_n q(n)$.

5.11   The asymptotic behavior of a sum

Problem 12153 (Proposed by O. Kouba, 127(1), 2020). For a real number x whose fractional part is not $1/2$, let $\langle x \rangle$ denote the nearest integer to x. For a positive integer n, let

$$a_n = \sum_{k=1}^{n} \frac{1}{\langle \sqrt{k} \rangle} - 2\sqrt{n}.$$

(a) Prove that the sequence $(a_n)_{n \geq 1}$ is convergent, and find its limit L.
(b) Prove that the set $\{\sqrt{n}(a_n - L) : n \geq 1\}$ is a dense subset of $[0, 1/4]$.

Discussion.
Note that $\langle \sqrt{k} \rangle = \lfloor \sqrt{k} + \frac{1}{2} \rfloor$ is equal to some positive integer j if and only if

$$j - \frac{1}{2} \leq \sqrt{k} < j + \frac{1}{2}$$

that is

$$j(j-1) + \frac{1}{4} \leq k < j(j+1) + \frac{1}{4}.$$

Hence $\langle \sqrt{k} \rangle = j$ for all $k \in (j(j-1), j(j+1)]$. This simplifies the sum in the definition of a_n to

$$\sum_{k=1}^{n} \frac{1}{\langle \sqrt{k} \rangle} = \sum_{j=1}^{m-1} 2 + \sum_{k=m(m-1)+1}^{n} \frac{1}{\langle \sqrt{k} \rangle} = \frac{n}{\langle \sqrt{n} \rangle} + \langle \sqrt{n} \rangle - 1$$

where $m = \langle \sqrt{n} \rangle$ is the smallest integer such that $n \leq m(m+1)$.

Solution.
We begin with

$$a_n = \sum_{k=1}^{n} \frac{1}{\langle \sqrt{k} \rangle} - 2\sqrt{n} = \frac{n}{\langle \sqrt{n} \rangle} + \langle \sqrt{n} \rangle - 1 - 2\sqrt{n} = \frac{(\sqrt{n} - \langle \sqrt{n} \rangle)^2}{\langle \sqrt{n} \rangle} - 1.$$

(a) As n goes to infinity, we have that $\langle\sqrt{n}\rangle \to +\infty$ and

$$0 \le a_n + 1 = \frac{(\sqrt{n} - \langle\sqrt{n}\rangle)^2}{\langle\sqrt{n}\rangle} = \frac{(\{\sqrt{n} + \frac{1}{2}\} - \frac{1}{2})^2}{\langle\sqrt{n}\rangle} \le \frac{1/4}{\langle\sqrt{n}\rangle} \to 0$$

where $\{x\} = x - \lfloor x \rfloor \in [0,1)$, and we conclude that $a_n \to L = -1$.

(b) Note that

$$\sqrt{n}(a_n - L) = \sqrt{n}(a_n + 1) = \frac{\sqrt{n}(\sqrt{n} - \langle\sqrt{n}\rangle)^2}{\langle\sqrt{n}\rangle}.$$

Let $j = \langle\sqrt{n}\rangle$. Then $n \in (j(j-1), j(j+1)]$ and

$$0 \le \sqrt{n}(a_n + 1) \le \frac{\sqrt{j(j+1)}(\sqrt{j(j+1)} - j)^2}{j} = \frac{1}{4} \cdot \frac{\sqrt{j(j+1)}}{\left(\frac{\sqrt{j+1}+\sqrt{j}}{2}\right)^2} < 1/4$$

where the AM-GM inequality is applied in the last step.
Let $\alpha \in [0, 1/4]$. Then there exist two sequences of integers $(p_k)_{k \ge 1}$ and $(q_k)_{k \ge 1}$ with $0 \le p_k \le q_k$ and $q_k \to +\infty$ such that $p_k/q_k \to 2\sqrt{\alpha} \in [0,1]$. Let $n_k = q_k^2 + p_k$ then

$$q_k^2 \le n_k \le q_k^2 + q_k < \left(q_k + \frac{1}{2}\right)^2$$

which implies that $\langle\sqrt{n_k}\rangle = q_k$. Therefore, as n goes to infinity,

$$\sqrt{n_k} - \langle\sqrt{n_k}\rangle = \sqrt{n_k} - q_k = \frac{n_k - q_k^2}{\sqrt{n_k} + q_k} = \frac{p_k}{\sqrt{q_k^2 + p_k} + q_k} \to \sqrt{\alpha}$$

and

$$\sqrt{n_k}(a_{n_k} + 1) = \frac{\sqrt{n_k}(\sqrt{n_k} - \langle\sqrt{n_k}\rangle)^2}{\langle\sqrt{n_k}\rangle} \to \alpha.$$

This proves that the set $\{\sqrt{n}(a_n + 1) : n \ge 1\}$ is a dense subset of $[0, 1/4]$. $\square$

Remark. The nearest integer function can be used to tidy up many otherwise complicated formulas. For example, let D_n be the number of derangements of a set of size n. It is well-known that $D_n = n! \sum_{k=0}^{n} (-1)^k/k!$. On the other hand, since the sum is a truncation of the series of e^{-1}, D_n is pretty closed to $n!/e$. Thus, we have a simpler formula for D_n namely,

$$D_n = \left\langle \frac{n!}{e} \right\rangle.$$

Similarly, we have a slightly simpler formula for the n-th Fibonacci number

$$F_n = \left\langle \frac{1}{\sqrt{5}} \left(\frac{1 + \sqrt{5}}{2}\right)^n \right\rangle.$$

Additional problems for practice.

1. **Putnam 2001-B3**. For any positive integer n, let $\langle n \rangle$ denote the closest integer to $\sqrt{n}$. Evaluate

$$\sum_{n=1}^{\infty} \frac{2^{\langle n \rangle} + 2^{-\langle n \rangle}}{2^n}.$$

2. **Problem 10212** (Proposed by S.-J. Ban, 99(4), 1992). For a real number x whose fractional part is not $1/2$, let $\langle x \rangle$ denote the nearest integer to x. Evaluate

$$\sum_{k=1}^{\infty} \frac{1}{\langle \sqrt[3]{k} \rangle^4}.$$

3. **Problem 11206** (Proposed by M. Ivan and A. Lupaş, 113(2), 2006). Find

$$\lim_{n \to \infty} \frac{1}{n} \sum_{k=1}^{n} \left\{ \frac{n}{k} \right\}^2$$

where $\{x\} = x - \lfloor x \rfloor$ denotes the fractional part of x.

4. **Problem E2813** (Proposed by A. D. Sands, 87(1), 1980). For $0 \le x < 1$, show that

$$\sum_{n=1}^{\infty} \frac{(-1)^{\lfloor 2^n x \rfloor}}{2^n} = 1 - 2x.$$

Also find the sum of the series for $x \ge 1$.

5.12   A family of sums with logarithmic powers

Problem 12203 (Proposed by R. Tauraso, 127(8), 2020). Let m be a nonnegative integer, and let μ be the Möbius function on $\mathbb{Z}^+$, defined by setting $\mu(k)$ equal to $(-1)^r$ if k is the product of r distinct primes and equal to 0 if k has a square prime factor. Evaluate

$$\lim_{n \to \infty} \frac{1}{\ln^m(n)} \sum_{k=1}^{n} \frac{\mu(k)}{k} \ln^{m+1} \left(\frac{n}{k} \right).$$

Discussion.
We will show that the limit is $m + 1$. The case $m = 0$ is known. See, for instance, see [57, p. 18] or [38, p. 95]:

$$\lim_{x \to \infty} \sum_{1 \le k \le x} \frac{\mu(k)}{k} \log \left(\frac{x}{k} \right) = \lim_{x \to \infty} \log(x) \sum_{1 \le k \le x} \frac{\mu(k)}{k} - \lim_{x \to \infty} \sum_{1 \le k \le x} \frac{\mu(k)}{k} \log(k)$$

$$= 0 - (-1) = 1.$$

We are going to show by induction that the statement holds for all positive integers m as well.

Solution.
It remains to show the induction step, i.e., if the statement holds for m, then it must also hold for the case $m + 1$. Recall that if f is a function defined on $[1, +\infty)$ and $F(x) := \sum_{1 \le k \le x} f\left(\frac{x}{k}\right)$ then the following Möbius inversion formula holds

$$\sum_{1 \le k \le x} \mu(k) F\left(\frac{x}{k}\right) = \sum_{1 \le k \le x} \mu(k) \sum_{1 \le j \le x/k} f\left(\frac{x/k}{j}\right)$$

$$= \sum_{1 \le n \le x} f\left(\frac{x}{n}\right) \sum_{k | n} \mu(k) = f(x) \qquad (5.13)$$

with $n = jk$. Now assume that $m \ge 1$ and let $f(x) = x \log^m(x)$, then

$$\frac{F(x)}{x} = \sum_{1 \le k \le x} \frac{1}{k} \log^m\left(\frac{x}{k}\right) = x \sum_{1 \le k \le x} \frac{1}{k} \sum_{j=0}^{m} \binom{m}{j} \log^{m-j}(x) \log^j(k)(-1)^j$$

$$= \sum_{j=0}^{m} \binom{m}{j} (-1)^j \log^{m-j}(x) \sum_{1 \le k \le x} \frac{\log^j(k)}{k}$$

$$= \sum_{j=0}^{m} \binom{m}{j} (-1)^j \log^{m-j}(x) \left(\frac{\log^{j+1}(x)}{j+1} + c_j + O\left(\frac{\log^j(x)}{x}\right)\right)$$

$$= \frac{1}{m+1} \sum_{j=0}^{m} \binom{m+1}{j+1} (-1)^j \log^{m+1}(x)$$

$$+ \sum_{j=0}^{m} \binom{m}{j} (-1)^j c_j \log^{m-j}(x) + O\left(\frac{\log^m(x)}{x}\right)$$

$$= \frac{\log^{m+1}(x)}{m+1} + \sum_{j=0}^{m} \binom{m}{j} (-1)^j c_j \log^{m-j}(x) + O\left(\frac{\log^m(x)}{x}\right),$$

where we have used the fact that for any $j \ge 0$ there exists $c_j \in \mathbb{R}$ such that for $x \to +\infty$,

$$\sum_{1 \le k \le x} \frac{\log^j(k)}{k} = \frac{\log^{j+1}(x)}{j+1} + c_j + O\left(\frac{\log^j(x)}{x}\right).$$

Therefore, by (5.13),

$$\log^m(x) = \frac{f(x)}{x} = \frac{1}{x} \sum_{1 \leq k \leq x} \mu(k) F\left(\frac{x}{k}\right) = \frac{1}{m+1} \sum_{1 \leq k \leq x} \frac{\mu(k)}{k} \log^{m+1}\left(\frac{x}{k}\right)$$

$$+ \sum_{j=0}^m \binom{m}{j}(-1)^j c_j \sum_{1 \leq k \leq x} \frac{\mu(k)}{k} \log^{m-j}\left(\frac{x}{k}\right) + O(1)$$

$$(5.14)$$

because for some constant C

$$\left| \sum_{1 \leq k \leq x} \mu(k) O\left(\log^m(x/k)\right) \right| \leq C \sum_{1 \leq k \leq x} \log^m(x/k)$$

$$\leq C \log^m(x) + C \int_1^x \log^m(x/t)\, dt$$

$$\leq C \log^m(x) + Cx \int_1^{+\infty} \frac{\log^m(y)}{y^2}\, dy = O(x).$$

Since $\sum_{1 \leq k \leq x} \frac{\mu(k)}{k} = O(1)$, by the inductive step, for $0 \leq j \leq m-1$, we have

$$\sum_{1 \leq k \leq x} \frac{\mu(k)}{k} \log^{m-j}\left(\frac{x}{k}\right) = O(\log^{m-j-1}(x)).$$

Dividing both sides of (5.14) by $\log^m(x)$, we get

$$1 = \frac{1}{m+1} \cdot \frac{1}{\log^m(x)} \sum_{1 \leq k \leq x} \frac{\mu(k)}{k} \log^{m+1}\left(\frac{x}{k}\right) + \sum_{j=0}^m \binom{m}{j}(-1)^j c_j \cdot o(1) + o(1).$$

Therefore,

$$\lim_{x \to \infty} \frac{1}{\log^m(x)} \sum_{1 \leq k \leq x} \frac{\mu(k)}{k} \log^{m+1}\left(\frac{x}{k}\right) = m+1.$$

This proves the induction step and so the limit as claimed. $\qquad \square$

Remark. We can shorten the induction step by directly invoking Landau's classical formula [63]: for $j \geq 1$, there exists a positive constant C such that

$$\sum_{k=1}^n \frac{\mu(k)}{k}(-1)^j (\ln k)^j = \left(\frac{1}{\zeta(s)}\right)^{(j)}\Bigg|_{s=1} + O\left(e^{-C\sqrt{\ln n}}\right),$$

where $\zeta(s) = \sum_{n=1}^\infty 1/n^s$ is the Riemann zeta function. To this end, by the binomial theorem, we have

$$\sum_{k=1}^{n} \frac{\mu(k)}{k} \ln^{m+1}\left(\frac{n}{k}\right) = \sum_{k=1}^{n} \frac{\mu(k)}{k}(\ln n - \ln k)^{m+1}$$

$$= \sum_{j=0}^{m+1} \binom{m+1}{j}(\ln n)^{m+1-j} \sum_{k=1}^{n} \frac{\mu(k)}{k}(-1)^j(\ln k)^j.$$

Applying Landau's formula yields

$$\sum_{k=1}^{n} \frac{\mu(k)}{k} \ln^{m+1}\left(\frac{n}{k}\right) = (m+1)(\ln n)^m + \sum_{k=1}^{m-1} C_k(m)(\ln n)^k + O(1),$$

from which the claimed limit follows immediately.

The function $M(x) := \sum_{n \le x} \mu(n)$ is known as Mertens function. As $|\mu(n)| \le 1$ holds for all $n \in \mathbb{N}$, this implies that $M(x) = O(x)$. The prime number theorem implies that

$$\lim_{x \to \infty} \frac{M(x)}{x} = 0$$

(see for example [15, Theorem 4.14]). For the true order of $M(x)$, Mertens conjectured that, for all $\epsilon > 0$ and $x > 0$,

$$M(x) = o(x^{1/2+\epsilon}). \tag{5.15}$$

Surprisedly, (5.15) is equivalent to the Riemann hypothesis. One part of this equivalence can be directly proved via

$$\frac{1}{\zeta(s)} = \sum_{n=1}^{\infty} \frac{\mu(n)}{n^s}.$$

See Problem 3 below. For the converse see [38, pp. 261-263].

Additional problems for practice.

1. Let p be a prime number. Show that

$$\sum_{p \le x} \frac{\ln(p)}{p} = \ln(x) + O(1).$$

2. It is well-known that

$$\zeta(s) = \frac{1}{s-1} + \sum_{k=1}^{\infty} \frac{(-1)^k}{k!}\gamma_k(s-1)^k,$$

where γ_k is the k-th Stieltjes' constant. Find the first few terms of the power series for $1/\zeta(s)$ at $s = 1$.

3. Prove that if (5.15) holds for every $\epsilon > 0$ then the Riemann hypothesis is true. *Hint.* First show that

$$\frac{1}{\zeta(s)} = s \int_0^\infty \frac{M(x)}{x^{s+1}} \, dx.$$

4. **Problem 12321** (Proposed by M. Sharifi, 129(5), 2022). Let p be a prime number, and let N be the number of perfect squares m such that the least nonnegative remainder of $p \pmod{m}$ is a perfect square. Prove that N is less than $2p^{1/3}$.

5. (Due to K. Rogers) Let $S(n)$ be the number of squarefree numbers less than or equal to n. Show that

$$S(n) \geq \frac{53}{88} n \quad \text{for all } n \in \mathbb{N}.$$

5.13 Another application of Bertrand's postulate

Problem 11761 (Proposed by B. Tomper, 121(3), 2014). For each positive integer n, determine the least integer m such that

$$\mathrm{lcm}\{1, 2, \ldots, m\} = \mathrm{lcm}\{n, n+1, \ldots, m\}.$$

Discussion.
First, observe that the proposed equality holds if and only if all $1, 2, n-1$ divide $\mathrm{lcm}\{n, \ldots, m\}$. Next, by the definition of least common multiple,

$$\mathrm{lcm}\{1, 2, \ldots, m\} = \prod_{p \text{ prime}} p^{\lfloor \ln(m)/\ln(p) \rfloor} \geq \mathrm{lcm}\{n, n+1, \ldots, m\}.$$

Note that if equality holds for some m_0 then it is verified also for all $m \geq m_0$. Let a_n be the least integer m such that the equality holds. It is easy to see that $a_1 = 1$, $a_2 = 2$. We claim that, for $n > 2$, a_n is twice the largest prime power less than n.

Solution.
Let $n > 2$, and let $a_n = 2p^\alpha$, then $p^\alpha \geq 2$, and, by Bertrand's postulate, there is a prime $q \in (p^\alpha, 2p^\alpha)$. It follows that $2p^\alpha \geq n$, otherwise $p^\alpha < q < 2p^\alpha < n$, contradicting the fact that p^α is largest power of a prime p^α less than n.
If $m < a_n = 2p^\alpha$, then $p^\alpha < n \leq m < 2p^\alpha$ implies that

$$p^\alpha \mid \mathrm{lcm}\{1, 2, \ldots, m\}, \quad \text{but} \quad p^\alpha \nmid \mathrm{lcm}\{n, n+1, \ldots, m\}$$

so $\mathrm{lcm}\{1, 2, \ldots, m\} > \mathrm{lcm}\{n, n+1, \ldots, m\}$.
It remains to show that if $m = a_n = 2p^\alpha$ then

$$\mathrm{lcm}\{1, 2, \ldots, 2p^\alpha\} = \mathrm{lcm}\{n, n+1, \ldots, 2p^\alpha\}.$$

By the pigeonhole principle, any number in $\{1, 2, \ldots, p^\alpha\}$ has at least one multiple in the set $\{p^\alpha + 1, p^\alpha + 2, \ldots, 2p^\alpha\}$ (which has p^α elements). Hence

$$\text{lcm}\{1, 2, \ldots, 2p^\alpha\} = \text{lcm}\{p^\alpha + 1, p^\alpha + 2, \ldots, 2p^\alpha\}.$$

In order to prove that

$$\text{lcm}\{p^\alpha + 1, p^\alpha + 2, \ldots, 2p^\alpha\} = \text{lcm}\{n, n + 1, \ldots, 2p^\alpha\}$$

it suffices to show that if q^β divides $\text{lcm}\{p^\alpha + 1, p^\alpha + 2, \ldots, n - 1\}$ for some prime q, then $q^\beta \leq 2p^\alpha - n + 1$ which implies, by the pigeonhole principle, that there is a multiple of q^β in the set $\{n, n + 1, \ldots, 2p^\alpha\}$.

It is easy to verify it holds for $n < 26$. If $n \geq 26$ then $p^\alpha \geq 25$ and by a stronger version of Bertrand's postulate proved in [75], there is a prime between p^α and $6p^\alpha/5$. As above, this implies that $6p^\alpha/5 \geq n$. Since p^α is largest power of a prime less than n, it follows that there is an integer $b \geq 2$ such that $bq^\beta \in \{p^\alpha + 1, p^\alpha + 2, \ldots, n - 1\}$. Thus $n > 2q^\beta$ and

$$2p^\alpha - n + 1 > 2p^\alpha - \frac{6p^\alpha}{5} = \frac{4p^\alpha}{5} \geq \frac{4}{5}\frac{5n}{6} = \frac{2n}{3} > \frac{n}{2} > q^\beta,$$

and the proof is complete.   □

Remark. Bertrand's postulate was first proved by P. Chebyshev in 1850. A much simpler argument was later offered by both S. Ramanujan and P. Erdős. For more on this, see [2, Chapter 2]. According to the prime number theorem, the number of primes between n and $2n$ grows as

$$\pi(2n) - \pi(n) \sim \frac{2n}{\ln(2n)} - \frac{n}{\ln(n)} \sim \frac{n}{\ln(n)}$$

which shows that, in the long run, there are far more primes in that range than Bertrand's postulate actually requires.

The sequence $\text{lcm}(1, 2, \ldots, n)$ is the sequence A003418 in the OEIS (https://oeis.org/A003418) and has many remarkable properties. For example in [39], it is shown this identity for any positive integer n,

$$\text{lcm}\{1, 2, \ldots, n\} = n \, \text{lcm}\left\{ \binom{n-1}{0}, \binom{n-1}{1}, \ldots, \binom{n-1}{n-1} \right\}.$$

This identity immediately yields a well-known nontrivial lower bound

$$\text{lcm}\{1, 2, \ldots, n\} \geq n \binom{n-1}{\lfloor (n-1)/2 \rfloor} \geq \sum_{k=0}^{n-1} \binom{n-1}{k} = 2^{n-1}.$$

Additional problems for practice.

1. **Putnam 2003-B3**. Show that for each positive integer n,

$$n! = \prod_{i=1}^{n} \mathrm{lcm}\{1, 2, \ldots, \lfloor n/i \rfloor\}.$$

2. **Putnam 2008-A3**. Start with a finite sequence $a_1, a_2, \ldots a_n$ of positive integers. If possible, choose two indices $j < k$ such that a_j does not divide a_k, and replace a_j and a_k by $\gcd(a_i, a_k)$ and $\mathrm{lcm}\{a_j, a_k\}$, respectively. Prove that if this process is repeated, it must eventually stop and the final sequence does not depend on the choices made.

3. **Problem 10797** (Proposed by P. Bateman and J. Kalb, 107(4), 2000). Let h and k be integers with $k > 0$, $h + k > 0$, and $\gcd(h, k) = 1$. For $n \geq 1$, prove that

$$\lim_{n \to \infty} \frac{1}{n} \log(\mathrm{lcm}(\{h+k, h+2k, \ldots, h+nk\})) = \frac{k}{\phi(k)} \sum_{\substack{1 \leq m \leq k \\ \gcd(m,k)=1}} \frac{1}{m}$$

where $\phi(k)$ is the Euler's totient function.

4. **Problem 11117** (Proposed by M. Nyblom, 111(10), 2004). An integer n is a *positive power* if there exist integers a and m such that $a > 1$, $m > 2$, and $n = a^m$. Let $N(x)$ denote the number of positive powers n such that $1 < n < x$. For real $x \geq 4$ and with $L = \lfloor \log_2(x) \rfloor$, show that

$$N(x) = \sum_{k=1}^{L-1} (-1)^{k+1} \sum_{2 \leq i_1 < \cdots < i_k \leq L} \lfloor x^{1/\mathrm{lcm}(\{i_1, \ldots, i_k\})} \rfloor.$$

5. **Problem 11346** (Proposed by C. Hillar, 115(2), 2008). Let n be an integer greater than 1, and let $S = \{2, \ldots, n\}$. For each nonempty subset A of S, let $\pi(A) = \prod_{j \in A} j$. Prove that when k is a positive integer and $k < n$,

$$\prod_{i=k}^{n} \mathrm{lcm}(\{1, \ldots, \lfloor n/i \rfloor\}) = \gcd(\{\pi(A) : |A| = n - k\}).$$

6. **Problem 11699** (Proposed by B. Farhi, 120(7), 2013). Let $(a_k)_{k \geq 1}$ be a strictly increasing sequence of positive integers such that $\sum_{k=2}^{\infty} \frac{1}{a_k \log(a_k)}$ diverges. Prove that

$$\mathrm{lcm}(a_1, \ldots, a_k) = \mathrm{lcm}(a_1, \ldots, a_{k+1})$$

for infinitely many positive integers k.

5.14   A property of the product of consecutive primes

Problem 12236 (Proposed by N. Safaei, 128(2), 2021). Let p_k be the k-th prime number, and let $a_n = \prod_{k=1}^{n} p_k$. Prove that for $n \in \mathbb{N}$ every positive integer less than a_n can be expressed as a sum of at most $2n$ distinct divisors of a_n.

Discussion.
Recall that a *squarefree number* is a positive integer which is divisible by no perfect square other than 1. Clearly, all divisors of a_n are squarefree numbers. First, we prove an important property on the squarefree numbers in the following lemma, which will play a key role for solving the proposed problem.

Lemma. *Any integer $N > 2$ can be written as the sum of two distinct square-free numbers.*

Proof. Let $(1, 2, 3, 5, 6, \dots) = (s_j)_{j \geq 1}$ be the strictly increasing sequence of the squarefree numbers and let $S(N)$ be the number of squarefree numbers less than or equal to N. By the inclusion-exclusion principle,

$$S(N) = \sum_{k \leq \lfloor \sqrt{N} \rfloor} \mu(k) \left\lfloor \frac{N}{k^2} \right\rfloor$$

$$= N \sum_{k=1}^{\infty} \frac{\mu(k)}{k^2} - N \sum_{k \geq \lfloor \sqrt{N} \rfloor + 1} \frac{\mu(k)}{k^2} - \sum_{k \leq \lfloor \sqrt{N} \rfloor} \mu(k) \left(\frac{N}{k^2} - \left\lfloor \frac{N}{k^2} \right\rfloor \right)$$

$$\geq \frac{6N}{\pi^2} - N \sum_{k \geq \lfloor \sqrt{N} \rfloor + 1} \frac{1}{k^2} - \sum_{k \leq \lfloor \sqrt{N} \rfloor} 1 \geq \frac{6N}{\pi^2} - N \int_{\lfloor \sqrt{N} \rfloor}^{\infty} \frac{dx}{x^2} - \lfloor \sqrt{N} \rfloor$$

$$\geq \frac{6N}{\pi^2} - \frac{N}{\lfloor \sqrt{N} \rfloor} - \lfloor \sqrt{N} \rfloor \geq \frac{6N}{\pi^2} - 2\sqrt{N} - 1.$$

We assume now that $N > 2$ is an integer which cannot be written as the sum of two distinct squarefree numbers. It follows that the set

$$\{s_1, s_2, \dots, s_{S(N)}, N - s_1, N - s_2, \dots, N - s_{S(N)}\} \subset [1, N - 1]$$

contains at least $2S(N) - 1$ distinct integers, and therefore

$$N - 1 \geq 2S(N) - 1 \geq \frac{12N}{\pi^2} - 4\sqrt{N} - 2 - 1$$

which implies that $N < 361.7$. On the other hand, by direct calculation, any integer $2 < N < 361.7$ can be written as the sum of two distinct square-free numbers. Hence we have a contradiction and the proof of the lemma is complete. $\qquad\square$

Solution.

We prove our statement by induction with respect to n.

The base case. For $n = 1$, the divisors of $a_2 = 2$ are 1 and 2. Then for $N = 1$ we just need the divisor 1 and we are done.

The induction step. Let $n \geq 2$ and let N be a positive integer less than a_n. We divide N by p_n getting a quotient q and a remainder r such that

$$0 \leq q \leq \frac{N}{p_n} < \frac{a_n}{p_n} = a_{n-1} \quad \text{and} \quad 0 \leq r < p_n.$$

If $q \geq 1$, by the induction hypothesis, there exist $d_1 < d_2 < \cdots < d_m$ distinct divisors of a_{n-1} with $1 \leq m \leq 2(n-1)$ such that $q = d_1 + d_2 + \cdots + d_m$. If $q = 0$ then we let $m = 0$.

Moreover, if $r > 2$ then, by the previous lemma, there are two distinct square-free numbers $t_1 < t_2$ such that $r = t_1 + t_2$. Notice that t_1 and t_2 are divisors of a_{n-1} and we let $l = 2$. If $r = 1$ or $r = 2$ then r divides a_{n-1} and we let $t_1 = r$ and $l = 1$. If $r = 0$ then we let $l = 0$.

Finally we may conclude that N can be written as sum of at most $m + l \leq 2(n-1) + 2 = 2n$ distinct divisors of a_n:

$$N = qp_n + r = \sum_{j=1}^{m}(d_j p_n) + \sum_{j=1}^{l} t_j.$$

□

Remark. The number $\prod_{k=1}^{n} p_k$ is also called the nth *primorial number*. Primorial numbers (see the sequence A002110 in the OEIS (https://oeis.org/A002110) display many interesting properties. Here, we just mention the so-called *Bonse's inequality*: if $n \geq 4$ then

$$\prod_{k=1}^{n} p_k > p_{n+1}^2.$$

A proof is easily obtained by using the Bertrand's postulate, which implies that $p_{n+1} < 2p_n < 4p_{n-1}$ for $n \geq 2$. Therefore, for $n \geq 5$,

$$\prod_{k=1}^{n} p_k \geq 2 \cdot 3 \cdot 5 \cdot p_{n-1} \cdot p_n > (4p_{n-1}) \cdot (2p_n) > p_{n+1}^2.$$

The Bonse's inequality trivially holds for $n = 4$: $2 \cdot 3 \cdot 5 \cdot 7 > 11^2$.

Additional problems for practice.

1. **MM Problem 798** (Proposed by P. A. Lindstrom, 44(3), 1971). Show that

$$\lim_{x \to \infty} \left(\prod_{p \leq x} p \right)^{1/x} = e$$

where the product is taken over all primes less than or equal to x.

2. **Problem 11045** (Proposed by M. P. Singh, 110(9), 2003). Prove that when n is a sufficiently large positive integer there exists a finite set S of prime numbers such that

$$\sum_{p \in S} \left\lfloor \frac{n}{p} \right\rfloor = n.$$

3. **Problem 11705** (Proposed by J. Loase, 120(5), 2013). Let $C(n)$ be the number of distinct multisets of two or more primes that sum to n. Prove that $C(n+1) \geq C(n)$ for all n.

4. **Problem 12179** (Proposed by N. MacKinnon, 127(4), 2020). A positive integer n is *good* if its prime factorization $2^{a_1} 3^{a_2} \cdots p_m^{a_m}$ has the property that a_i/a_{i+1} is an integer whenever $1 \leq i < m$. Find all n greater than 2 such that $n!$ is good.

5. **Problem 12493** (Proposed by Z. Franco, 131(9), 2024). Compute $\lim_{n \to \infty} \sqrt[n]{d(1! \cdot 2! \cdots n!)}$, where $d(k)$ is the number of positive divisors of k.

5.15   A congruence for the integer part of a power of a cosine

Problem 12292 (Proposed by N. Osipov, 128(10), 2021). Let p be a prime number, and let $r = 1/(2\cos(4\pi/7))$. Evaluate $\lfloor r^{p+2} \rfloor$ modulo p.

Discussion.
Let $f(z) = \sum_{k \geq 0} a_k z^k$. By the *series multisection formula*, we find

$$\sum_{\substack{k \geq 0 \\ k \equiv d \, (\mathrm{mod} \, m)}} a_k z^k = \sum_{k \geq 0} \left(\frac{1}{m} \sum_{j=0}^{m-1} \omega_m^{(d-k)j} \right) a_k z^k = \frac{1}{m} \sum_{j=0}^{m-1} \omega_m^{dj} f(\omega_m^{-j} z)$$

where $\omega_m = e^{\frac{2\pi i}{m}}$. In particular, if $f(z) = (1+z)^{2n}$ then $a_k = \binom{2n}{k}$ and it follows that

$$\sum_{\substack{k=0 \\ k \equiv d \, (\mathrm{mod} \, m)}}^{2n} \binom{2n}{k} = \frac{1}{m} \sum_{j=0}^{m-1} \omega_m^{dj} (1 + \omega_m^{-j})^{2n}$$

$$= \frac{2^{2n}}{m} + \frac{1}{m} \sum_{j=1}^{m-1} \omega_m^{(d-n)j} (\omega_m^{\frac{j}{2}} + \omega_m^{-\frac{j}{2}})^{2n}$$

$$= \frac{2^{2n}}{m} + \frac{1}{m} \sum_{j=1}^{\frac{m-1}{2}} 2 \cos \left(\frac{\pi(2d-2n)j}{m} \right) \left(2 \cos \left(\frac{\pi j}{m} \right) \right)^{2n}.$$

Hence, by letting $m = 7$ and $d = n$, we obtain

$$\sum_{\substack{k=0 \\ k \equiv n \,(\text{mod } 7)}}^{2n} \binom{2n}{k} = \frac{2^{2n}}{7} + \frac{2}{7}\sum_{j=1}^{3}\left(2\cos\left(\frac{\pi j}{7}\right)\right)^{2n} = \frac{2^{2n}}{7} + \frac{2c_{2n}}{7} \qquad (5.16)$$

where

$$c_n = (2\cos(2\pi/7))^n + (2\cos(4\pi/7))^n + (2\cos(6\pi/7))^n.$$

Moreover, for $m = 7$ and $d = n + 3$, we get

$$\sum_{\substack{k=0 \\ k \equiv n+3 \,(\text{mod } 7)}}^{2n} \binom{2n}{k} = \frac{2^{2n}}{7} + \frac{1}{7}\sum_{j=1}^{3}(-1)^j\left(2\cos\left(\frac{\pi j}{7}\right)\right)^{2n+1} = \frac{2^{2n}}{7} + \frac{c_{2n+1}}{7}.$$

$$(5.17)$$

These identities (5.16) and (5.17) will be used to determine $\lfloor(2\cos(4\pi/7))^{-(p+2)}\rfloor$ modulo p.

Solution.

We show that for any prime p,

$$\left\lfloor(2\cos(4\pi/7))^{-(p+2)}\right\rfloor \equiv \begin{cases} 2 & \text{if } p \equiv 0,3,4 \quad (\text{mod } 7) \\ -5 & \text{if } p \equiv 2,5 \quad\quad (\text{mod } 7) \\ -12 & \text{if } p \equiv 1,6 \quad\quad (\text{mod } 7) \end{cases} \quad (\text{mod } p). \quad (5.18)$$

We divide the proof in several steps.

1) From the discussion above, for any integer $n \geq 0$,

$$c_{2n} = \frac{7}{2}\sum_{\substack{k=0 \\ k \equiv n \,(\text{mod } 7)}}^{2n} \binom{2n}{k} - 2^{2n-1} \quad \text{and} \quad c_{2n+1} = 7\sum_{\substack{k=0 \\ k \equiv n+3 \,(\text{mod } 7)}}^{2n} \binom{2n}{k} - 2^{2n}.$$

2) For $n \geq 0$,

$$c_{-n} = \frac{c_n^2 - c_{2n}}{2}.$$

Let $\alpha = 2\cos(2\pi/7)$, $\beta = 2\cos(4\pi/7)$, $\gamma = 2\cos(6\pi/7)$. Since $\alpha\beta\gamma = 1$ then

$$2c_{-n} = 2(\beta\gamma)^n + 2(\alpha\gamma)^n + 2(\alpha\beta)^n$$
$$= (\alpha^n + \beta^n + \gamma^n)^2 - (\alpha^{2n} + \beta^{2n} + \gamma^{2n}) = c_n^2 - c_{2n}.$$

3) For $n \geq 2$,

$$c_{-n} = \lfloor(2\cos(4\pi/7))^{-n}\rfloor + 1.$$

Note that $|(2\cos(4\pi/7))^{-1}| > 1$, and

$$-0.56 < (2\cos(6\pi/7))^{-1} < 0 < (2\cos(2\pi/7))^{-1} < 0.81.$$

Then the claim follows because for $n \geq 2$,

$$0 < (2\cos(2\pi/7))^{-n} + (2\cos(6\pi/7))^{-n} < (0.81)^2 + (0.56)^2 < 1.$$

4) For any prime $p \neq 2, 7$,

$$c_{p+2} \equiv \begin{cases} 3 & \text{if } p \equiv 2, 5 \pmod 7 \\ -4 & \text{if } p \equiv 1, 3, 4, 6 \pmod 7 \end{cases} \pmod p.$$

By Lucas' theorem, $\binom{p+1}{k} \equiv \binom{1}{k_1}\binom{1}{k_0}$ modulo p where $k = pk_1 + k_0$ in base p, and therefore

$$S := \sum_{\substack{k=0 \\ k \equiv \frac{p+1}{2}+3 \,(\text{mod } 7)}}^{p+1} \binom{p+1}{k} \equiv \sum_{\substack{k=0 \\ 2k \equiv p \,(\text{mod } 7)}}^{1} 1 + \sum_{\substack{k=p \\ 2k \equiv p \,(\text{mod } 7)}}^{p+1} 1$$

$$\equiv \begin{cases} 1 & \text{if } p \equiv 2, 5 \pmod 7 \\ 0 & \text{if } p \equiv 1, 3, 4, 6 \pmod 7 \end{cases} \pmod p.$$

Hence, since $p + 2$ is odd, it follows that $c_{p+2} = 7S - 2^{p+1} \equiv 7S - 4 \pmod p$, and we are done.

5) For any prime $p \neq 2, 7$,

$$c_{2p+4} \equiv \begin{cases} 10 & \text{if } p \equiv 3, 4 \pmod 7 \\ 17 & \text{if } p \equiv 2, 5 \pmod 7 \\ 38 & \text{if } p \equiv 1, 6 \pmod 7 \end{cases} \pmod p.$$

By Lucas' theorem, $\binom{2p+4}{k} \equiv \binom{2}{k_1}\binom{4}{k_0}$ modulo p, and therefore

$$S := \sum_{\substack{k=0 \\ k \equiv p+2 \,(\text{mod } 7)}}^{2p+4} \binom{2p+4}{k} \equiv 2\binom{4}{2} + 2 \sum_{\substack{k=0 \\ k \equiv p+2 \,(\text{mod } 7)}}^{4} \binom{4}{k}$$

$$\equiv \begin{cases} 12 & \text{if } p \equiv 3, 4 \pmod 7 \\ 14 & \text{if } p \equiv 2, 5 \pmod 7 \\ 20 & \text{if } p \equiv 1, 6 \pmod 7 \end{cases} \pmod p.$$

Hence, since $2p + 4$ is even, it follows that $c_{2p+4} = \frac{7}{2}S - 2^{2p+3} \equiv \frac{7}{2}S - 32$ (mod p) and we are done.

6) We directly check that (5.18) holds for $p = 2, 7$. For any other prime p, by the previous steps, we may conclude that

$$\left\lfloor (2\cos(4\pi/7))^{-(p+2)} \right\rfloor = c_{-(p+2)} - 1 = \frac{c_{p+2}^2 - c_{2p+4}}{2} - 1$$

$$\equiv \begin{cases} \frac{(-4)^2 - 10}{2} - 1 & \text{if } p \equiv 3, 4 \pmod 7 \\ \frac{3^2 - 17}{2} - 1 & \text{if } p \equiv 2, 5 \pmod 7 \\ \frac{(-4)^2 - 38}{2} - 1 & \text{if } p \equiv 1, 6 \pmod 7 \end{cases} \pmod p.$$

and the formula (5.18) is completely confirmed. □

Remark. $\{c_n\}_{n\geq 0}$ is the sequence A094648 in the OEIS (https://oeis.org/A094648):

$$3, -1, 5, -4, 13, -16, 38, -57, 117, -193, 370, -639, 1186, -2094, \ldots$$

and it satisfies the recurrence formula $c_{n+3} = -c_{n+2} + 2c_{n+1} + c_n$. By using a similar approach, we can show that for any prime p,

$$c_p \equiv -1 \pmod{p}, \quad c_{2p} \equiv 5 \pmod{p}, \quad \lfloor (2\cos(4\pi/7))^{-p} \rfloor \equiv -3 \pmod{p}.$$

Additional problems for practice.

1. **Putnam 1986-A2.** What is the units (i.e., rightmost) digit of $\left\lfloor \frac{10^{20000}}{10^{100}+3} \right\rfloor$?

2. **Problem E2766** (Proposed by I. Borosh and D. Hensley, 86(3), 1979). Let r be a positive rational number not an integer. Prove that there are infinitely many positive integers n such that $\lfloor nr \rfloor$ is a prime.

3. **Problem 12006** (Proposed by J. D. Lee, 124(10), 2017). When n is an integer and $n \geq 2$, let $a_n = \lceil n/\pi \rceil$ and $b_n = \lceil \csc(\pi/n) \rceil$. The sequences $a_2, a_3, \ldots$ and $b_2, b_3, \ldots$ are, respectively,

$$1, 1, 2, 2, 2, 3, 3, 3, 4, 4, 4, 5, 5, 5, 6, 6, 6, 7, 7, 7, 8, 8, 8, 8, 9, \ldots$$

and

$$1, 2, 2, 2, 2, 3, 3, 3, 4, 4, 4, 5, 5, 5, 6, 6, 6, 7, 7, 7, 8, 8, 8, 8, 9, \ldots.$$

They differ when $n = 3$. Are they equal for all larger n?

4. **Problem 10743** (Proposed by C. Popescu, 106(8), 1999). Let $R = \sum(-1)^i \binom{n}{i}$, where the sum is taken over all $i \in \{0, 1, \ldots, n-1\}$ such that $i+1$ is a quadratic residue modulo p, and let $N = \sum(-1)^j \binom{n}{j}$, where the sum is taken over all $j \in \{0, 1, \ldots, n-1\}$ such that $j+1$ is a quadratic nonresidue modulo p. Prove that exactly one of R and N is divisible by p.

5. **Problem 10852** (Proposed by C. Popescu, 108(2), 2001). Given a prime number $p \equiv 7 \pmod{8}$ evaluate

$$\sum_{k=1}^{(p-1)/2} \left\lfloor \frac{k^2}{p} + \frac{1}{2} \right\rfloor.$$

6. **Problem 11333** (Proposed by R. Tauraso, 114(10), 2007). Let α and β be positive irrational numbers. Show that for any positive integer n,

$$\sum_{k=0}^{\lfloor n/\alpha \rfloor - 1} \left\lfloor \frac{\lceil (k + \{n/\alpha\})\alpha \rceil}{\beta} \right\rfloor = \sum_{k=0}^{\lfloor n/\beta \rfloor - 1} \left\lfloor \frac{\lceil (k + \{n/\beta\})\beta \rceil}{\alpha} \right\rfloor$$

where $\{x\}$ denotes the fractional part of x.

7. **Problem 11428** (Proposed by W. Blumberg, 116(4), 2009). Let p be a prime that is congruent to 3 mod 4, and let a and q be integers, with $p \nmid q$. Show that

$$\sum_{k=1}^{p} \left\lfloor \frac{qk^2 + a}{p} \right\rfloor = 2a + 1 + \sum_{k=1}^{p} \left\lfloor \frac{qk^2 - a - 1}{p} \right\rfloor.$$

8. **Problem 11728** (Proposed by W. Blumberg, 120(8), 2013). Let p be a prime congruent to 7 modulo 8. Prove that

$$\sum_{k=1}^{p} \left\lfloor \frac{k^2 + k}{p} \right\rfloor = \frac{2p^2 + 3p + 7}{6}.$$

9. **Problem 12252** (Proposed by N. Q. Minh, 128(5), 2021). Let b, m, and n be positive integers with $b \geq 2$. Prove

$$\sum_{\substack{k=1 \\ b \nmid k}}^{b^n - 1} \left\lceil \frac{\lfloor n - \log_b(k) \rfloor}{m} \right\rceil = 1 + \left\lfloor \frac{b^{m-1}(b^{n-1} - 1)(b - 1)}{b^m - 1} \right\rfloor.$$

5.16   A variant of the Collatz map

Problem 12426 (Proposed by R. Kaufman and J. Lagarias, 130(10), 2023).
(a) The $3n + 1$ function takes n to $n/2$ if n is even and to $(3n + 1)/2$ if n is odd. Show that for every positive integer m there exists a positive integer a such that am reaches 1 upon iteration of the $3n + 1$ function.
(b) Show the same result for the $5n + 1$ function, defined by replacing $3n + 1$ by $5n + 1$ in the definition in (a).

Discussion.
Let us consider a more general function

$$T_b(n) = \begin{cases} \dfrac{n}{2} & \text{if } n \text{ is even,} \\[2ex] \dfrac{bn + 1}{2} & \text{if } n \text{ is odd,} \end{cases}$$

where $b > 1$ is an odd integer, and we try to solve the problem for T_b. Then (a) and (b) follow immediately by setting $b = 3$ and $b = 5$.

Solution.
Given the positive integer m, we rewrite it as the product of a power of two 2^k and a positive odd integer d. Since $\gcd(2, bd) = 1$, by Euler's theorem, it follows that

$$2^{\varphi(bd)} \equiv 1 \pmod{bd}$$

where φ denotes the totient function. Hence

$$a := \frac{2^{\varphi(bd)} - 1}{bd}$$

is an odd integer, and, for $k \geq 1$,

$$T_b^{(k+\varphi(bd))}(am) = T^{(k+\varphi(bd))}(2^k ad) = T_b^{(\varphi(bd))}(ad)$$

$$= T_b^{(\varphi(bd)-1)}\left(\frac{bad + 1}{2}\right)$$

$$= T_b^{(\varphi(bd)-1)}\left(2^{\varphi(bd)-1}\right) = 1.$$

□

Remark. The Collatz conjecture is one of the long-standing open problems of mathematics. It states that starting from any positive integer n, iterations of the function $T_3(n)$ will eventually reach the number 1. The conjecture has been verified for all $n < 2^{70}$ (last update 2024), but a fool proof remains unapproachable at present. Regarding the Collatz conjecture, P. Erdős is quoted as saying: "Mathematics is not yet ready for such problems". As a survey on this subject, we recommend the reading of the collection of papers [60].

Additional problems for practice.

1. **MM Problem 1343** (Proposed by R. L. Graham, 63(2), 1990). What is the behavior of the recursive sequence defined by $x_{n+2} = (1 + x_{n+1})/x_n$ with x_0, x_1 arbitrary positive numbers? Answer the same question for $x_{n+3} = (1 + x_{n+1} + x_{n+2})/x_n$.

2. **MM Problem 1459** (Proposed by D. M. Bloom, 67(5), 1994). The now notorious *Newman-Conway sequence* $1, 1, 2, 2, 3, 4, 4, 4, 5, 6, 7, 7, \ldots$ is defined by the recurrence $P(1) = P(2) = 1$, $P(n) = P(P(n-1)) + P(n - P(n-1))$ for $n \geq 3$. Richard Guy wrote: "I have an earlier manuscript of Conway in which he has written $P(2^k) = 2^{k-1}$ (easy), $P(2n) \leq 2P(n)$ (hard)". Prove Conway's "hard" inequality: $P(2n) \leq 2P(n)$.

3. **Problem 10568** (Proposed by D. E. Knuth, 104(1), 1997). Let n be a nonnegative integer. The sequence defined by $x_0 = n$ and $x_{k+1} = x_k - \lceil \sqrt{x_k} \rceil$ for $k \geq 0$ converges to 0. Let $f(n)$ be the number of steps required; i.e., $x_{f(n)} = 0$ but $x_{f(n)-1} > 0$. Find a closed form for $f(n)$.

4. **Problem 10927** (Proposed by J. C. Lagarias, E. M. Rains, and N. J. A. Sloane, 109(2), 2002). Define a sequence $(a_n)_{n \geq 1}$ by letting $a_1 = 1$, $a_2 = 2$, and $a_3 = 3$, and for $n > 3$ letting a_n be the smallest integer among those not already used such that $\gcd(a_{n-1}, a_n) \geq 3$. The sequence begins $1, 2, 3, 6, 9, 12, 4, 8, 16, 20, 5, 10, 15, \ldots$. Prove that it is a permutation of $\mathbb{N}$.

5. **Problem 10974** (Proposed by S. Marivani, 109(9), 2002). The digital root $\rho(n)$ of a positive integer n is the eventual image of n under the mapping that carries an integer n to the sum of its base-ten digits. Thus $\rho(10974) = \rho(21) = 3$. Find $\rho(F_n)$, where F_n is the nth Fibonacci number.

6. **Problem 12237** (Proposed by D. E. Knuth, 128(3), 2021). Let $x_0 = 1$ and $x_{n+1} = x_n + \lfloor x_n^{3/10} \rfloor$ for $n \geq 0$. What are the first 40 decimal digits of x_n when $n = 10^{100}$?

Chapter 6

Potpourri

In this chapter, we collect 14 problems which range from geometry and number theory to combinatorics and probability. The solutions touch on various topics, including asymptotic estimates of sums involving the floor function, characterization of palindrome numbers, test for existence of real zeros, enumerations of rationals, algorithmic puzzles and Markov chains. For the solutions that involve more sophisticated techniques, we include references and informative backgrounds.

6.1 Fixed point of the distance to the boundary

Problem 10998 (Proposed by R. Tauraso, 110(3), 2003). Let D be a nonempty, open, connected, and relatively compact set in a metric space X with metric d. Prove that if f is a continuous map from D into D such that $f(D)$ is open, then there exists a point $x_0 \in D$ such that

$$d(x_0, \partial D) = d(f(x_0), \partial D).$$

Discussion.
In a metric space (X, d), the distance to a nonempty set $S \subset X$,

$$d(x, S) = \inf\{d(x, s) \ : \ s \in S\}$$

is a continuous function. Indeed, by the triangle inequality, $d(x, s) \leq d(x, y) + d(y, s)$, and by taking infimum for $s \in S$ we find $d(x, S) \leq d(x, y) + d(y, S)$. Similarly, $d(y, S) \leq d(y, x) + d(x, S)$. Thus,

$$|d(x, S) - d(y, S)| \leq d(x, y).$$

Let us consider the function

$$F(x) = d(f(x), \partial D) - d(x, \partial D). \tag{6.1}$$

DOI: 10.1201/9781003607809-6

Then also F is continuous in X. Combining this fact with the property of connectedness, we give two proofs. The first proof relies on the nonexistence of separation of D, and the second proof applies the intermediate value theorem to F in $\overline{D}$.

Solution I.

We prove by contradiction. Assume that there is no such point x_0 exists. Let $F(x)$ be given by (6.1). Define

$$D^+ := \{x \in D : F(x) > 0\} \quad \text{and} \quad D^- := \{x \in D : F(x) < 0\}.$$

Then D^+ and D^- are open, disjoint, and $D^+ \cup D^- = D$. Since D is connected, one of these two sets must be empty. By the assumption that D is relatively compact, so $\overline{D}$ is compact, hence there is a point $c \in \overline{D}$ (it is not unique in general) such that

$$d(c, \partial D) = d_M := \max_{x \in \overline{D}} d(x, \partial D).$$

Moreover, $d_M > 0$ because D is open and non-empty. Thus c stays inside D and since it reaches the maximal distance from the boundary, it belongs to D^-. Therefore the set D^+ has to be empty and $D = D^-$. This fact gives that for all $r > 0$, $f^{-1}(D_r) \subset D_r$, where D_r is the compact set $\{x \in D : d(x, \partial D) \geq r\}$.

We now claim that the range set $f(D)$ is closed in D. Let $\{y_n\}_{n \geq 0}$ be a sequence in $f(D)$ which converges to $y \in D$ and let r be the distance of the compact set $\{y_n\}_{n \geq 0} \cup \{y\}$ from the boundary ∂D. Therefore $\{y_n\}_{n \geq 0} \cup \{y\} \subset D_r$ and the inclusion $f^{-1}(D_r) \subset D_r$ implies that if $\{x_n\}_{n \geq 0}$ is a sequence in D such that $f(x_n) = y_n$ then it is contained in D_r and, since D_r is compact, it converges, up to a subsequence, to some $x \in D$. Hence, by the continuity of f, $y = f(x) \in f(D)$. So $f(D)$ is a non-empty open and closed subset of the connected set D that is $f(D) = D$. In particular there exists $x \in D$ such that $f(x) = c$ and

$$d_M = d(c, \partial D) = d(f(x), \partial D) < d(x, \partial D)$$

which is against the definition of d_M. $\qquad\square$

Solution II.

Let $F(x)$ be defined by (6.1). We show that there exist $y, z \in D$ such that $F(y) \leq 0$ and $F(z) \geq 0$, respectively. Since D is connected, the existence of x_0 follows from the intermediate value theorem.

In view of that $d(x, \partial D)$ is continuous on X and $\overline{D}$ is compact, there exists a $y \in \overline{D}$ with $d(y, \partial D)$ maximal. Since D is nonempty and open, and point in D will have some positive distance to the boundary, and thus $y \in D$. Since $f(y) \in D$, we have $d(f(y), \partial D) \leq d(y, \partial D)$. Therefor $F(y) = d(f(y), \partial D) - d(y, \partial D) \leq 0$.

On the other hand, for the existence of z with $F(z) \geq 0$ we consider two cases. If $f(D) = D$, then there is some $z \in D$ with $f(z) = y$ and

$$F(z) = d(f(z), \partial D) - d(z, \partial D) = d(y, \partial D) - d(z, \partial D) \geq 0.$$

If $F(D) \neq D$, since D is connected and $f(D)$ is open, then there exists $u \in D \cap \partial f(D)$ such that $u \notin f(D)$. Take a sequence $z_n \in D$ such that $f(z_n) \to u$ as $n \to \infty$. Since $\overline{D}$ is compact, without loss of generality (otherwise we go to a subsequence), we may assume that $z_n \to v \in \overline{D}$. Since $u \notin f(D)$, v must be in ∂D. Thus $d(z_n, \partial D) \to d(v, \partial D) = 0$. However, $d(f(z_n), \partial D) \to d(u, \partial D) > 0$. This implies that for large enough n we have $F(z_n) > 0$. $\square$

Remark. When D is the open unit disc centered in the origin and d is the Euclidean distance in $\mathbb{R}^2$ then the statement is: *if f is a continuous self-map of D such that $f(D)$ is open in D then there exists a rotation r around the origin such that $r \circ f$ has a fixed point in D.* If $f(D)$ is not open then the statement is false. In this case, $f(x_1, x_2) = (1 + x_1^2 + x_2^2)/2$ gives a counterexample.

Additional problems for practice.

1. **Problem E3428** (Proposed by A. B. Boghossian, 98(3), 1991). Let S be a non-empty interval on the real line. Let $f : S \to S$ be a continuous function having the property that for each $x \in S$ there exists a positive integer $n = n(x)$ with $f^n(x) = x$, where f^n denotes the nth iterate of f. For given S characterize all such functions.

2. **Problem 10442** (Proposed by R. Bielawski, 102(3), 1995). Let f be a continuous function from the unit disc D in $\mathbb{R}^2$ to itself such that $f \circ f$ is the identity on D and f is the identity on the unit circle ∂D. Show that f is the identity on D.

3. **Problem 10991** (Proposed by R. Mortini, 110(2), 2003). For complex $a, z \in D = \{s : |s| < 1\}$, let $F(a, z) = \frac{a+z}{1+\bar{a}z}$ be a map of D onto D. Let ρ denote the pseudohyperbolic distance, defined by $\rho(a, b) = \left| \frac{a-b}{1-\bar{a}b} \right|$. (a) Prove that there exists a function $C : D \to \mathbb{R}^+$ so that

$$\rho(F(a, z), F(b, z)) \leq C(z)\rho(a, b)$$

for every $a, b, z \in D$.

(b) Find the minimal value of $C(z)$ for which this bound holds.

4. **Problem 11070** (Proposed by R. Tauraso, 111(3), 2004). Let f and g be two commuting analytic maps from a non-empty open connected set $D \subset \mathbb{C}$ into D. Suppose that $z_0 \in D$ be a fixed point of both f and g, and that neither $f'(z_0)$ nor $g'(z_0)$ is a root of unity. Suppose also there exists an integer $N \geq 1$ such that $f^{(k)}(z_0) = g^{(k)}(z_0) = 0$ for $1 \leq k \leq N-1$, while $f^{(N)}(z_0) = g^{(N)}(z_0) \neq 0$. Prove that the restriction of f and g to D are equal.

6.2   A variant alternating series with the floor function

Problem 11809 (Proposed by O. Kouba, 121(10), 2014). Let $\{a_n\}_{n\geq 1}$ be a sequence of real numbers.
(a) Suppose $\{a_n\}_{n\geq 1}$ consists of nonnegative numbers and is nonincreasing, and $\sum_{n=1}^{\infty} a_n/\sqrt{n}$ converges. Prove that

$$\sum_{n=1}^{\infty} (-1)^{\lfloor \sqrt{n} \rfloor} a_n$$

converges.
(b) Find a nonincreasing sequence $\{a_n\}_{n\geq 1}$ of positive numbers such that $\sqrt{n}a_n \to 0$ and

$$\sum_{n=1}^{\infty} (-1)^{\lfloor \sqrt{n} \rfloor} a_n$$

diverges.

Discussion.
Since the proposed series is a variant alternating series, we naturally convert it into an alternating series. In fact, for $m^2 \leq n < (m+1)^2$, we have $(-1)^{\lfloor \sqrt{n} \rfloor} = (-1)^m$. Then

$$\sum_{n=1}^{\infty} (-1)^{\lfloor \sqrt{n} \rfloor} a_n = \sum_{m=1}^{\infty} (-1)^m \left(\sum_{k=m^2}^{(m+1)^2 - 1} a_k \right).$$

However, the assumption that $\sum_{n=1}^{\infty} a_n/\sqrt{n}$ converges seems irrelevant to justifying use of the alternating series test.
On the other hand, applying Abel's summation formula yields

$$\sum_{k=1}^{n} \frac{a_k}{\sqrt{k}} = \sum_{k=1}^{n-1} \left(\sum_{i=1}^{k} \frac{1}{\sqrt{i}} \right) (a_k - a_{k+1}) + \left(\sum_{k=1}^{n} \frac{1}{\sqrt{k}} \right) a_n,$$

and

$$\sum_{k=1}^{n} (-1)^{\lfloor \sqrt{k} \rfloor} a_k = \sum_{k=1}^{n-1} S_k(a_k - a_{k+1}) + S_n\, a_n. \qquad (6.2)$$

where $S_n := \sum_{k=1}^{n} (-1)^{\lfloor \sqrt{k} \rfloor}$. It is well-known that

$$\sum_{k=1}^{n} \frac{1}{\sqrt{k}} = O(\sqrt{n}).$$

Thus, if we establish that $S_n = O(\sqrt{n})$, combining with (6.2), then we can show that $\sum_{k=1}^{n} (-1)^{\lfloor \sqrt{k} \rfloor} a_k$ indeed is a Cauchy sequence. So it converges.

Solution.

(a) We first note that $\lim_{n\to\infty} \sqrt{n}a_n = 0$. Otherwise there exist $\varepsilon > 0$ and $n_0 \geq 1$, such that for all $n \geq n_0$, $\sqrt{n}a_n > \varepsilon$ and

$$\sum_{k=n_0}^{n} \frac{a_k}{\sqrt{k}} > \varepsilon \sum_{k=n_0}^{n} \frac{1}{k},$$

which contradicts the fact that $\sum_{n=1}^{\infty} a_n/\sqrt{n}$ converges.

Next, observe that, for any integer $l \geq 0$,

$$(-1)^{\lfloor \sqrt{k} \rfloor} = \begin{cases} 1, & \text{for } (2l)^2 \leq k < (2j+1)^2, \\ -1, & \text{for } (2l+1)^2 \leq k < (2j+2)^2. \end{cases}$$

Let $a = \lfloor \sqrt{n} \rfloor$. Then

$$S_n = \sum_{k=1}^{n} (-1)^{\lfloor \sqrt{k} \rfloor} = \sum_{k=1}^{a-1} (-1)^k ((k+1)^2 - k^2) + (-1)^a (n+1-a^2)$$

$$= -(-1)^a a - 1 + (-1)^a (n+1-a^2),$$

and so $|S_n| \leq |n - a^2 - a| + 2 \leq a + 2 \leq \sqrt{n} + 2$. For $1 \leq n \leq m$, by (6.2), we find

$$\sum_{k=n}^{m} (-1)^{\lfloor \sqrt{k} \rfloor} a_k = \sum_{k=n}^{m-1} S_k(a_k - a_{k+1}) - S_{n-1}a_n + S_m a_m. \tag{6.3}$$

Since $\{a_n\}_{n\geq 1}$ is a nonincreasing sequence, we have

$$\sum_{k=n}^{m-1} |S_k(a_k - a_{k+1})| \leq \sum_{k=n}^{m-1} (\sqrt{k}+2)(a_k - a_{k+1})$$

$$= \sum_{k=n}^{m} (\sqrt{k} - \sqrt{k-1})a_k + (\sqrt{n-1}+2)a_n - (\sqrt{m}+2)a_m$$

$$\leq \sum_{k=n}^{m} \frac{a_k}{\sqrt{k}} + (\sqrt{n-1}+2)a_n - (\sqrt{m}+2)a_m.$$

By (6.3), this yields

$$\left| \sum_{k=n}^{m} (-1)^{\lfloor \sqrt{k} \rfloor} a_k \right| \leq \sum_{k=n}^{m-1} |S_k(a_k - a_{k+1})| + |S_{n-1}|a_n + |S_m|a_m$$

$$\leq \sum_{k=n}^{m} \frac{a_k}{\sqrt{k}} + 2(\sqrt{n-1}+2)a_n,$$

which implies that $\{\sum_{k=1}^{n}(-1)^{\lfloor\sqrt{k}\rfloor}a_k\}_{n\geq1}$ is a Cauchy sequence because each term on the right hand side goes to zero as n and m go to infinity.
(b) For $n \geq 1$, let

$$a_n = \frac{1}{x_n \ln(1 + x_n)} \qquad \text{with} \quad x_n = \lfloor(\sqrt{n}+1)/2\rfloor.$$

Then $\{a_n\}_{n\geq1}$ is a nonincreasing sequence of positive numbers such that for $n > 1$,

$$0 < \sqrt{n}a_n < \frac{2\sqrt{n}}{(\sqrt{n}-1)\ln(\sqrt{n}+1)/2)}$$

because $x - 1 < \lfloor x \rfloor$. Hence $\lim_{n\to\infty}\sqrt{n}a_n = 0$. Moreover

$$(-1)^{\lfloor\sqrt{n}\rfloor} = -1 \quad \text{and} \quad x_n = m \quad \text{for } (2m-1)^2 \leq n < (2m)^2,$$
$$(-1)^{\lfloor\sqrt{n}\rfloor} = +1 \quad \text{and} \quad x_n = m \quad \text{for } (2m)^2 \leq n < (2m+1)^2,$$

which implies that for $N < (2m+1)^2$,

$$\sum_{n=1}^{N}(-1)^{\lfloor\sqrt{n}\rfloor}a_n \geq \sum_{n=1}^{m-1}\frac{2}{n\ln(1+n)} - \frac{(4m-1)}{m\ln(1+m)}.$$

Therefore the series diverges because

$$\sum_{n=1}^{\infty}(-1)^{\lfloor\sqrt{n}\rfloor}a_n \geq \sum_{n=1}^{\infty}\frac{2}{n\ln(1+n)} - \lim_{m\to\infty}\frac{(4m-1)}{m\ln(1+m)} = +\infty.$$

$\square$

Remark. Stimulated by Hardy's counterexample $\sum_{n=2}^{\infty}(-1)^n/(\sqrt{n}+(-1)^n)$ to the alternating series test, we give another example for part (b). Let

$$a_n = f(\lfloor\sqrt{n}\rfloor) \quad \text{where} \quad f(n) = \frac{1}{(2n+1-(-1)^n)\ln(n)}.$$

Then

$$\sum_{n=(2k)^2}^{(2k+1)^2-1}(-1)^{\lfloor\sqrt{n}\rfloor}a_n = \frac{4k+1}{4k\ln(2k)} = \frac{1}{\ln(2k)} + \frac{1}{4k\ln(2k)},$$

$$\sum_{n=(2k+1)^2}^{(2k+2)^2-1}(-1)^{\lfloor\sqrt{n}\rfloor}a_n = -\frac{4k+3}{4(k+1)\ln(2k+1)}$$

$$= -\frac{1}{\ln(2k+1)} + \frac{1}{4(k+1)\ln(2k+1)}.$$

Adding these two sums yields

$$\sum_{n=(2k)^2}^{(2k+2)^2-1} (-1)^{\lfloor\sqrt{n}\rfloor} a_n = \frac{1}{\ln(2k)} + \frac{1}{4k\ln(2k)} - \frac{1}{\ln(2k+1)} + \frac{1}{4(k+1)\ln(2k+1)}$$

$$\geq \frac{1}{2(k+1)\ln(2k+1)}.$$

The divergence of the series $\sum_{k=1}^{\infty} \frac{1}{2(k+1)\ln(2k+1)}$ implies that

$$\lim_{k\to\infty} \sum_{n=1}^{4k^2-1} (-1)^{\lfloor\sqrt{n}\rfloor} a_n = \infty.$$

So the series $\sum_{n=1}^{\infty} (-1)^{\lfloor\sqrt{n}\rfloor} a_n$ diverges.

Additional problems for practice.

1. For $n \geq 2$, let

$$a_n = \frac{1}{\sqrt{n}\ln n}.$$

show that $\sum_{n=2}^{\infty} (-1)^{\lfloor\sqrt{n}\rfloor} a_n$ diverges.

2. Show the following series

$$\frac{1}{\sqrt{1}} - \frac{1}{\sqrt{2}} - \frac{1}{\sqrt{3}} + \frac{1}{\sqrt{4}} + \frac{1}{\sqrt{5}} + \frac{1}{\sqrt{6}} - \frac{1}{\sqrt{7}} - \frac{1}{\sqrt{8}} - \frac{1}{\sqrt{9}} - \frac{1}{\sqrt{10}} + \cdots$$

has a bounded partial sums, but it does not converge. Show the associated pair parenthesized series

$$\left(\frac{1}{\sqrt{1}} - \frac{1}{\sqrt{2}} - \frac{1}{\sqrt{3}}\right) + \left(\frac{1}{\sqrt{4}} + \frac{1}{\sqrt{5}} + \frac{1}{\sqrt{6}} - \frac{1}{\sqrt{7}} - \frac{1}{\sqrt{8}} - \frac{1}{\sqrt{9}} - \frac{1}{\sqrt{10}}\right) + \cdots$$

converges.

3. **Problem 6015** (Proposed by H. D. Ruderman, 83(6), 1976). Prove that the following series converges

$$\sum_{n=1}^{\infty} \frac{(-1)^{\lfloor n\sqrt{2}\rfloor}}{n}.$$

Estimate its value.

4. **Open Problem.** Let α be an irrational number. Define $S_n(\alpha) = \sum_{k=1}^{n} (-1)^{\lfloor k\alpha\rfloor}$. In the featured solution of Problem 6015, D. Borwein obtained $S_n(\sqrt{2}) = O(\ln n)$.

(a) Is there an irrational number α such that $\sum_{n=1}^{\infty} \frac{(-1)^{\lfloor n\alpha\rfloor}}{n}$ diverges?

    (b) Estimate $S_n(e)$.

    (c) Does $\sum_{n=1}^{\infty} \frac{(-1)^{\lfloor ne \rfloor}}{n}$ converge?

5. **Problem 11384** (Proposed by M. Omarjee, 115(8), 2008). Let p_n denote the nth prime. Show that

$$\sum_{n=1}^{\infty} \frac{(-1)^{\lfloor \sqrt{n} \rfloor}}{p_n}$$

converges. *Remark.* The published solution [2010, 745] has a minor defection. You can solve this problem by using the result in Part (a) and the elementary bound $p_n > \frac{1}{4} n \ln n$.

6. **Problem 11999** (Proposed by O. Kouba, 124(8), 2017). Evaluate

$$\sum_{k=1}^{\infty} \frac{(-1)^{\lfloor \sqrt{k} + \sqrt{k+1} \rfloor}}{k(k+1)}.$$

7. **Problem 12486** (Proposed by Z. Franco and H. Ohtsuka, 131(8), 2024). Let $\phi = (1 + \sqrt{5})/2$, and let n be a positive integer. Let $f^1(x) = \lceil x\phi^n \rceil$ and $g^1(x) = \lfloor x\phi^n \rfloor$, and let $f^m(x) = f^{m-1}(f^1(x))$ and $g^m(x) = g^{m-1}(g^1(x))$ for $m > 1$. Evaluate $f^m(1)$ and $g^m(1)$.

6.3 A real analytic function with no zeros

Problem 12000 (Proposed by M. Sawhney, 124(8), 2017). Let $H_k = \sum_{i=1}^{k} 1/i$. Prove that the function $f : \mathbb{R} \to \mathbb{R}$ defined by

$$f(x) = 1 + \sum_{n=1}^{\infty} \frac{x^n}{\prod_{k=1}^{n} H_k}$$

has no real zeroes.

Discussion.
Note that the radius of convergence for the power series $f(x) = 1 + \sum_{n=1}^{\infty} a_n x^n$ is

$$R = \lim_{n \to \infty} \left| \frac{a_n}{a_{n+1}} \right| = \lim_{n \to \infty} H_{n+1} = +\infty.$$

Therefore f is well-defined and continuous in $\mathbb{R}$. Clearly, $f(x) > 0$ for all $x \geq 0$. To prove $f(x)$ has no real zeros, it suffices to show that $f(x) > 0$ for all $x < 0$ as well.

Solution.
We prove by contradiction. Assume that f has at least a real zero and let

$$t := \sup\{x \in \mathbb{R} \,:\, f(x) = 0\}.$$

Then $t < 0$ is well-defined. Moreover, by continuity, $f(t) = 0$ and $f(x) > 0$ for $x \in (t, 0]$. Hence

$$I := \int_t^0 \frac{f(x) - f(t)}{x - t}\, dx = \int_t^0 \frac{f(x)}{x - t}\, dx > 0. \tag{6.4}$$

On the other hand, since power series can be integrated term-by-term inside the interval of convergence, we have

$$I = \int_t^0 \sum_{n=1}^\infty \frac{x^n - t^n}{(x - t) \prod_{k=1}^n H_k}\, dx = \sum_{n=1}^\infty \frac{1}{\prod_{k=1}^n H_k} \int_t^0 \frac{x^n - t^n}{x - t}\, dx \quad (s := x/t)$$

$$= -\sum_{n=1}^\infty \frac{t^n}{\prod_{k=1}^n H_k} \int_0^1 \frac{s^n - 1}{s - 1}\, ds = -\sum_{n=1}^\infty \frac{t^n}{\prod_{k=1}^n H_k} \sum_{k=0}^{n-1} \int_0^1 s^k\, ds$$

$$= -\sum_{n=1}^\infty \frac{t^n H_n}{\prod_{k=1}^n H_k} = -t\left(1 + \sum_{n=2}^\infty \frac{t^{n-1}}{\prod_{k=1}^{n-1} H_k}\right) = -tf(t) = 0$$

which contradicts (6.4). This proves that f has no real zeros. $\qquad\square$

Remark. Determine the existence of real zeros for a function is pretty challenge. The classical Sturm's theorem relates the number of real zeros of a polynomial in an interval to the number of sign alternations in a sequence of polynomial division-like calculations (see for example [70, p. 171]). This enables us to count the total number of real zeros and to find these zeros numerically. It is worth noting that Vieta's formula and the power sum symmetric functions of zeros can be used to find *necessary* conditions for all the zeros to be real. For example, let $f(x) = x^3 + bx^2 + cx + d$ with distinct zeros r_1, r_2, r_3. Define

$$\Delta(f) = \prod_{i<j} (r_i - r_j)^2$$

and the Vandermonde matrix

$$V = \begin{bmatrix} 1 & r_1 & r_1^2 \\ 1 & r_2 & r_2^2 \\ 1 & r_3 & r_3^2 \end{bmatrix}.$$

It is well-known that $\det(V) = \prod_{1 \le i < j \le 3} (r_i - r_j)$. Define $s_k = r_1^k + r_2^k + r_3^k$, the k-th power sums. Then

$$V^T V = \begin{bmatrix} 3 & s_1 & s_2 \\ s_1 & s_2 & s_3 \\ s_2 & s_3 & s_4 \end{bmatrix}$$

is positive definite. Note that $\det(V^T V) = \Delta(f) > 0$. Moreover, the principal leading minor

$$\det \begin{bmatrix} 3 & s_1 \\ s_1 & s_2 \end{bmatrix} > 0. \tag{6.5}$$

By Vieta's formula, we have

$$s_1 = -b, \; s_2 = b^2 - 2c, \; s_3 = -b^3 + 3bc - 3d, \; s_4 = b^4 - 4b^2 c + 4bd + 2c^2.$$

Thus (6.5) yields $b^2 > 3c$, and $\det(V^T V) > 0$ implies that

$$\Delta(f) = b^2 c^2 - 4ac^3 - 4b^3 d - 27a^2 d^2 + 18abcd > 0.$$

These two inequalities are necessary for f having only real zeros. Along the same lines, the principal leading minors of $V^T V$ will yield $n - 1$ inequalities for n-th degree polynomials $f(x)$ having only real zeros.

Additional problems for practice.

1. Let $f(x) = x^n + a_1 x^{n-1} + a_2 x^{n-2} + \cdots + a_{n-1} x + a_n$. Show that
   (a) If $f(x)$ has only real zeros, then $(n-1)a_1^2 \geq 2na_2$.
   (b) If $f(x)$ has only positive zeros, then $na_n \leq a_1 a_{n-1}$.

2. **Putnam 1990-B5.** Is there an infinite sequence $a_0, a_1, a_2, \ldots$ of nonzero real numbers such that for $n = 1, 2, 3, \ldots$ the polynomial

   $$p_n(x) = a_0 + a_1 x + a_2 x^2 + \cdots + a_n x^n$$

   has exactly n distinct real roots?

3. **Putnam 2014-B4.** Show that for each positive integer n, all the roots of the polynomial

   $$\sum_{k=0}^{n} 2^{k(n-k)} x^k$$

   are real numbers.

4. **Putnam 2018-B2.** Let n be a positive integer, and let

   $$f_n(z) = n + (n-1)z + (n-2)z^2 + \cdots + z^{n-1}.$$

   Prove that f_n has no roots in the closed unit disk $\{z \in \mathbb{C} : |z| \leq 1\}$.

5. **Putnam 2022-A6.** Let n be a positive integer. Determine, in terms of n, the largest integer m with the following property: There exist real numbers $x_1, \ldots, x_{2n}$ with $-1 < x_1 < x_2 < \cdots < x_{2n} < 1$ such that the sum of lengths of the n intervals

   $$[x_1^{2k-1}, x_2^{2k-1}], [x_3^{2k-1}, x_4^{2k-1}], \cdots, [x_{2n-1}^{2k-1}, x_{2n}^{2k-1}]$$

is equal to 1 for all integers k with $1 \le k \le m$.

Hint: (Due to Evan Dummit) Define the polynomial

$$p(x) = (x - x_1)(x + x_2) \cdots (x - x_{2n-1})(x + x_{2n})(x + 1).$$

Let s_k and e_k denote the k-th power sum and the k-th elementary symmetric function of the roots of $p(x)$ respectively. Then the desired property becomes $s_{2k-1} = 0$ for $k = 1, \ldots, m$. Show that $e_{2k-1} = 0$ for $k = 1, \ldots, m$.

6. **Problem 6654** (Proposed by W. O. Egerland and C. E. Hansen, 98(2), 1991). Suppose ω is real, n is a positive integer greater than 1, and $a_1, a_2, \ldots, a_n$ are complex number with $|a_k| < 1$ for $k = 1, 2, \ldots, n$. Prove that the equation

$$e^{i\omega}(z - a_1)(z - a_2) \cdots (z - a_n) = z(1 - \bar{a}_1 z)(1 - \bar{a}_2 z) \cdots (1 - \bar{a}_n z)$$

has at least $n - 1$ roots on the unit circle.

7. **Problem 11226** (Proposed by F. Beaucoup, 113(5), 2006). Let $a_1, \ldots, a_n$ be real numbers, each greater than 1. If $n \ge 2$, show that there is exactly one solution in the interval $(0, 1)$ to

$$\prod_{j=1}^{n} (1 - x^{a_j}) = 1 - x.$$

8. **Problem 11801** (Proposed by D. Carter, 121(8), 2014). Let f be a polynomial in one variable with rational coefficients that has no nonnegative real root. Show that there is a nonzero polynomial g with rational coefficients such that the coefficients of $f \cdot g$ are positive.

9. **Problem 12354** (Proposed by S. Filipovski, 129(9), 2022). Let n and k be positive integers with $n \ge 3$, and let

$$p(x) = x^n + x^{n-1} + \cdots + x - k.$$

(a) Prove that the roots of $p(x)$ in the complex plane are simple.

(b) Prove that if $k > n$, then $p(x)$ has at least one root with negative real part and nonzero imaginary part.

10. **Problem 12420** (Proposed by K. H. Kim and K. Stolarsky, 130(9), 2023). For a polynomial P of positive degree, let P^* be P with its leading term deleted. Show that there are arbitrarily large integers N for which there is a polynomial P with integer coefficients such that P has exactly N zeros on the unit circle and P^* has at least $2N$ zeros on the unit circle.

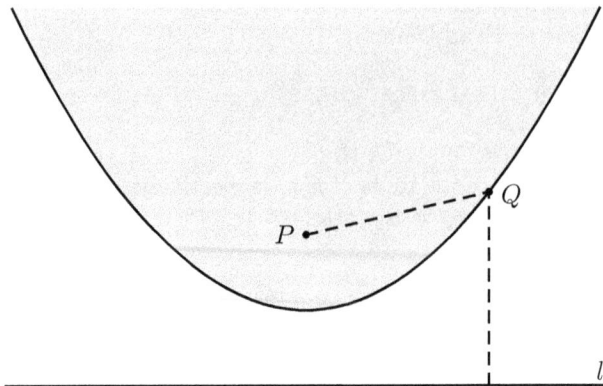

Figure 6.1: A parabola with focus P and directrix l

11. **Problem 12523** (Proposed by A. Stenger, 132(3), 2025). Let $x^n + a_{n-1}x^{n-1} + \cdots + a_1 x + a_0$ be a monic polynomial of degree n, with $n \geq 2$, all of whose zeros are real. Show that the difference between the largest and smallest zeros is at most

$$2\sqrt{\frac{n-1}{2n}a_{n-1}^2 - a_{n-2}},$$

and exhibit, for each $n \geq 2$, a polynomial of degree n for which there is a difference equal to this bound.

6.4   Finding a special half-line

Problem 11854 (Proposed by R. Tauraso, 122(7), 2015). In the Euclidean plane, given a finite number of points $P_1, \ldots, P_n$, and a finite number of lines $l_1, \ldots, l_m$, prove that there is a half-line h such that for any point $Q \in h$, for any $k \in \{1, \ldots, m\}$ and for any $j \in \{1, \ldots, n\}$, $d(Q, l_k)$, the distance from Q to the line l_k, is less than $d(Q, P_j)$, the distance from Q to the point P_j.

Discussion.
Recall that a parabola is the set of all points Q that are equidistant from a fixed point P, the focus, and a specific line l, the directrix.
As shown in Figure 6.1, $d(Q, P) < d(Q, l)$ indicates that Q lies in the interior (shaded region) of the parabola; and $d(Q, P) > d(Q, l)$ implies that Q lies outside of the parabola.

Given a line which is not parallel to the axis of symmetry of parabolas, we see that the line intersects the parabola at most twice. In the case intersecting the parabola twice, only a finite segment of the line lies in the interior of the parabola. This observation will guide us to single out the required half-line.

Solution.

The set

$$S_{jk} := \{Q \in \mathbb{R}^2 \; : \; d(Q, l_k) \geq d(Q, P_j)\}$$

is the closed interior of the parabola with focus P_j and directrix l_k (if $P_j \in l_k$ then S_{ij} degenerates to the line which intersect orthogonally l_k at P_j). Hence the problem is equivalent to showing that the set

$$\mathbb{R}^2 \setminus \bigcup_{(j,k)} S_{jk}$$

contains a half-line. Let r be a line which is not orthogonal to l_k for any $k \in \{1, \ldots, m\}$ (we can make such a choice because the set of lines is finite). Since the line r is not parallel to the axis of any parabola ∂S_{jk}, it follows that the intersection of r with S_{jk}, is a segment (possibly empty or just a single point). Therefore the line r certainly contains a half-line which is disjoint from the finite collection of those segments. □

Remark. Points and lines are the most basic objects in Euclidean geometry. The first of Euclid's five axioms for plane geometry is about the fundamental relationship between points and lines: any two points define a unique line. Given an arrangement of points, a line containing just two of them is called an *ordinary line*. The famous *Sylvester-Gallai theorem* states that every finite set of non-collinear points in the plane has an ordinary line. A proof from the BOOK can be found in [2, Chapter 11]. In 1948, Bruijn and Erdös improved this theorem to: Any $n \geq 3$ non-collinear points in the plane define at least n lines (not necessarily ordinary lines). But it does not say how many ordinary lines among these lines. Since then, to find the minimum number of ordinary lines determined over every set of n non-collinear points is still an open problem. The best result so far is duo to B. Green and T. Tao [51]: For sufficiently large point sets, the number of ordinary lines is at least $n/2$.

Additional problems for practice.

1. Given a set of $2n$ points in the plane with no three collinear, and then arbitrarily color each point red or blue. Prove that it is always possible to pair up the red points with the blue points by drawing line segments connecting them so that no two of the line segments intersect.

2. Let $P_1, \ldots, P_n$ points in the plane. Show that there is a point P such that every line through P has at least $n/3$ points P_i in each of the two closed half-planes it determines. A proof is given in [21, p. 93].

3. Let $D_1, \ldots, D_n$ be unit discs (in the plane) with centers $c_1, \ldots, c_n$, $n \geq 3$, such that no line meets more than two of them. Then

$$\sum_{1 \leq i < j \leq n} \frac{1}{d(c_i, c_j)} < \frac{n\pi}{4}.$$

A proof can be found in [22, p. 249].

4. **Putnam 2001-A6.** Can an arc of a parabola inside a circle of radius 1 have a length greater than 4?

5. **Putnam 2006-B3.** Let S be a finite set of points in the plane. A *linear partition* of S is an unordered pair $\{A, B\}$ of subsets of S such that $A \cup B = S, A \cap B = \emptyset$, and A and B lie on opposite sides of some straight line disjoint from S (A or B may be empty). Let L_S be the number of linear partitions of S. For each positive integer n, find the maximum of L_S over all sets S of n points.

6. **Problem E2598** (Proposed by E. Just, 83(5), 1976). Does there exist a set of rational points which is dense in the plane such that the distance between each pair of points in the set is irrational?

7. **Problem 11703** (Proposed by R. Bagby, 120(4), 2013). For $\lambda > 0$, let

$$\Gamma(\lambda) = \{(x, y, z) \in \mathbb{R}^3 : z \geq \lambda\sqrt{x^2 + y^2}\},$$

and let $C(\lambda)$ be the (half-cone) boundary of $\Gamma(\lambda)$. Prove that every point in the interior of $\Gamma(\lambda)$ is the focus of at least one ellipse in $C(\lambda)$, and find the largest μ such that every ellipse in $C(\lambda)$ has at least one focus in $\Gamma(\mu)$.

6.5 Squares of palindromes

Problem 11922 (Proposed by M. Alekseyev, 123(7), 2016). Find every positive integer n such that both n and n^2 are palindromes when written in the binary numeral system (and with no leading zeros).

Discussion.
A palindrome number is a positive integer which remains unchanged when its digits are reversed. For example, $313 = 100111001_2$ is both decimal and binary palindrome. It is easy to see that there are infinitely many positive integers n such that both n and n^2 are decimal palindromes. In fact, for every integer $k \geq 0$, let $n = 10^k + 1$. Then both n and $n^2 = 10^{2k} + 2 \cdot 10^k + 1$ are decimal palindromes. Indeed, even n^3 and n^4 are also decimal palindromes.
    Let

$$n = \sum_{i=0}^{k} a_i 2^i \quad \text{with } a_k = 1, 0 \leq a_i \leq 1 \text{ for } 0 \leq i \leq k - 1.$$

Then n is binary palindrome if and only if $a_i = a_{k-i}$ for all i. For $n \leq 100$, clearly $n = 1 = 1_2$ is palindrome. By *Mathematica*, we only find $3 = 11_2$ and $3^2 = 1001_2$ are palindromes. In the following, we show that $n = \{1, 3\}$ are the unique pair such that both n and n^2 are binary palindromes.

Solution.
We first state and prove the following general fact. If $n \equiv a \pmod{2^m}$ then

$$n^2 = (a + q2^m)^2 = a^2 + 2a2^m + 2^{2m} \equiv a^2 \pmod{2^{m+1}}.$$

Hence the m least significant bits of the binary representation of n determine exactly the $m + 1$ least significant bits of the binary representation of n^2.

Let n be a binary palindrome. We divide the proof in several cases with respect to d i.e., the number of 1s in the binary representation of n.

1) If $d = 1$ then $n = 1 = 1_2$ whose square is trivially a binary palindrome.

2) If $d = 2$ then $n = 2^k + 1 = 10^{k-1}1_2$ with $k \geq 1$. If $k = 1$ then $3 = 11_2$ whose square $3^2 = 1001_2$ is a binary palindrome. If $k \geq 2$ then $2k > k + 1$ and

$$n^2 = 2^{2k} + 2^{k+1} + 1 = 10^{k-2}10^k1_2$$

is not a binary palindrome.

3) If $d = 3$ then

$$n = 2^{2k} + 2^k + 1 = 10^{k-1}10^{k-1}1_2$$

with $k \geq 1$. If $k = 1$ then $n = 7$ whose square is not a binary palindrome. If $k \geq 2$ then $4k > 3k + 1 > 2k + 1 > 2k > k + 1 > 0$, and

$$n^2 = 2^{4k} + 2^{3k+1} + 2^{2k+1} + 2^{2k} + 2^{k+1} + 1 = 10^{k-2}10^{k-1}110^{k-2}10^k1_2$$

is not a binary palindrome.

4) Let $d \geq 4$ then

$$n = 2^j + 2^{j-k} + \cdots + 2^k + 1 = 10^{k-1}1 * 10^{k-1}1_2$$

with $j - k > k \geq 1$.

4.1) If $d \geq 4$ and $k \geq 2$ then $j \geq 5$ and $2^j + 2^{j-k} < n < 2^j + 2^{j-k+1}$ implies

$$2^{2j} + 2^{2j+1-k} < n^2 < 2^{2j} + 2^{2j-k+2} + 2^{2j-2k+2} < 2^{2j+2}.$$

Since $n = *10^{k-1}1_2$, by the fact stated above, it follows that $n^2 = *10^k1_2$ and if n^2 is a binary palindrome then $n^2 = 10^k1*_2$. Thus $n^2 < 2^{2j} + 2^{2j-k}$ or $n^2 > 2^{2j+1} + 2^{2j-k}$ which contradict the previous inequalities.

4.2) If $d = 4$ and $k = 1$ then

$$n = 2^j + 2^{j-1} + 2 + 1 = 110^{j-3}11_2$$

with $j \geq 3$. For $j = 3, 4, 5$ we have respectively $n = 15$, $n = 27$ and $n = 51$ whose squares are not binary palindromes. For $j \geq 6$,

$$n^2 = 2^{2j+1} + 2^{2j-2} + 2^{j+3} + 2^j + 9 = 10010^{j-6}10010^{j-4}1001$$

which is not a binary palindrome.

4.3) If $d = 5$ and $k = 1$ then

$$n = 2^{2j} + 2^{2j-1} + 2^j + 2 + 1 = 110^{j-2}10^{j-2}11_2$$

with $j \geq 2$. For $j = 2, 3$ we have respectively $n = 31$, and $n = 107$ whose squares are not binary palindrome. For $j \geq 4$,

$$n^2 = 2^{4j+1} + 2^{4j-2} + 2^{3j+1} + 2^{3j} + 2^{2j+3} + 2^{2j+1} + 2^{j+2} + 2^{j+1} + 9$$
$$= 10010^{j-4}110^{j-4}1010^{j-2}110^{j-3}1001$$

which is not a binary palindrome.

4.4) Let $d \geq 6$ and $k = 1$ then

$$n = 2^j + 2^{j-1} + 2^{j-i} + \ldots + 2^i + 2 + 1 = 110^{i-2}1 * 10^{i-2}11_2$$

with $j - i > i \geq 2$.

4.4.1) If $d \geq 6$ and $i = 2$ then $n = 111 * 111_2$ with $j \geq 5$. Now $7 \cdot 2^{j-2} < n < 2^{j+1}$ implies

$$2^{2j+1} < 49 \cdot 2^{2j-4} < n^2 < 2^{2j+2}.$$

Since $n^2 = *0001_2$, it follows that if n^2 is a binary palindrome then $n^2 = 1000*_2$ and

$$n^2 < 2^{2j+1} + 2^{2j-2} < \left(1 + 2^{-1} + 2^{-2}\right)^2 2^{2j}$$

which implies $n < 2^j + 2^{j-1} + 2^{j-2}$ that is against the fact that $n > 2^j + 2^{j-1} + 2^{j-2}$.

4.4.2) If $d \geq 6$, $i \geq 3$ then $n = 110^{i-2}1 * 10^{i-2}11_2$ with $j > 2i$. Now $3 \cdot 2^{j-1} < n < 2^{j+1}$ implies

$$2^{2j+1} < 9 \cdot 2^{2j-2} < n^2 < 2^{2j+2}.$$

Since $n^2 = *0^{i-3}1001_2$ it follows that if n^2 is a binary palindrome then $n^2 = 10010^{i-3}*_2$ and

$$n^2 < 2^{2j+1} + 2^{2j-2} + 2^{2j+1-i} = \left(2 + 2^{-2} + 2^{1-i}\right)2^{2j} < \left(1 + 2^{-1} + 2^{-i}\right)^2 2^{2j}$$

which implies $n < 2^j + 2^{j-1} + 2^{j-i}$ that is against the fact that $n > 2^j + 2^{j-1} + 2^{j-i}$. $\qquad\square$

Remark. Palindrome numbers have received much attention in recreational mathematics. The palindrome numbers and the palindrome squares are given by the sequences A002133 (https://oeis.org/A002113) and A002779 (https://oeis.org/A002779) in the OEIS, respectively. In A251673, we see that the numbers that are not palindromes, but whose squares are palindromes are rare, yet they are infinite. For instance, each number of the form $111 \cdot 100^k + 91 \cdot 10^k + 111$ with $k \geq 3$ has this property. On the other hand, the only known non-palindromic number whose cube is a palindrome is 2201.

Additional problems for practice.

1. If both n and n^2 are decimal palindromes, show that n^2 has an odd number of digits.

2. If $n = \sum_{i=0}^{k} a_i \cdot 10^i$ with $k > 0$ is decimal palindrome, show that n^2 is palindromic if and only if $\sum_{i=1}^{k} a_i^2 \le 9$.

3. If both $n = \sum_{i=0}^{k} a_i \cdot 10^i$ with $k > 0$ and n^2 are decimal palindromes, show that $a_i \in \{0, 1, 2\}$ for each $0 \le i \le k$.

4. Let $w_0, w_1, \ldots$ be the sequence of the *Fibonacci words*, defined by $w_0 = 0$, $w_1 = 1$, and, for $n \ge 2$, $w_n = w_{n-2}w_{n-1}$, the concatenation of w_{n-2} and w_{n-1}. The first few terms of the sequence are

   $$0, 1, 01, 101, 01101, 10101101, 0110110101101, 101011010110110101101, \ldots.$$

   Show that, for $n \ge 3$, removing the first two symbols from w_n yields a palindrome.

5. **Putnam 2002-B5**. A palindrome in base b is a positive integer whose base-b digits read the same backwards and forwards; for example, 2002 is a 4-digit palindrome in base 10. Note that 200 is not a palindrome in base 10, but it is the 3-digit palindrome 242 in base 9, and 404 in base 7. Prove that there is an integer which is 3-digit palindrome in base b for at least 2002 different values of b.

6. **Putnam 2007-A4**. A *repunit* is a positive integer whose digits in base 10 are all ones. Find all polynomials f with real coefficients such that if n is a repunit, then so is $f(n)$.

7. **Problem E3156** (Proposed by R. M. Robinson, 93(6), 1986). Suppose that r, s, t are integers with $r \ge 0$, $s \ge 0$, $t = r + s \ge 2$. Is there a word W of length t in the alphabet $\{a, b\}$ such that $W = AB = Cab$, where A, B, C are palindromes, and the lengths of A and B are r and s? Show that such a word W exists if and only if $r + 2$ is prime to $s - 2$, and that in this case it is unique.

6.6 Counting some strange mappings

Problem 11901 (Proposed by D. E. Knuth, 123(4), 2016). For $n \in \mathbb{Z}^+$, let $[n] = \{1, \ldots, n\}$. Define the functions $\uparrow$ and $\downarrow$ on $[n]$ by $\uparrow x = \min\{x + 1, n\}$ and $\downarrow x = \min\{x - 1, 1\}$. How many distinct mappings from $[n]$ to $[n]$ occur as compositions of $\uparrow$ and $\downarrow$?

Discussion.

Let's start with a few basic remarks. Since the functions ↑ and ↓ on $[n]$ are non-decreasing, any composition will also be non-decreasing. The identity function arises as an empty composition. Additionally, constant functions can be easily obtained: Given $x \in [n]$, for all $k \in [n]$,

$$\underbrace{\uparrow \cdots \uparrow}_{x-1}\underbrace{\downarrow \cdots \downarrow}_{n-1}(k) = \underbrace{\uparrow \cdots \uparrow}_{x-1}(1) = x.$$

As we will see, not all the non-decreasing functions can be derived this way, and in fact, their number is actually much smaller.

Solution.

We will show that the mappings from $[n]$ to $[n]$ which can be written as compositions (possibly empty) of ↑ and ↓ are those with the following forms:

$$(f(1), f(2), \ldots, f(n)) = (\underbrace{x, \ldots, x}_{a}, \underbrace{x+1, \ldots, x+c}_{c}, \underbrace{x+c+1, \ldots, x+c+1}_{b}).$$

$$(6.6)$$

where $a + b + c = n$ with $a \geq 1$, $b \geq 0$, $c \geq 0$ and $1 \leq x \leq n - c$. Let $\mathcal{F}_n$ be the set of such mappings.

It is easy to verify that composition with both ↑ and ↓ preserves this form. On the other hand, if $f \in \mathcal{F}_n$ then it occurs as a composition of ↑ and ↓. In fact, as we already noted above, we have the n constant maps, where $b = c = 0$, and for $b \geq 1$,

$$\underbrace{\uparrow \cdots \uparrow}_{x-1}\underbrace{\downarrow \cdots \downarrow}_{a+b-2}\underbrace{\uparrow \cdots \uparrow}_{b-1}(k) = \underbrace{\uparrow \cdots \uparrow}_{x-1}\underbrace{\downarrow \cdots \downarrow}_{a+b-2} \begin{cases} k+b-1 & \text{if } 1 \leq k \leq n-b, \\ n & \text{if } n-b < k \leq n, \end{cases}$$

$$= \underbrace{\uparrow \cdots \uparrow}_{x-1} \begin{cases} 1 & \text{if } 1 \leq k \leq a, \\ k-a+1 & \text{if } a < k \leq n-b, \\ n-a-b+2 & \text{if } n-b < k \leq n, \end{cases}$$

$$= \begin{cases} x & \text{if } 1 \leq k \leq a, \\ x+k-a & \text{if } a < k \leq n-b, \\ x+c+1 & \text{if } n-b < k \leq n. \end{cases}$$

which gives a map of the form (6.6).

Thus, it remains to evaluate the cardinality of $\mathcal{F}_n$. In view of our discussion, we conclude, all the functions in $\mathcal{F}_n$ can be counted as

$$n + \sum_{a=1}^{n-1}\sum_{b=1}^{n-a}\sum_{x=1}^{a+b-1} 1 = n + \binom{n}{2} + 2\binom{n}{3} = \frac{n(2n^2 - 3n + 7)}{6}.$$

The first few terms in the sequence are 1, 3, 8, 18, 35, 61, 98, 148, 213, 295, which is the entry A081489 in the OEIS (https://oeis.org/A081489). □

Remark. The total number of non-decreasing functions from $[n]$ to $[m]$ can be obtained as an application of a combinatorial counting technique known as *stars and bars*. A non-decreasing function from $[n]$ to $[m]$ is completely determined by the number x_k of times it assumes the value k for any $k \in [m]$. So it suffices to enumerate the nonnegative integer solutions of the equation

$$x_1 + x_2 + \cdots + x_m = n.$$

Counting all such solutions is like placing n *stars* along a line and separate them into m groups by using $m - 1$ *bars*. For example, if $n = 10$ and $m = 4$, then the solution $1 + 3 + 0 + 6 = 10$ can be represented as the configuration

$$\star \mid \star \star \star \mid\mid \star \star \star \star \star \star \,.$$

Since the $m-1$ separating bars can be placed into $n+m-1$ possible positions, we find that the number of configurations is the binomial coefficient

$$\binom{n+m-1}{m-1}$$

which is precisely the number of non-decreasing function from $[n]$ to $[m]$ as we desired.

Additional problems for practice.

1. **MM Problem 1684** (Proposed by E. S. Brown and C. J. Hillar, 76(5), 2003). Let S be the set of all n-letter words in two letters, say a and b. Define an equivalence relation on S as follows: given a word W, the reverse of W, the complement of W (that is, change all as to bs and all bs to as) and the reverse of the complement are all equivalent to W. Find the number of equivalence classes of S that do not contain any palindromes.

2. **Problem 10298** (Proposed by D. E. Knuth, 100(4), 1993). Let $\left\{ {m+n \atop n} \right\}$ denote the number of ways to partition a set of $m + n$ elements into n nonempty subsets. Prove that

$$\frac{2^m 3^{\lfloor m/2 \rfloor} 4^{\lfloor m/3 \rfloor} 5^{\lfloor m/4 \rfloor} \cdots}{(n+1)(n+2)\cdots(n+m)} \left\{ {m+n \atop n} \right\}.$$

3. **Problem 11183** (Proposed by D. Beckwith, 112(9), 2005). The left and right pillars of a triumphal arch are each built of blocks of height 1 or 2. Blocks of height 2 may not sit upon blocks of height 1. How many designs are feasible if the lintel must sit upon the pillars and if exactly n blocks must be used in the construction of the pillars?

4. **Problem 11249** (Proposed by D. Beckwith, 113(8), 2006). A *node-labeled rooted tree* is a tree such that any parent with label k has $k+1$ children, labeled $1, 2, \ldots, k+1$, and such that the root vertex (generation 0) has label k. Find the population of generation n.

5. **Problem 11424** (Proposed by E. Deutsch, 116(3), 2009). Find the number x_n of bit strings of length n in which the number of 00 substrings is equal to the number of 11 substrings.

6. **Problem 12069** (Proposed by D. E. Knuth, 125(9), 2018). Place n nonattacking rooks on an n-by-n chessboard in such a way as to maximize the sum of the Euclidean distances from the rooks to the center of the chessboard. (Regard a rook as a point positioned at the center of its square.) How many placements attain this maximum?

6.7 Comparing the coefficients of two generating functions

Problem 12113 (Proposed by R. P. Stanley, 126(5), 2019). Define $f(n)$ and $g(n)$ for $n \geq 0$ by

$$\sum_{n\geq 0} f(n)x^n = \sum_{j\geq 0} x^{2^j} \prod_{k=0}^{j-1}\left(1 + x^{2^k} + x^{3\cdot 2^k}\right)$$

and

$$\sum_{n\geq 0} g(n)x^n = \prod_{i\geq 0}\left(1 + x^{2^i} + x^{3\cdot 2^i}\right).$$

Find all values of n for which $f(n) = g(n)$, and find $f(n)$ for these values.

Discussion.
To gain insight into the behavior of the sequences $f(n)$ and $g(n)$, by *Mathematica*, we compute the first 24 terms of each one:

n	0	1	2	3	4	5	6	7	8	9	10	11
$f(n)$	0	1	1	1	1	2	1	2	1	2	2	3
$g(n)$	1	1	1	2	1	2	2	3	1	3	2	3
n	12	13	14	15	16	17	18	19	20	21	22	23
$f(n)$	1	3	2	3	1	3	2	3	2	4	3	5
$g(n)$	2	4	3	5	1	4	3	4	2	5	3	5

In the OEIS, $f(n)$ and $g(n)$ appear as sequences A082498 (https://oeis.org/A082498) and A120562 (https://oeis.org/A120562), respectively. Based on the above table, we observe that

(a) $f(n) = g(n)$ occurs at $n = 1, 2, 4, 5, 8, 10, 11, 16, 20, 22, 23$, which is the sequence A191986 (https://oeis.org/A191986) in the OEIS. This monotonically ordered sequence is given by $2^i(3 \cdot 2^j - 1)$ for $i, j \geq 0$.

(b) The equal values are $1, 1, 1, 2, 1, 2, 3, 1, 2, 3, 5$, respectively. All entries are Fibonacci numbers!

By using the generating functions and recursive relations, we now prove the properties (a) and (b) hold in general.

Solution.

By shifting the index, we can easily obtain

$$F(x) := \sum_{n \geq 0} f(n)x^n = x + (1 + x + x^3)\, F(x^2)$$

$$G(x) := \sum_{n \geq 0} g(n)x^n = (1 + x + x^3)\, G(x^2)$$

Rewrite them as

$$\sum_{n \geq 0} f(n)x^n = x + \sum_{n \geq 1} f(n)x^{2n} + \sum_{n \geq 1} f(n)x^{2n+1} + \sum_{n \geq 1} f(n)x^{2n+3}$$

$$= x + \sum_{n \geq 1} f(n)x^{2n} + \sum_{n \geq 1}(f(n) + f(n-1))x^{2n+1};$$

$$\sum_{n \geq 0} g(n)x^n = \sum_{n \geq 0} g(n)x^{2n} + \sum_{n \geq 0} g(n)x^{2n+1} + \sum_{n \geq 0} g(n)x^{2n+3}$$

$$= 1 + x + \sum_{n \geq 1} g(n)x^{2n} + \sum_{n \geq 1}(g(n) + g(n-1))x^{2n+1}.$$

Therefore, both sequences $f(n)$ and $g(n)$ satisfy the same recurrence relation: $a_1 = 1$ and for $n \geq 2$,

$$a_{2n} = a_n, \qquad a_{2n+1} = a_n + a_{n-1}, \tag{6.7}$$

but with different initial conditions, i.e., $f(0) = 0$ and $g(0) = 1$.

By (6.7), we find that $f(n) \leq g(n)$, and equality holds if and only if n is 1 or it is of the form $2^i(3 \cdot 2^j - 1)$ with $i, j \geq 0$. Moreover

$$f(n) = g(n) = \begin{cases} 1, & \text{if } n = 1, \\ F_{j+2}, & \text{if } n = 2^i(3 \cdot 2^j - 1) \text{ with } i, j \geq 0, \end{cases}$$

where F_k is the k-th Fibonacci number, which is defined by $F_0 = 0$, $F_1 = 1$, $F_k = F_{k-1} + F_{k-2}$ for $k \geq 2$.

This can be asserted by induction on j. Indeed, if $j = 0$, applying (6.7) proves the base case:

$$f(2^i(3 \cdot 2^0 - 1)) = f(2^{i+1} \cdot 1) = f(1) = 1 = F_2.$$

If $n = 2^i(3 \cdot 2^{j+1}) - 1$, since $3 \cdot 2^{j+1} - 1 = 2(3 \cdot 2^j - 1) + 1$, applying (6.7) and the induction hypothesis, we obtain

$$f(2^i \cdot (3 \cdot 2^{j+1} - 1)) = f(3 \cdot 2^{j+1} - 1) = f(3 \cdot 2^j - 1) + f(3 \cdot 2^j - 2)$$

$$= F_{j+2} + f(2(3 \cdot 2^{j-1} - 1)) = F_{j+2} + f(3 \cdot 2^{j-1} - 1)$$

$$= F_{j+2} + F_{j+1} = F_{j+3},$$

as claimed. $\qquad \square$

Remark. In view of the form of the generating function of g, $g(n)$ is the number of partitions of n into distinct parts, in which each part has the form 2^j or $3 \cdot 2^j$ with $j \geq 0$ and 2^j or $3 \cdot 2^j$ are just used once for same j. i.e.,

$$n = \sum_{j=0}^{m} a_j 2^j \quad \text{for some positive integer } m, \text{ with } a_j \in \{0, 1, 3\} \text{ and } a_m \neq 0.$$

Clearly, $f(n)$ counts such n with $a_m = 1$. The featured solution by H. Kwong and the editors [91] is based on the fact that if $a_m = 1$ then $n = 3 \cdot 2^p - 2^q$ with $0 \leq q \leq p$.

It is curious to observe that the recurrence (6.7) satisfied by both $f(n)$ and $g(n)$ is quite similar to the one which defines the celebrated *Stern's diatomic sequence*: let $a_0 = 0, a_1 = 1$, and for $n \geq 1$,

$$a_{2n} = a_n, \qquad a_{2n+1} = a_n + a_{n+1},$$

which is the sequence A002487 (`https://oeis.org/A002487`) in the OEIS. The first few terms are

$$0, 1, 1, 2, 1, 3, 2, 3, 1, 4, 3, 5, 2, 5, 3, 4, 1, 5, 4, 7, 3, 8, 5, 7, 2, 7, 5, 8, 3, 7, 4, 5, 1, \ldots,$$

and its generating function is

$$\sum_{n \geq 0} a_n x^n = x \prod_{i \geq 0} \left(1 + x^{2^i} + x^{2 \cdot 2^i} \right).$$

The Stern's diatomic sequence has a long history and exhibits many charming properties, among which we find:

(1) For $n \geq 1$, a_n and a_{n+1} are relatively prime.

(2) a_{n+1} is the number of partitions of the integer n into powers of 2 for which each part has multiplicity 1 or 2. i.e.,

$$n = \sum_{j=0}^{m} a_j 2^j \quad \text{for some positive integer } m \text{ and } a_j \in \{0, 1, 2\}.$$

(3) In the Stern's diatomic array:

```
1
1  2
1  3  2  3
1  4  3  5  2  5  3  4  ...
```

the row maxima are $1, 2, 3, 5, \ldots$, the Fibonacci sequence.

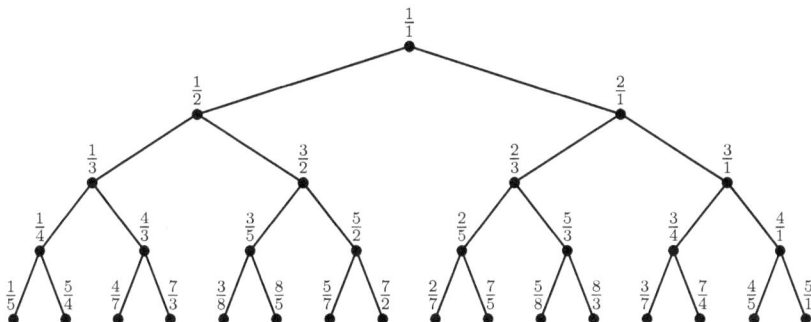

Figure 6.2: Calkin-Wilf tree

For details, the interested reader can refer S. Northshield's excellent tour on this sequence [77]. There are 5 challenge problems for future research listed at the end of this paper.

It is worth pointing out that the sequence $\{a_n/a_{n+1}\}_{n\geq1}$ enumerates every positive rational number and no number appears twice. This property can be illustrated through a binary tree as proposed by N. Calkin and H. Wilf in [27]. The binary tree is constructed inductively in this way: we start with $\frac{1}{1}$ at the top of the tree, and then each vertex $\frac{p}{q}$ has two children, namely its left child $\frac{p}{p+q}$ and its right child $\frac{p+q}{q}$.

Reading the fractions in order by row, we obtain exactly the sequence $\{a_n/a_{n+1}\}_{n\geq1}$:

$$\frac{1}{1},\frac{1}{2},\frac{2}{1},\frac{1}{3},\frac{3}{2},\frac{2}{3},\frac{3}{1},\frac{1}{4},\frac{4}{3},\frac{3}{5},\frac{5}{2},\frac{2}{5},\frac{5}{3},\frac{3}{4},\frac{4}{1},\frac{1}{5},\frac{5}{4},\frac{4}{7},\frac{7}{3},\frac{3}{8},\frac{8}{5},\frac{5}{7},\frac{7}{2},\cdots \qquad (6.8)$$

Another enumeration of the positive rational numbers is given by the closely related Stern-Brocot tree (see [47, p. 116]).

Additional problems for practice.

1. Let a_n be the Stern diatomic sequence defined above. Show that

$$\sum_{n=1}^{\infty}\frac{2}{a_{2n}a_{2n+1}^2a_{2n+2}}=1 \quad \text{and} \quad \sum_{n \text{ odd}}\frac{1}{x^{a_n}-1}=\frac{x}{(x-1)^2} \qquad \text{for } x>1.$$

2. Let the sequence a_n be defined by $a_1=1, a_{2n}=a_n$, and $a_{2n+1}=(-1)^na_n$. Show that

$$a_n=(-1)^{b_0b_1+b_1b_2+\cdots b_{k-1}b_k}, \qquad \text{where } n=(b_kb_{k-1}\ldots b_0)_2$$

in binary representation

3. Modify Pascal's triangle as follows: row 0 is 1; in each subsequent row, 1 is at the beginning and end, beside add adjacent number just as in Pascal's triangle, also bring down the number in row n directly below it in row $(n+1)$. Here are the first 4 rows of the new triangle

$$
\begin{array}{ccccccccccccc}
&&&&&& 1 &&&&&& \\
&&&& 1 && 1 && 1 &&&& \\
&& 1 && 1 && 2 && 1 && 2 && 1 && 1 \\
1 && 1 && 2 && 1 && 3 && 2 && 3 && 1 && 3 && 2 && 3 && 1 && 2 && 1 && 1
\end{array}
$$

R. P. Stanley called it as *Stern's triangle* and he discusses some of its properties in [90, Chapter 6].

Show that

(a) The number of entries in row n is $2^{n+1} - 1$, and the sum of its entries is 3^n.

(b) The largest entry in row n is the Fibonacci number F_{n+1}.

(c) Let the kth entry in row n be $C(n, k)$. Define

$$
G_n(x) = \sum_{k=0}^{2^{n+1}-2} C(n, k) x^k.
$$

Show that $G_n(x) = \prod_{i=0}^{n-1} \left(1 + x^{2^i} + x^{2 \cdot 2^i}\right)$.

(d) Let $b_n = \sum_{k=0}^{2^{n+1}-2} C^2(n, k)$. Show that

$$
\sum_{n=0}^{\infty} b_n x^n = \frac{1 - 2x}{1 - 5x + 2x^2}.
$$

4. **Putnam 2002-A5.** Define a sequence by $a_0 = 1$, together with the rules $a_{2n+1} = a_n$ and $a_{2n+2} = a_n + a_{n+1}$ for each integer $n \geq 0$. Prove that every positive rational number appears in the set

$$
\left\{ \frac{a_{n-1}}{a_n} : n \geq 1 \right\} = \left\{ \frac{1}{1}, \frac{1}{2}, \frac{2}{1}, \frac{1}{3}, \frac{3}{2}, \cdots \right\}.
$$

5. **Putnam 2013-B1.** For positive integer n, let the numbers $c(n)$ be determined by the rules: $c(1) = 1$, $c(2n) = c(n)$, and $c(2n + 1) = (-1)^n c(n)$. Find the value of

$$
\sum_{n=1}^{2013} c(n)c(n + 2).
$$

6. **Problem E3415** (Proposed by P. Flajolet and D. E. Knuth, 98(1), 1991). Find the coefficient of $x_1^{k_1} x_2^{k_2} \cdots x_n^{k_n}$ in

$$
(1 - x_1)^{-a_1} (1 - x_1 - x_2)^{-a_2} \cdots (1 - x_1 - x_2 - \cdots - x_n)^{-a_n}.
$$

7. **Problem 10357** (Proposed by I. Gessel, 101(1), 1994). Define integers $a_{m,n}$ by

$$\frac{1}{1 - u - v + 2uv} = \sum_{m,n=0}^{\infty} a_{m,n} u^m v^n.$$

Show that $(-1)^k a_{2k,2k+2}$ is the Catalan number $\frac{1}{k+1}\binom{2k}{k}$.

8. **Problem 10906** (Proposed by D. E. Knuth, 108(9), 2001). For $n \geq 1$, let $\nu_2(n) = k$ if n is divisible by 2^k but not by 2^{k+1}. Let $x_0 = 0$, and define x_n for $n \geq 1$ recursively by

$$\frac{1}{x_n} = 1 + 2\nu_2(n) - x_{n-1}.$$

Prove that every nonnegative rational number occurs exactly once in the sequence $x_0, x_1, x_2, \ldots$.

6.8 A vector sum of modulus at least 1

Problem 12202 (Proposed by K. T.-L. Koo, 127(8), 2020). Let V be a finite set of vectors in $\mathbb{R}^2$ such that $\sum_{v \in V} |v| = \pi$. Prove that there exists a subset U of V such that

$$\left| \sum_{v \in U} v \right| \geq 1.$$

Discussion.
We rewrite the proposed inequality as

$$\left| \sum_{v \in U \subset V} v \right| \geq \frac{1}{\pi} \sum_{v \in V} |v|.$$

This indicates that there exists some subset with a sum whose modulus is a large portion of the sum of all the moduli. To find such required subset, we single out two approaches. First, we use the fact that maximum value is greater than or equal the expected value from probability. Second, we construct a convex polygon and then apply the geometric inequality: If p and d are the perimeter and diameter of a convex polygon, then $p < \pi d$ [105, p. 257].

Solution I.
Choose a unit vector w randomly, define

$$P_w(v) = \begin{cases} v \cdot w = |v| \cos(\theta), & \text{if } \cos(\theta) > 0, \\ 0, & \text{otherwise,} \end{cases}$$

where θ is the angle between v and w. This yields $|v| \geq P_w(v)$.

Let $U_w = \{v \in V : P_w(v) > 0\}$, i.e., the angle between w and v is acute. Then

$$\left| \sum_{v \in U_w} v \right| \geq P_w \left(\sum_{v \in U_w} v \right) = \sum_{v \in U_w} P_w(v) = \sum_{v \in V} P_w(v).$$

On the other hand, since the expected value of $P_w(v)$ is

$$E[P_w(v)] = \frac{1}{2\pi} \int_{-\pi}^{\pi} P_w(v)\, d\theta = \frac{1}{2\pi} \int_{-\pi/2}^{\pi/2} |v| \cos(\theta)\, d\theta = \frac{|v|}{\pi},$$

we obtain

$$E\left[\sum_{v \in V} P_w(v) \right] = \sum_{v \in V} E[P_w(v)] = \frac{1}{\pi} \sum_{v \in V} |v| = 1.$$

This shows that there exists a unit vector w such that U_w is the required subset. $\square$

Solution II
Let $V = \{v_1, v_2, \ldots, v_n\}$. We choose a vector w such that $v_1 + v_2 + \cdots + v_n + w = 0$. For any vector v, let θ_v be the angle from the positive x-axis to v with $0 \leq \theta_v < 2\pi$. Let $u_1, u_2, \ldots, u_{n+1}$ be a permutation of $v_1, v_2, \ldots, v_n, w$ such that

$$\theta_{u_1} \leq \theta_{u_2} \leq \cdots \leq \theta_{u_{n+1}}.$$

Thus, the endpoints of the partial sums $\left\{ \sum_{i=1}^{k} u_i \right\}_{k=1}^{n+1}$ form the vertices of a convex polygon (possibly degenerate). Let p and d be the perimeter and diameter of this polygon, respectively. Then

$$\pi = \sum_{v \in V} |v| \leq \sum_{v \in V} |v| + |w| = p < \pi d,$$

so $d > 1$. The set U can be chosen to be a collection of vectors (exclude w) whose sum gives a diameter of the polygon. $\square$

Remark. The lower bound 1 is the best possible. Consider the set of evenly distributed two-dimensional vectors (with complex number notation)

$$V_n = \left\{ v_k = \frac{\pi}{n} e^{i \frac{2\pi k}{n}} \; : \; k = 0, \ldots, n-1 \right\}$$

where n is an even positive integer. Let $U_n = \{v_k \; : \; k = 0, \ldots, \frac{n}{2} - 1\} \subset V_n$. Then $\sum_{v \in V_n} |v| = \pi$ and

$$\sum_{v \in U_n} |v| = \frac{\pi}{n} \left| \sum_{k=0}^{\frac{n}{2}-1} e^{i \frac{2\pi k}{n}} \right| = \frac{\pi}{n} \left| \frac{1 - e^{i\pi}}{1 - e^{i \frac{2\pi}{n}}} \right|$$

$$= \frac{\pi}{n} \frac{2}{\left| e^{-i \frac{\pi}{n}} - e^{i \frac{\pi}{n}} \right|} = \frac{\pi/n}{\sin(\pi/n)} \to 1 \quad \text{as } n \to \infty.$$

Additional problems for practice.

1. Let V be a finite set of vectors in $\mathbb{R}^2$, and let $U_\theta = \{v \in V : \mathrm{Re}(ve^{-i\theta}) \geq 0\}$. Show that there is a $\theta \in (0, 2\pi)$ such that

$$\left| \sum_{v \in U_\theta} v \right| \geq \frac{1}{\pi} \sum_{v \in V} |v|.$$

2. (*An extension to higher dimensions*). Show that every finite set of vectors $\{v_1, v_2, \ldots, v_n\} \subset \mathbb{R}^k$ has a subset U such that

$$\left| \sum_{v \in U} v \right| \geq c(k) \sum_{j=1}^{n} |v_j|,$$

where

$$c(k) = \frac{\Gamma(k/2)}{2\sqrt{\pi}\,\Gamma((k+1)/2)}.$$

3. **Putnam 2006-B2**. Prove that, for every set $X = \{x_1, x_2, \ldots, x_n\}$ of n real numbers, there exists a nonempty subset S of X and an integer m such that

$$\left| m + \sum_{s \in S} s \right| \leq \frac{1}{n+1}.$$

4. **Problem 11825** (Proposed by M. Dincă and S. Radulescu, 122(3), 2015). Let E be a normed linear space. Given $x_1, \ldots x_n \in E$ (with $n \geq 2$) such that $\|x_k\| = 1$ for $1 \leq k \leq n$ and the origin of E is in the convex hull of $\{x_1, \ldots, x_n\}$, prove that $\|x_1 + \cdots + x_n\| \leq n - 2$.

5. **Problem 12163** (Proposed by T. Speckhofer, 127(2), 2020). Let $\mathbb{R}^n$ have the usual dot product and norm. When $v = (x_1, \ldots, x_n) \in \mathbb{R}^n$, let $\Sigma v = x_1 + \cdots + x_n$. Prove

$$\|v\|^2 \|w\|^2 \geq (v \cdot w)^2 + \frac{(\|v\|\,|\Sigma w| - \|w\|\,|\Sigma v|)^2}{n}$$

for all $v, w \in \mathbb{R}^n$.

6. **Problem 12329** (Proposed by L. Giugiuc, 129(6), 2022). Let n be a positive integer with $n \geq 3$. For each positive integer m with $m \geq 2$, find all real values λ_m such that there are m distinct unit vectors $v_1, \ldots, v_m$ in $\mathbb{R}^n$ satisfying $v_i \cdot v_j = \lambda_m$ for all i, j with $1 \leq i < j \leq m$.

7. **Problem 12342** (Proposed by G. Stoica, 129(8), 2022). Let $v_1, \ldots, v_n$ be unit vectors in $\mathbb{R}^d$. Prove that if u maximizes $\prod_{i=1}^{n} |v_i \cdot u|$ over all unit vectors $u \in \mathbb{R}^d$, then for all i,

$$|v_i \cdot u| \geq \sin\left(\frac{\pi}{2n}\right).$$

6.9 Counting equilateral triangles in hypercubes

Problem 12261 (Proposed by A. Stadler, 128(6), 2021). Let a_n be the number of equilateral triangles whose vertices are chosen from the vertices of the n-dimensional cube. Compute

$$\lim_{n\to\infty} \frac{na_n}{8^n}.$$

Discussion.
To search for patterns of a_n, we begin by examining the case $n = 3$. Let the 3D unit cube be in the first octant with one vertex at the origin. Then $a_3 = 8$. The vertices of these equilateral triangles are

$$\{(0,0,0),\ (1,1,0),\ (1,0,1)\},\quad \{(0,0,0),\ (1,1,0),\ (0,1,1)\},$$
$$\{(0,0,0),\ (1,0,1),\ (0,1,1)\},\quad \{(1,0,0),\ (0,1,0),\ (0,0,1)\},$$
$$\{(1,0,0),\ (0,1,0),\ (1,1,1)\},\quad \{(1,0,0),\ (0,0,1),\ (1,1,1)\},$$
$$\{(0,1,0),\ (0,0,1),\ (1,1,1)\},\quad \{(1,1,0),\ (1,0,1),\ (0,1,1)\}.$$

This leads to the following observations:

(1) No two vertices of an equilateral triangles lies on the same edge of the cube.

(2) For each vertex v, the number of equilateral triangles with v as a vertex are same.

(3) In an equilateral triangle $\triangle AOB$ with the origin O, the equalities $d(A,O) = d(B,O) = d(A,B)$ imply that both A and B have same number of 1s. Moreover, the number of places where A has a 1 and B has a 0 must be equal to the number of places where A has a 0 and B has a 1.

Based on these observations, we are able to find a closed form for a_n by counting arguments.

Solution.
We show the limit is $1/(3\sqrt{3}\pi)$. To this end, we introduce the n-dimensional coordinates to the vertices of the hypercube: one vertex at the origin O and the rest with coordinates 0's and 1's.
Let N be the number of equilateral triangles with the origin as a vertex. First, we prove that

$$N = \frac{1}{2} \sum_{k=1}^{\lfloor n/3 \rfloor} \binom{n}{k}\binom{n-k}{k}\binom{n-2k}{k}. \tag{6.9}$$

Once we establish (6.9), by symmetry, for each vertex v in the cube, the number of equilateral triangles with v as a vertex are same. Thus,

$$a_n = 2^n \cdot \frac{N}{3} = \frac{2^n}{6} \sum_{k=1}^{\lfloor n/3 \rfloor} \binom{n}{k}\binom{n-k}{k}\binom{n-2k}{k}.$$

To find N, we map each vertex coordinate to a set based on the following rules:

(R_1) delete all zeros in v;

(R_2) replace the 1 in v by its ordered coordinate component number.

For example, let the map be P. Then

$$P((1,1,1,1,0,0)) = \{1,2,3,4\}, \quad P((1,1,0,0,1,1)) = \{1,2,5,6\}.$$

Clearly, this map is bijective. For an equilateral triangle $\triangle AOB$, since the number of places where A has a 1 and B has a 0 is same as the number of places where A has a 0 and B has a 1, we have $|P(A)| = |P(B)|$. Let $|P(A)| = m$. Then

$$|P(A) \cap P(B)| = \frac{m}{2} := k.$$

Moreover, we have

$$|P(A) \cup P(B)| = |P(A)| + |P(B)| - |P(A) \cap P(B)| = 3k \le n.$$

Thus, we find that the number of ways to choose A and B is the same as the number of ways to choose $P(A)$ and $P(B)$. At the level k, there are $\binom{n}{k}$ ways to choose $P(A) \cap P(B)$, so we have that the number of ways to choose $P(A)$ and $P(B)$ is equal to

$$\binom{n}{k} \cdot \frac{\binom{n-k}{k}\binom{n-2k}{k}}{2}.$$

Hence,

$$N = \sum_{k=1}^{\lfloor n/3 \rfloor} \binom{n}{k} \cdot \frac{\binom{n-k}{k}\binom{n-2k}{k}}{2}.$$

This proves (6.9).

Applying $\binom{n}{k} = \frac{n!}{k!(n-k)!}$, we can condense a_n as

$$a_n = \frac{2^n}{6} \sum_{k=1}^{\lfloor n/3 \rfloor} \binom{n}{3k} \frac{(3k)!}{(k!)^3}.$$

Next, we show that, for large n,

$$a_n \sim \frac{8^n}{3\sqrt{3\pi n}}. \tag{6.10}$$

By the Stirling approximation, we find

$$a_n \sim \frac{2^n}{6} \sum_{k=1}^{\lfloor n/3 \rfloor} \binom{n}{3k} \frac{(6\pi k)^{1/2}(3k)^{3k}e^{-3k}}{(2\pi k)^{3/2}k^{3k}e^{-3k}} = \frac{2^n}{4\sqrt{3\pi}} \sum_{k=1}^{\lfloor n/3 \rfloor} \binom{n}{3k} \frac{3^{3k}}{k}$$

$$\sim \frac{2^n}{4\sqrt{3n\pi}} \sum_{k=1}^{\lfloor n/3 \rfloor} \binom{n+1}{3k+1} 3^{3k+1}$$

where we noted that

$$\binom{n+1}{3k+1} = \frac{n+1}{3k+1}\binom{n}{3k} \sim \frac{n}{3k}\binom{n}{3k}.$$

Let $\omega = e^{2\pi i/3}$. Applying the *multisection formula*, we have

$$\sum_{k=1}^{\lfloor n/3 \rfloor}\binom{n+1}{3k+1}3^{3k+1} = \frac{1}{3}\left((1+3)^{n+1} + \omega^{-1}(1+3\omega)^{n+1} + \omega(1+3\omega^{-1})^{n+1}\right).$$

Since $|1+3\omega| = |1+3\omega^{-1}| = \sqrt{7} < 3$, it follows that

$$\sum_{k=1}^{\lfloor n/3 \rfloor}\binom{n+1}{3k+1}3^{3k+1} \sim \frac{1}{3}4^{n+1},$$

which proves (6.10) and the requested limit

$$\lim_{n\to\infty}\frac{na_n}{8^n} = \frac{1}{3\sqrt{3}\,\pi},$$

as claimed. $\square$

Remark. Here is another counting approach to find a_n. Let $\mathbf{v}_1, \mathbf{v}_2, \mathbf{v}_3 \in \{0,1\}^n$ be vertices of n-cube, and let

$$S_1 = \{i \in [1,n] : (\mathbf{v}_2)_i \neq (\mathbf{v}_3)_i\},$$
$$S_2 = \{i \in [1,n] : (\mathbf{v}_1)_i \neq (\mathbf{v}_3)_i\},$$
$$S_3 = \{i \in [1,n] : (\mathbf{v}_1)_i \neq (\mathbf{v}_2)_i\}.$$

Note that $S_1 = (S_2 \setminus S_3) \cup (S_3 \setminus S_2)$ and therefore $|S_1| = |S_2| + |S_3| - 2|S_2 \cap S_3|$.
The triangle with vertices $\mathbf{v}_1, \mathbf{v}_2, \mathbf{v}_3$ is equilateral if and only if $|S_1| = |S_2| = |S_3| > 0$, that is

$$|S_2| = |S_3| = 2|S_2 \cap S_3| = 2k$$

for some positive integer k and its side length is $\sqrt{2k}$.
In order to count such triangles, we first choose the vertex $\mathbf{v}_1$ in 2^n ways, then we select the vertex $\mathbf{v}_2$ by choosing the set of indices S_3 in $\binom{n}{2k}$ ways. Finally the vertex $\mathbf{v}_3$ is determined as soon as we choose the set $S_2 \cap S_3$ in $\binom{2k}{k}$ ways, and the set $S_2 \setminus S_3$ in $\binom{n-2k}{k}$ ways. Since each triple $(\mathbf{v}_1, \mathbf{v}_2, \mathbf{v}_3)$ has been counted 3! times, we divide the final product by 6. Therefore, the number of equilateral triangles of side length $\sqrt{2k}$ is

$$\frac{2^n}{3!}\binom{n}{2k}\binom{2k}{k}\binom{n-2k}{k} = \frac{2^n}{6}\binom{n}{3k}\frac{(3k)!}{(k!)^3},$$

and by summing over all $1 \leq k \leq n/3$ we find the total number of equilateral triangles in the n-cube:

$$a_n = \frac{2^n}{6} \sum_{k=1}^{n/3} \binom{n}{3k} \frac{3k!}{(k!)^3}.$$

Starting from $n = 3$, the first few terms of the sequence are:

$$8, 64, 320, 2240, 17920, 121856, 831488, \ldots$$

, which is the entry A344854 (https://oeis.org/A344854) in the OEIS.

Additional problems for practice.

1. Let $t(n)$ be the number of triangles with integer sides and perimeter n. Show that for any integer $n \geq 3$,

$$t(n) = \begin{cases} \text{round}(n^2/48) & \text{if } n \text{ is even,} \\ \text{round}((n+3)^2/48) & \text{if } n \text{ is odd.} \end{cases}$$

2. **Problem E3157** (Proposed by L. I. Nicolaescu, 93(6), 1986). How many sets of four distinct points forming the vertices of a trapezoid are there if the points are chosen from the vertices of a regular n-gon with $n > 4$?

3. **Putnam 2002-A3**. Let $n \geq 2$ be an integer and T_n be the number of nonempty subsets S of $\{1, 2, 3, \ldots, n\}$ with the property that the average of the elements of S is an integer. Prove that $T_n - n$ is always even.

4. **Putnam 2006-B4**. Let Z denote the set of points in $\mathbb{R}^n$ whose coordinates are 0 and 1. (Thus Z has 2^n elements, which are the vertices of a unit hypercube in $\mathbb{R}^n$.) Given a vector subspace V of $\mathbb{R}^n$, let $Z(V)$ denote the number of members of Z that lie in V. Let k be given, $0 \leq k \leq n$. Find the maximum, over all vector subspaces $V \subset \mathbb{R}^n$ of dimension k, of the number of points in $V \cap Z$.

5. **Putnam 2015-B4**. Let T be the set of all triples (a, b, c) of positive integers for which there exist triangles with side lengths a, b, c. Express

$$\sum_{(a,b,c) \in T} \frac{2^a}{3^b 5^c}$$

as a rational number in lowest terms.

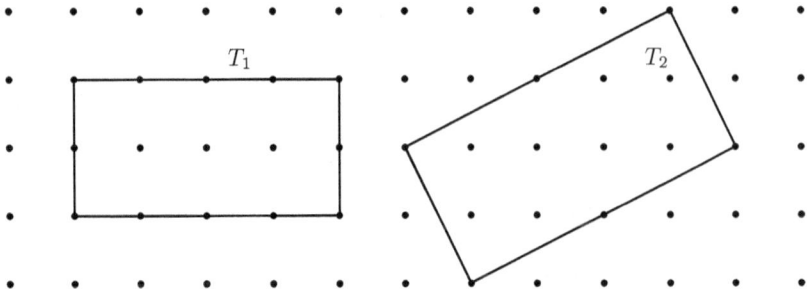

Figure 6.3: Two types of rectangles

6.10 Counting rectangles with prime area

Problem 12320 (Proposed by E. Treviño, 129(4), 2022). Consider the grid of n^2 lattice points $\{1, \ldots, n\}^2$. Let $S_1(n)$ be the number of rectangles with corners in the grid that have area equal to a prime integer congruent to 1 (mod 4). Define $S_3(n)$ similarly using primes congruent to 3 (mod 4). Prove that there is a value n_0 such that $S_1(n) > S_3(n)$ for $n \geq n_0$.

Discussion.
There are two types of rectangles with corners in the grid: the rectangles with horizontal and vertical sides, say T_1-*type* and the rectangles with tilted sides, say T_2-*type*. (See Figure 6.3.)
For a T_1-type rectangle with prime area p, its side lengths must be 1 and p, respectively. For a T_2-type rectangle, each side has length $\sqrt{l}$ for some integer $l \geq 2$. i.e., $p = \sqrt{l_1} \cdot \sqrt{l_2} = \sqrt{l_1 \cdot l_2}$. The primary of p implies that $l_1 = l_2$. So the rectangle must be a square and $p = a^2 + b^2$ for some positive integers a and b. By Fermat's theorem: *An odd prime p is expressible as a sum of two squares if and only if $p \equiv 1$ (mod 4)*, we see that all T_2-type rectangles are excluded from the counting of S_3. With these facts in hand, we are ready to count $S_1(n)$ and $S_3(n)$.

Solution.
We count the number of T_1-type rectangles first. In the grid $n \times n$, for $p < n$, there are $2(n-1)(n-p)$ copies of rectangles with area p. Thus, the total number of T_1 type rectangles with prime area is

$$\sum_{\substack{p \text{ prime} \\ p < n}} 2(n-1)(n-p).$$

Now we count the number of T_2-type rectangles with area p. In this case, a rectangle with tilted sides is defined by two orthogonal vectors $s(a, b)$ and $t(-b, a)$ at the bottom vertex, with s, t, a, b positive integers such that $\gcd(a, b) = 1$. Its area is equal to $s \cdot t \cdot (a^2 + b^2)$. Note that this kind of rectangle is inscribed in a square with horizontal and vertical sides of length $a + b$. Hence the area is a prime number p if and only if $s = t = 1$ and $a^2 + b^2 = p$. In the grid $n \times n$, for $a + b < n$, there are $(n - a - b)^2$ copies of such rectangle.

Based on Discussion above, applying Fermat's theorem yields

$$S_1(n) = \sum_{\substack{p \equiv 1 \,(\text{mod } 4) \\ p < n}} 2(n-1)(n-p) + \sum_{\substack{p \equiv 1 \,(\text{mod } 4) \\ a+b<n,\, a^2+b^2=p}} (n-a-b)^2,$$

$$S_3(n) = \sum_{\substack{p \equiv 3 \,(\text{mod } 4) \\ p < n}} 2(n-1)(n-p).$$

By the prime number theorem, for $\alpha \geq 0$,

$$T_\alpha(x) := \sum_{p \leq x} p^\alpha \sim \frac{x^{\alpha+1}}{(\alpha+1)\log(x)},$$

and if the sum is restricted to the primes congruent to 1 modulo 4 or to 3 modulo 4, then, by the prime number theorem for arithmetic progressions, there is an extra factor $1/\varphi(4) = 1/2$ in the asymptotic estimate. Therefore

$$S_3(n) \sim \sum_{\substack{p \equiv 3 \,(\text{mod } 4) \\ p < n}} (2n^2 - 2np) \sim n^2 T_0(n) - n T_1(n)$$

$$\sim n^2 \frac{n}{\log(n)} - n\frac{n^2}{2\log(n)} = \frac{n^3}{2\log(n)}.$$

As regards $S_1(n)$, since

$$(a+b)^2 \leq (1^2 + 1^2)(a^2 + b^2) = 2p \implies a + b \leq \sqrt{2p},$$

we have that

$$S_1(n) \geq \sum_{\substack{p \equiv 1 \,(\text{mod } 4) \\ a+b<n,\, a^2+b^2=p}} (n-a-b)^2 \geq \sum_{\substack{p \equiv 1 \,(\text{mod } 4) \\ \sqrt{2p}<n}} (n-\sqrt{2p})^2.$$

Moreover

$$\sum_{\substack{p \equiv 1 \,(\text{mod } 4) \\ \sqrt{2p}<n}} (n-\sqrt{2p})^2 = \sum_{\substack{p \equiv 1 \,(\text{mod } 4) \\ p<n^2/2}} (n^2 - 2\sqrt{2}np^{1/2} + 2p)$$

$$\sim \frac{n^2}{2}T_0(n^2/2) - \sqrt{2}nT_{1/2}(n^2/2) + T_1(n^2/2)$$

$$\sim \frac{n^2}{2}\frac{(n^2/2)}{\log(n^2/2)} - \sqrt{2}n\frac{2(n^2/2)^{3/2}}{3\log(n^2/2)} + \frac{(n^2/2)^2}{2\log(n^2/2)}$$

$$\sim \frac{n^4}{\log(n)}\left(\frac{1}{8} - \frac{1}{6} + \frac{1}{16}\right) = \frac{n^4}{48\log(n)}.$$

Hence, there is some positive constant c such that, for every sufficiently large n,

$$\frac{S_1(n)}{S_3(n)} \geq c \frac{n^4/\log(n)}{n^3/\log(n)} = cn$$

which implies that there is a value n_0 such that $S_1(n) > S_3(n)$ for $n \geq n_0$. $\square$

Remark. Let q, r be integers with $q \geq 2$ and $\gcd(q, r) = 1$. We define $\pi(x; q, r)$ as the number of primes $p \leq x$ with $p \equiv q \pmod{r}$. The prime number theorem for arithmetic progressions claims that

$$\pi(x; q, r) \sim \frac{1}{\varphi(q)} \cdot \frac{x}{\ln(x)} \qquad \text{as } x \to \infty,$$

where $\varphi(n)$ is Euler's phi-function (see, for instance, [15, Theorem 7.10]).

By restricting p to $p \leq n^2/8$, we can give a short proof. In this case, we have $a + b \leq \sqrt{2p} \leq n/2$. Thus there are at least $n^2/2$ squares in the grid, and so

$$S_1(n) \geq \sum \pi(n^2/8; 4, 1) \cdot \frac{n^2}{2}.$$

Since $\varphi(4) = 2$ and $\pi(n^2/8; 4, 1) \sim x/(2\ln x)$, for sufficiently large n, we have

$$S_1(n) \geq \frac{n^2/8}{2\ln(n^2/8)} \cdot \frac{n^2}{2} = \frac{n^4}{32\ln(n^2/8)} > n^3.$$

On the other hand, we have

$$S_3 \leq \sum_{k=1}^{n-1} 2(n-1)(n-k) = n(n-1)^2 < n^3.$$

Therefore, for sufficiently large n, we obtain $S_1(n) > S_3(n)$.

Additional problems for practice.

1. **Problem E3450** (Proposed by D. Bowman, 98(6), 1991). Let $T(n)$ be the number of triangles lying in the subset $[0, n] \times [0, n]$ of the Cartesian plane whose sides lie on lines of slope 0, ∞, 1, or -1 passing through points with integer coordinates. Derive a closed formula for $T(n)$.

2. **Problem 11871** (Proposed by C. Lupu and S. Spataru, 122(9), 2015). Let ABC be a triangle in the Cartesian plane with vertices in $\mathbb{Z}^2$. Show that, if P is an interior lattice point of ABC, then at least one of the angles PAB, PBC, and PCA has a radian measure that is not a rational multiple of π.

3. **Problem 11894** (Proposed by E. J. Ionascu, 123(3), 2016). Let a, b, c, and d be integers such that $a^2 + b^2 + c^2 = d^2$ and $d \neq 0$. Let x, y, and z be three integers such that $ax + by + cz = 0$.

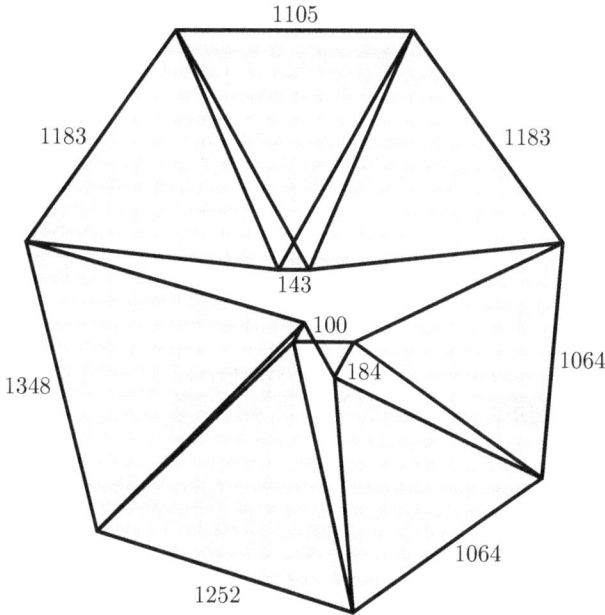

Figure 6.4: Ten saturated equilateral triangles with integer sides

(a) Prove that $x^2 + y^2 + z^2$ can be written as the sum of two squares.

(b) Let $ABCD$ be a square in $\mathbb{R}^3$ with integer vertices A, B, C, and D. Show that the side lengths of $ABCD$ have the form $\sqrt{l}$, where l is the sum of two squares.

4. **Problem 12257** (Proposed by E. Friedman and J. Tilley, 128(5), 2021). An arrangement of equilateral triangles in the plane is called *saturated* if the intersection of any two is either empty or is a common vertex and every vertex is shared by exactly two triangles. What is the smallest positive integer n such that there exists a saturated arrangement of n equilateral triangles with integer length sides? *Hint*: Show that smallest integer is 10. Below there is an example of an arrangement of 10 saturated equilateral triangles with integer sides (the lengths are specified beside each triangle). It remains to prove that there is no arrangement with less than 10 triangles.

6.11   Expected number of throws of an n-sided die

Problem 12299 (Proposed by E. Vigren, 129(1), 2022). Let $x_{0,n} = x_{1,n} = 1$ and, for integers k with $2 \le k \le n-1$, let

$$x_{k,n} = \frac{1}{k}\left(nx_{k-1,n} - \sum_{j=1}^{k-1} x_{j,n}\right).$$

Let $T_n = n^2 x_{n-1,n} - n + 1$. The first few values of T_n are

$$1, 3, 7, 47/3, 427/12, 416/5.$$

Prove that T_n is the expected number of throws of an n-sided die until the last n throws contain all possible face values.

Discussion.
It is well-known that for a fair n-side die, the expected number of throws required to get all possible face values is nH_n, where H_n is the n-th harmonic number. It is often called the *coupon collector problem* (see [102, p. 157]). For example, if $n = 6$, then $E = 14\frac{7}{10}$. The given value $T_6 = 416/5 \ne E$ implicitly suggests that the proposed problem is differ from the coupon collector problem. The use of a Markov chain is helpful here. We illustrate the idea for $n = 6$, the same argument goes through the general case.

We calculate the expected number by creating a set of linear recurrences. Let E_i be the expected number of rolls from a point where the last i rolls were distinct. We have $E_0 = 1 + E_1$, since there must be a roll, and that takes us to the state where the last 1 roll is distinct. Then another roll occurs; at this point, with probability $1/6$ the roll is the same as the last roll, and so we remain in the same state, or, with probability $5/6$, a different face appears, and then the last two rolls are distinct. The pattern continues this way. Therefore,

$$E_0 = 1 + E_1,$$
$$E_1 = 1 + \frac{1}{6}E_1 + \frac{5}{6}E_2,$$
$$E_2 = 1 + \frac{1}{6}E_1 + \frac{1}{6}E_2 + \frac{4}{6}E_3,$$
$$E_3 = 1 + \frac{1}{6}E_1 + \frac{1}{6}E_2 + \frac{1}{6}E_3 + \frac{3}{6}E_4,$$
$$E_4 = 1 + \frac{1}{6}E_1 + \frac{1}{6}E_2 + \frac{1}{6}E_3 + \frac{1}{6}E_4 + \frac{2}{6}E_5,$$
$$E_5 = 1 + \frac{1}{6}E_1 + \frac{1}{6}E_2 + \frac{1}{6}E_3 + \frac{1}{6}E_4 + \frac{1}{6}E_5 + \frac{1}{6}E_6,$$
$$E_6 = 0.$$

By the back substitution, solving this system yields $E_0 = 416/5$.

On the other hand, to get a feel for the sequence $x_{k,n}$, we compute the first few terms:

$$x_{2,n} = \frac{1}{2!}(n-1),$$

$$x_{3,n} = \frac{1}{3!}\left(n^2 - (2n+1)\right),$$

$$x_{4,n} = \frac{1}{4!}\left(n^3 - (3n^2 + 2n + 2)\right),$$

$$x_{5,n} = \frac{1}{5!}\left(n^4 - (4n^3 + 3n^2 + 4n + 6)\right),$$

$$x_{6,n} = \frac{1}{6!}\left(n^5 - (5n^4 + 4n^3 + 6n^2 + 12n + 12)\right).$$

Based on the emerged pattern, for $n \geq 2, 1 \leq k \leq n-1$, an easy induction argument confirms

$$x_{k,n} = \frac{1}{k!}\left(n^{k-1} - \sum_{j=0}^{k-2}(k-1-j)j!n^{k-2-j}\right). \tag{6.11}$$

Now, we are ready to show $T_n = E_0$ in general.

Solution.

By (6.11), we find that

$$T_n = n^2 x_{n-1,n} - n + 1 = \frac{1}{(n-1)!}\left(n^n - \sum_{j=0}^{n-3}(n-2-j)j!n^{n-1-j}\right) - n + 1$$

$$= \frac{1}{(n-1)!}\left(n^n + \sum_{j=0}^{n-3}j!n^{n-1-j} + \sum_{j=0}^{n-3}((j+1)!n^{n-(j+1)} - j!n^{n-j})\right) - n + 1$$

$$= \frac{1}{(n-1)!}\left(n^n + \sum_{j=0}^{n-3}j!n^{n-1-j} + (n-2)!n^2 - n^n\right) - n + 1$$

$$= \frac{1}{n!}\sum_{j=0}^{n-3}j!n^{n-j} + \frac{n^2}{n-1} - n + 1 = \frac{1}{n!}\sum_{j=0}^{n-1}j!n^{n-j}.$$

On the other hand, we consider the random process of throwing the n-sided die as Markov chain with states $s_1, \ldots, s_n$. The state s_j is the set of all sequences of throws where the last j values are all distinct, but the last $j+1$ values are not. Let $E(s_j)$ be the expected number of throws to reach state s_n, starting from s_j. Then trivially $E(s_n) = 0$. Furthermore, for $1 \leq j \leq n-1$,

$$E(s_j) = \frac{1}{n}\sum_{i=1}^{j}(E(s_i) + 1) + \left(1 - \frac{j}{n}\right)(E(s_{j+1}) + 1)$$

$$= \frac{1}{n}\sum_{i=1}^{j}E(s_i) + \left(1 - \frac{j}{n}\right)E(s_{j+1}) + 1 \tag{6.12}$$

because from state s_j there is probability $1/n$ to go to any state $s_1, \ldots, s_j$, and there is probability $1 - j/n$ to go to state s_{j+1}. Hence it remains to verify that $T_n = E(s_1) + 1$ (at the beginning we need one extra-step to reach state s_1).

By subtraction, for $2 \leq j \leq n - 1$,

$$E(s_{j-1}) - E(s_j) = \left(1 - \frac{j-1}{n}\right)E(s_j) - \frac{E(s_j)}{n} - \left(1 - \frac{j}{n}\right)E(s_{j+1})$$

$$= \frac{n-j}{n}(E(s_j) - E(s_{j+1}))$$

which implies that

$$E(s_{j-1}) - E(s_j) = \frac{n-j}{n}(E(s_j) - E(s_{j+1}))$$

$$= \frac{(n-j)(n-j-1)}{n^2}(E(s_{j+1}) - E(s_{j+2}))$$

$$= \cdots = \frac{(n-j)!}{n^{n-j}}(E(s_{n-1}) - E(s_n)) = \frac{(n-j)!}{n^{n-j}}E(s_{n-1}).$$

Therefore, for $1 \leq i \leq n - 2$,

$$E(s_i) = \sum_{j=i+1}^{n-1} (E(s_{j-1}) - E(s_j)) + E(s_{n-1})$$

$$= E(s_{n-1}) \sum_{j=i+1}^{n} \frac{(n-j)!}{n^{n-j}} = E(s_{n-1}) \sum_{j=0}^{n-i-1} \frac{j!}{n^j}.$$

Letting $j = n - 1$ in (6.12), we get

$$E(s_{n-1}) = \frac{1}{n}\sum_{i=1}^{n-2} E(s_i) + \frac{E(s_{n-1})}{n} + 1 = \frac{E(s_{n-1})}{n}\sum_{i=1}^{n-1}\sum_{j=0}^{n-i-1} \frac{j!}{n^j} + 1$$

$$= E(s_{n-1}) \sum_{j=0}^{n-2} \frac{j!(n-j-1)}{n^{j+1}} + 1 = E(s_{n-1}) \sum_{j=0}^{n-2} \left(\frac{j!}{n^j} - \frac{(j+1)!}{n^{j+1}}\right) + 1$$

$$= E(s_{n-1})\left(1 - \frac{n!}{n^n}\right) + 1$$

and we find that $E(s_{n-1}) = n^n/n!$. Finally,

$$E(s_1) + 1 = E(s_{n-1}) \sum_{j=0}^{n-2} \frac{j!}{n^j} + 1 = \frac{1}{n!}\sum_{j=0}^{n-2} j!n^{n-j} + 1 = \frac{1}{n!}\sum_{j=0}^{n-1} j!n^{n-j}$$

which is the same formula given above for T_n. This completes the proof. $\square$

Remark. There is an alternative way to find E_0. Let $n = 6$ again. The transition matrix A is given by

$$
A = \begin{pmatrix}
0 & 1 & 0 & 0 & 0 & 0 & 0 \\
0 & 1/6 & 5/6 & 0 & 0 & 0 & 0 \\
0 & 1/6 & 1/6 & 4/6 & 0 & 0 & 0 \\
0 & 1/6 & 1/6 & 1/6 & 3/6 & 0 & 0 \\
0 & 1/6 & 1/6 & 1/6 & 1/6 & 2/6 & 0 \\
0 & 1/6 & 1/6 & 1/6 & 1/6 & 1/6 & 1/6 \\
0 & 0 & 0 & 0 & 0 & 0 & 1
\end{pmatrix}
$$

The last entry in the first row of A^n is the probability of throwing die until the last n throw contain all six face values. Let $B = \lim_{k \to \infty} A^k$, $C = (I - A - B)^{-1}$, and let C' be the diagonal matrix whose diagonal entries are the same as C, J be the matrix of all 1's and D be a diagonal matrix with $D_{ii} = 1/B_{ii}$. Then applying the *Mean first passage theorem* (see [52, 11.5] leads to

$$E = (I - C + JC')D,$$

where $E = (e_{ij})$, in which, for $i \neq j$, e_{ij} is the expected number of steps before the process state j for the first time after starting in state i, and e_{ii} is the expected number of steps before the chain reenters state i.

Additional problems for practice.

1. If we throw a 6-sided die n times, what is the probability that all faces have appeared in order, in some six consecutive throws (i.e., what is the probability that the subsequence 123456 appears among the n throws)?

2. **Putnam 2020-A4.** Consider a horizontal strip of $N + 2$ squares in which the first and the last square are black and the remaining N squares are all white. Choose a white square uniformly at random, choose one of its two neighbors with equal probability, and color this neighboring square black if it is not already black. Repeat this process until all the remaining white squares have only black neighbors. Let $w(N)$ be the expected number of white squares remaining. Find

$$\lim_{N \to \infty} \frac{w(N)}{N}.$$

3. **Problem 11477** (Proposed by A. Gonzalez and J. H. Nieto, 117(1), 2010). Several boxes sit in a row, numbered from 0 on the left to n on the right. A frog hops from box to box, starting at time 0 in box 0. If at time t, the frog is in box k, it hops one box to the left with probability k/n and one box to the right with probability $1 - k/n$. Let $p_i(k)$ be the probability that the frog launches its $(t + 1)$th hop from box k. Find $\lim_{i \to \infty} p_{2i}(k)$ and $\lim_{i \to \infty} p_{2i+1}(k)$.

4. **Problem 11623** (Proposed by A. Gabhe and M. N. Deshpande, 119(2), 2012). A fair coin is tossed n times and the results recorded as a bit string. A run is a maximal subsequence of (possibly just one) identical tosses. Let the random variable X_n be the number of runs in the bit string not immediately followed by a longer run. (For instance, with bit string 1001101110, there are six runs, of lengths 1, 2, 2, 1, 3, and 1. Of these, the 2nd, 3rd, 5th, and 6th are not followed by a longer run, so $X_{10} = 4$.) Find $E(X_n)$.

5. **Problem 11672** (Proposed by J. L. Palacios, 119(9), 2012). A random walk starts at the origin and moves up-right or down-right with equal probability. What is the expected value of the first time that the walk is k steps below its then-current all time high? (Thus, for instance, with the walk $UDDUUUUDDUDD\cdots$, the walk is three steps below its maximum-so-far on step 12.)

6. **Problem 11707** (Proposed by J. L. Palacios, 120(5), 2013). For $N \geq 1$, consider the following random walk on the $(N + 1)$-cycle with vertices labeled $0, 1, \ldots, N$. The walk begins at vertex 0 and continues until every vertex has been visited and the walk returns to vertex 0. Prove that the expected number of visits to any vertex other than 0 is $(2N + 1)/3$.

7. **Problem 12450** (Proposed by E. Vigren, 131(3), 2024). Let n be an odd positive integer, and suppose that $x_1, \ldots, x_n$ are chosen randomly and uniformly from the interval $[0, 1]$. For $1 \leq i \leq n$, let $y_i = x_i - x_i^2$. What is the expected value of the median of $\{y_1, \ldots, y_n\}$?

6.12 Removing tiles

Problem 12309 (Proposed by J. DeVincentis, T. C. Occhipinti, and D. J. Velleman, 129(3), 2022). Consider a square grid that is infinite in all directions, with tiles placed on finitely many squares of the grid. Two grid squares are called adjacent if they share an edge. There are two types of legal moves:

(a) If two tiles are on adjacent squares, then they can both be removed.

(b)] If a tile is on a square and all adjacent squares are unoccupied, then the tile can be removed with four new tiles then placed on the four adjacent squares.

For which initial configurations is it possible to eliminate all tiles from the grid?

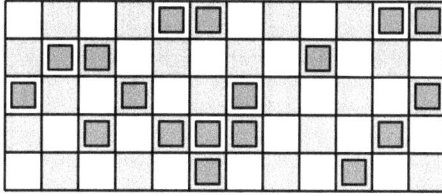

Figure 6.5: The colored grid with some tiles

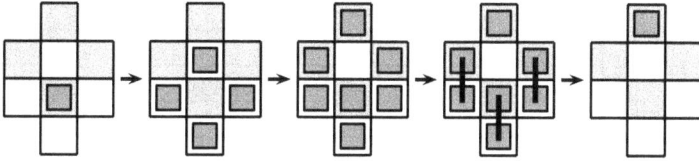

Figure 6.6: Moving a tile upward by using move (c)

Discussion.

As in a chessboard, we color the squares of the infinite grid alternating black and white.

Let W and B be the number of tiles placed on white and black squares, respectively. After a few trials, we observe that the difference $W - B$ follows a favorable pattern with respect to the moves given in the problem. Indeed, for move (a), $W - B = (W - 1) - (B - 1)$ does not change, and, for move (b), $W - B$ either increases or decreases by 5.

Hence, the difference $W - B$ is invariant modulo 5. If we succeed in eliminating all the tiles, the final difference will be zero which means that $W \equiv B \pmod 5$ is a necessary condition for all the tiles to be removed. In the following, we show that the condition $W \equiv B \pmod 5$ is also sufficient.

Solution.

From the above Discussion, we assume that initial configuration satisfies $W \equiv B \pmod 5$ and we prove that all the tiles can be eliminated.

We first remove all the tiles on white squares: if a tile on a white square is adjacent to another tile then it can be eliminated by making move (a), otherwise we remove it by applying move (b). After this first step, the tiles are all on black squares, and, due to the condition, the total number of tiles is a multiple of 5.

Now we notice that any tile on a black square can be moved by two squares horizontally or vertically to an unoccupied black square in any direction. We denote this composite move by (c). For instance, in order to move a tile upward we may apply the moves pictured in Figure 6.5.

It is easy to verify that move (c) is not affected by the presence of other tiles on black squares.

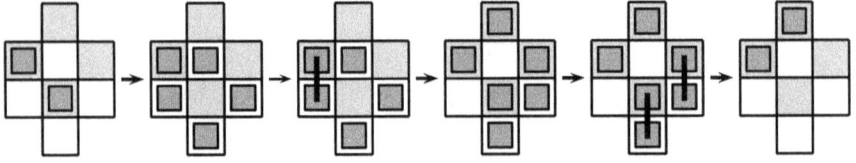

Figure 6.7: Moving a tile upward by using move (c) with another tile on a black square

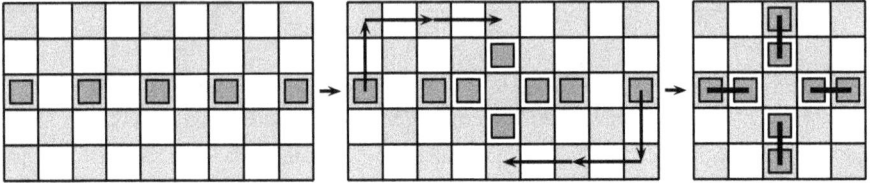

Figure 6.8: (i) 5 tiles in a single row

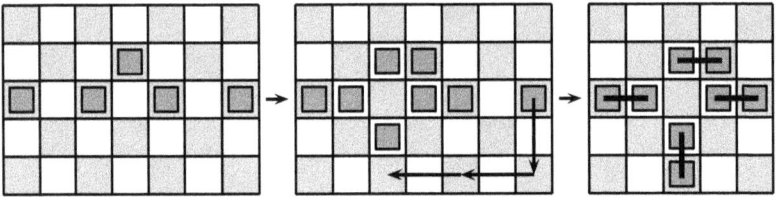

Figure 6.9: (ii) 4 tiles in a row and 1 in the another

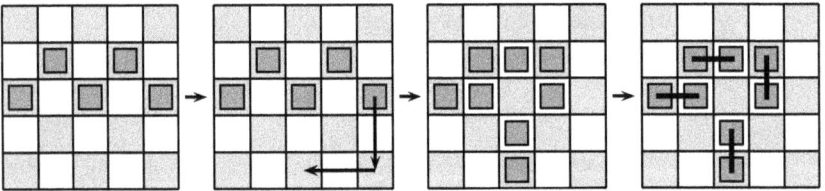

Figure 6.10: (iii) 3 tiles in a row and 2 in the another

As a second step, by applying move (c) repeatedly, we are able to arrange the tiles, in groups of 5, all along two adjacent horizontal lines (moving the tiles by a 2-step they can't change their parity). Each group can be reduced to one of these three cases: i) 5 tiles in a single row, ii) 4 tiles in a row and 1 in the another, iii) 3 tiles in a row and 2 in the another.

For each case we give a picture that illustrate how to remove all the tiles by using moves (a), (b), and (c).

Since we are able to eliminate all the tiles in each group, we are done.   □

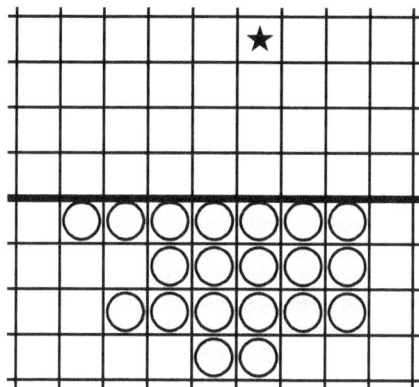

Figure 6.11: Show that an army of 20 pegs can send a peg into the fourth row

Remark. The proposed problem is a typical example of *algorithmic puzzle*, that is a problem where one is required to identify a procedure that meets the given requirements. If you liked it, we suggest [66] which is a collection of one hundred and fifty puzzles. The book includes many old classics like the eight queens puzzle, knight's tours, and the Tower of Hanoi. Another source is the first book of mathematical puzzles and games of Martin Gardner [41] that features a chapter dedicated to the game of Nim.

A variant of peg-solitaire which has interesting algorithmic aspect, is the so-called *Conway's soldiers*, a puzzle devised by J. H. Conway, a renowned mathematician of legendary creativity. Imagine drawing a horizontal line on an infinite grid of squares. On one side of this line stands a finite army, with each square occupied by at most one soldier (peg), whereas the other side is empty. A move consists of a peg jumping over an adjacent peg onto an unoccupied square, either horizontally or vertically, with the jumped-over peg being removed. The question is: how many soldiers are required to move a scout forward by n steps into the empty halfplane?

The surprising answer is that, with a finite army of pegs, it is impossible to go beyond the fourth row, meaning that $n \leq 4$. This limitation can be proven using a potential function associated with the positions of the pegs (see [17, p. 715] and [1]). For extensions of the problem to other shapes of the front line separating the army from the unexplored region, please consult [35, 53].

Additional problems for practice.

1. **MM Problem 1350** (Proposed by H. Noland, 63(3), 1990). In the well-known Tower of Hanoi puzzle one starts with three pegs, two of which are empty and one of which contains n disks, no two of the same size, stacked in order of size, with the smallest on top. It is required to move all the disks to one of the empty pegs, by moving one disk at a time,

subject to the condition that no disk ever rest on a smaller one. It is easy to show that the number of moves required is $2^n - 1$. Suppose instead that one has $2n$ disks, numbered according to size, with the smallest numbered 1. If all the odd-numbered disks occupy one peg and all the even numbered disks another, stacked according to size, how many moves are required to move all the disks onto the empty peg, the requirement again being that no disk ever rest on a smaller?

2. **MM Problem 2017** (Proposed by K. S. Soo, 90(2), 2017). Consider the following modification of the classical game of Nim. Initially, there are one or more heaps, each consisting of one or more stones. Players Alice and Bob take turns (Alice plays first). A player's move consists of choosing one or more heaps and removing exactly one stone from each of them. The player who takes the last stone loses. Determine all initial states for which Alice has a winning strategy.

3. **Problem 10287** (Proposed by A. K. Austin, 100(2), 1993). We have a doubly-infinite (i.e. indexed by $\mathbb{Z}$) row of squares and we start with counters in those squares to the left of some point (e. g. those with negative index). For a fixed positive integer k, the allowable moves consist of selecting k consecutive squares, discarding one of the counters in those squares, and rearranging the remaining counters within the k selected squares (with at most one counter in a square). Prove or disprove that there is an integer $N = N(k)$ such that no sequence of moves will allow a counter to be placed N squares into the region which originally contained no counters.

4. **Problem 10916** (Proposed by G. Ehrlich, 109(1), 2002). Available are two beakers A and B, having volumes a liters and b liters, respectively, a source of water, and a drain. Water may be poured into the beakers from the source or from each other, either filling the receiving beaker or emptying the source beaker, and beakers may be emptied into the drain. Using only these operations, show that if a and b are relatively prime positive integers, then for every integer m with $1 < m < b$ it is possible to reach a state in which beaker B contains m liters.

5. **Problem 12348** (Proposed by E. Vigren and H. Rullgård, 129(8), 2022). We have n people in a circle, numbered from 1 to n clockwise. They are removed one at a time as follows, until just one remains. At each step, remove the nth person among those remaining, where the count starts at the lowest numbered person remaining and proceeds clockwise. Let $W(n)$ be the number of the last person remaining.

(a) What is $W(10^{12})$?

(b) For $n \geq 5$, show that $W(n) = n - 4$ if and only if $n/2$ is a Sophie Germain prime (i.e., $n/2$ and $n + 1$ are prime).

(c) Find the smallest even number that does not equal $W(n)$ for any n.

6. **Problem 12314** (Proposed by G. Galperin and Y. J. Ionin, 129(4), 2022). Let n, m, and k be positive integers with $k \leq n - 1$. Consider n devices each of which can be in any of m states denoted $0, 1, \ldots, m - 1$. A move consists of selecting a set of k devices and adding 1 (mod m) to each of their states. Prove that for any n, m, k as specified and any initial states of the n devices, there exists a sequence of moves that leaves each device in the state 0 or 1.

7. **Problem 11570** (Proposed by K. Bresniker and S. Wagon, 118(4), 2011). Alice and Bob play a number game. Starting with a positive integer n, they take turns changing the number; Alice goes first. Each player in turn may change the current number k to either $k - 1$ or $\lceil k/2 \rceil$. The person who changes 1 to 0 wins. For instance, when $n = 3$, the players have no choice, k proceeds from 3 to 2 to 1 to 0, and Alice wins. When $n = 4$, Alice wins if and only if her first move is to 2. For which initial n does Alice have a winning strategy?

6.13 Writing a Gaussian integer as sum of powers of $1 + i$

Problem 12335 (Proposed by T. Karzes, S. Lucas, J. Madison and J. Propp, 129(7), 2022). A Gaussian integer is a complex number $z = a + ib$ for integers a and b. Show that every Gaussian integer can be written in at most one way as a sum of distinct powers of $1 + i$, and that the Gaussian integer z can be expressed as such a sum if and only if $i - z$ cannot.

Discussion.
Let $\mathbb{Z}[i] = \{z = a + ib : a, b \in \mathbb{Z}\}$ be the set of Gaussian integers. We work backward by examining an example. If $z = -1 + 5i$ is a sum of distinct powers of $1 + i$, we perform Euclidean type of division

$$\frac{z}{1+i} = 2 + 3i \in \mathbb{Z}[i], \qquad \frac{2 + 3i - 1}{1+i} = 2 + i \in \mathbb{Z}[i],$$

and

$$\frac{2 + i - 1}{1 + i} = 1 \in \mathbb{Z}[i], \qquad \frac{1 - 1}{1 + i} = 0 \in \mathbb{Z}[i].$$

This yields that

$$z = (1 + i)(2 + 3i) = (1 + i)[1 + (1 + 3i)] = (1 + i)[1 + (1 + i)(2 + i)]$$
$$= (1 + i)\{1 + (1 + i)[1 + (1 + i)]\} = (1 + i) + (1 + i)^2 + (1 + i)^3.$$

In general, let $z = a + ib \in \mathbb{Z}[i]$. Since

$$\frac{z}{1+i} = \frac{1}{2}(1-i)z = \frac{a+b}{2} + i\frac{b-a}{2},$$

we see that $z/(1+i) \in \mathbb{Z}[i]$ if and only if both $a+b$ and $a-b$ are even. This holds if and only if $a^2 + b^2$ is even because $2(a^2 + b^2) = (a+b)^2 + (a-b)^2$. On the other hand, if $a+b$ is odd, then $(a-1)+b$ is even, so $(z-1)/(1+i) \in \mathbb{Z}[i]$. This motivates us to introduce the following function

$$T(z) = \begin{cases} z/(1+i), & \text{if } a^2 + b^2 \text{ is even,} \\ (z-1)/(1+i), & \text{if } a^2 + b^2 \text{ is odd.} \end{cases}$$

Solution.
We define a function $T : \mathbb{Z}[i] \to \mathbb{Z}[i]$ by

$$T(z) = \frac{z - (N(z))_2}{1+i} = \frac{(a + ib - r)(1-i)}{2} = \frac{a+b-r}{2} + i\frac{b-a+r}{2} \quad (6.13)$$

where $N(z) = a^2 + b^2$ is the norm of z, and $r = (N(z))_2 = (a+b)_2 \in \{0, 1\}$ is the remainder of the division of $N(z)$ by 2.

Now we show that for all $z \in \mathbb{Z}[i]$ the sequence of iterates $\{T^n(z)\}_{n\geq 0}$ is eventually constant and such constant, say $L(z)$, can be 0 or i. Indeed, by (6.13),

$$N(T(z)) = \frac{(a+b-r)^2 + (b-a+r)^2}{4} = \frac{(a-r)^2 + b^2}{2} < N(z) = a^2 + b^2$$

if and only if $a^2 + b^2 > 0$ and $r = 0$, or $(a+1)^2 + b^2 > 2$ and $r = 1$, that is

$$z \notin S = \{0, i, -i, -1, -2+i, -2-i\}.$$

Hence $N(T^n(z))$ is strictly decreasing until $T^n(z) \in S$, thereafter

$$-2 - i \xrightarrow{T} -2 + i \xrightarrow{T} -1 + 2i \xrightarrow{T} 2i \xrightarrow{T} 1 + i \xrightarrow{T} 1 \xrightarrow{T} 0 \circlearrowleft$$
$$-i \xrightarrow{T} -1 \xrightarrow{T} -1 + i \xrightarrow{T} i \circlearrowleft$$

and we are done.
Moreover $L(z) = 0$ if and only if $L(i-z) = i$: if $(N(z))_2 = r$ then $(N(i-z))_2 = 1 - r$ and by (6.13),

$$T(i-z) = T(-a + i(1-b)) = \frac{-a + (1-b) - (1-r)}{2} + i\frac{(1-b) + a + (1-r)}{2}$$
$$= -\frac{a+b-r}{2} + i\left(1 - \frac{b-a+r}{2}\right) = i - T(z).$$

Hence, for all $n \geq 0$,

$$T^n(i - z) = T^{n-1}(i - T(z)) = T^{n-2}(i - T(T(z))) = \cdots = i - T^n(z)$$

which implies that $L(i - z) = i - L(z)$.

So if $L(z) = 0$ then $T^n(z) = 0$ for some positive integer n and

$$z = T(z)(1 + i) + r_0 = T(T(z))(1 + i)^2 + r_1(1 + i) + r_0$$

$$= \cdots = T^n(z)(1 + i)^n + \sum_{k=0}^{n-1} r_k(1 + i)^k = \sum_{k=0}^{n-1} r_k(1 + i)^k$$

that is $z \in \mathbb{Z}[i]$ can be represented as a sum of distinct powers of $1 + i$. On the other hand, if z can be represented as a sum of distinct powers of $1 + i$, that is $z = \sum_{k=0}^{n-1} a_k(1 + i)^k$ with $a_0, \ldots, a_{n-1} \in \{0, 1\}$ then, due to the fact that $(1 + i)^k = 2i(1 + i)^{k-2}$ for $k \geq 2$,

$$z \equiv a_1(1 + i) + a_0 \pmod{2}$$

implies

$$N(z) \equiv (a_0 + a_1)^2 + a_1^2 \equiv a_0 \pmod{2}$$

and therefore $r_0 = (N(z))_2 = a_0$. By applying the same argument, we obtain $T^n(z) = 0$, $L(z) = 0$, and $r_k = a_k$ for $k = 0, \ldots, n-1$. Hence, we may conclude that $z \in \mathbb{Z}[i]$ can be written as a sum of distinct powers of $1 + i$ if and only if $L(z) = 0$. Moreover, it follows also that such representation is unique. Since $L(z) = 0$ if and only if $L(i - z) = i$, then z has such representation if and only if $i - z$ has not. □

Remark. As noted by W. Gilbert in [43], if we draw a grid in the complex plane where each unit square is centered at a Gaussian integer and we shade all the squares whose center can be written as a sum of distinct powers of $1 + i$, then we obtain a fascinating snowflake spiral shown in Figure 6.12. The property that $L(z) = 0$ if and only if $L(i - z) = i$ says that the black spiral and the white spiral are symmetric with respect to the point $i/2$.

Additional problems for practice.

1. Let $z = a + ib \neq 0$ be a Gaussian integer. There is a natural number n such that z^n is real if and only if $a = 0$ or $b = 0$ or $a = b$ or $a = -b$.

2. Define the residue class of α modulo μ by

$$[\alpha]_\mu = \{z \in \mathbb{Z}[i] : z \equiv \alpha(\text{mod } \mu)\}.$$

Illustrate $[\alpha]_{1+i}$ with a two color diagram inside the square $[-30, 30] \times [-30, 30]$

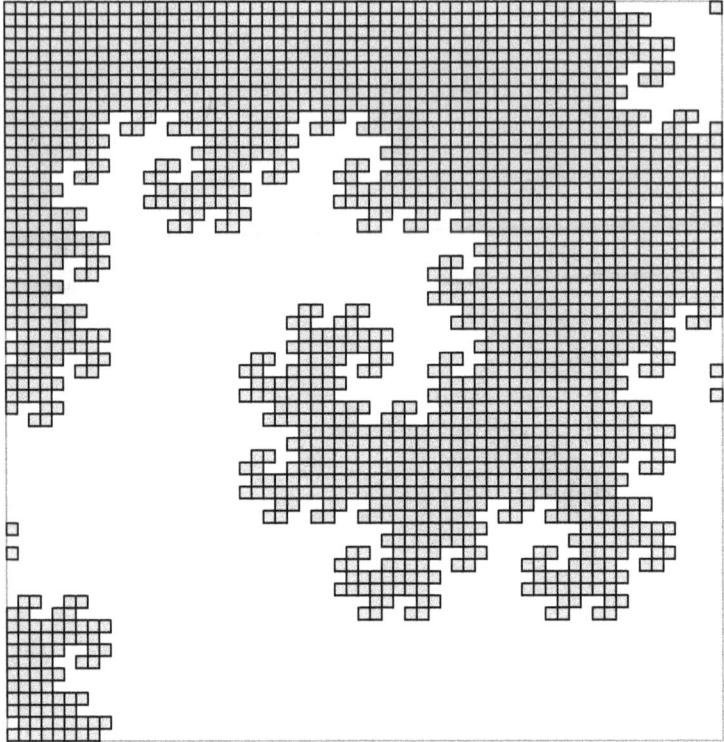

Figure 6.12: The snowflake spiral inside the square $[-30, 30] \times [-30, 30]$

3. Show that a Gaussian integer is expressible as a sum of squares of Gaussian integers if and only if its imaginary part is even (see [76]).

4. **Problem 10940** (Proposed by Y. Nievergelt, 109(4), 2002). Prove that for each real number r with $0 < r < 1$, and for any complex numbers $z_1, \ldots, z_k$ in the closed disk of radius r about the origin, there is a complex number z_0 in that disk such that

$$\prod_{j=1}^{k} (1 + z_j) = (1 + z_0)^k.$$

5. **Problem 11700** (Proposed by E. O'Dorney, 120(4), 2013). Let n be an integer greater than 1. Let a, b, and c be complex numbers with

$$a + b + c = a^n + b^n + c^n = 0$$

Prove that the absolute values of a, b, and c cannot be distinct.

6. **Problem 11840** (Proposed by G. Stoica, 122(5), 2015). Let $z_1, \ldots, z_n$ be complex number. Prove that

$$\left(\sum_{k=1}^{n} |z_k| \right)^2 - \left| \sum_{k=1}^{n} z_k \right|^2 \geq \left(\sum_{k=1}^{n} |\operatorname{Re}(z_k)| - \left| \sum_{k=1}^{n} \operatorname{Re}(z_k) \right| \right)^2.$$

7. **Problem 12074** (Proposed by E. P. Goldenberg, 125(9), 2018). Start with an equilateral triangle of area 1. Attach externally three equilateral triangles to the vertices of the original triangle, so that the altitude of each new triangle is an extension of one side of the original triangle and half its length. Always use the side that is counterclockwise from the vertex. Continue this process, producing each new generation by attaching three triangles to each triangle of the previous generation. Let T_n be the union of all triangles (and their interiors) produced through generation n. What is the area of $\bigcup_{n=1}^{\infty} T_n$.

6.14   Enumerating the positive rationals

Problem 12391 (Proposed by R. Tauraso, 130(5), 2023). Does there exist an enumeration $\{q_n\}_{n \geq 1}$ of the positive rationals such that $\{|q_{n+1} - q_n|\}_{n \geq 1}$ is another enumeration of the positive rationals?

Discussion.
An enumeration which meets the required property does exist and we will construct such a sequence $\{q_n\}_{n \geq 1}$ in a recursive way. All we need are two auxiliary lists of positive rationals which provide us the necessary elements for this construction.

Solution.
Let $R = \{r_k\}_{k \geq 1}$ and $D = \{d_j\}_{j \geq 1}$ be two enumerations of $\mathbb{Q}^+$: R is the *list of all positive rational numbers* and D is the *list of all positive rational distances*.
We set $q_1 = r_1$. Now let $n \geq 1$, assuming that we already set $q_1, q_2, \ldots, q_n$, we show how to define

$$q_{n+1}, q_{n+2}, q_{n+3}, q_{n+4}.$$

so that one of them is r_k, the element of the list R with the smallest index which has not been assigned yet, and one of the distances

$$|q_{n+1} - q_n|, |q_{n+2} - q_{n+1}|, |q_{n+3} - q_{n+2}|, |q_{n+4} - q_{n+3}|$$

is d_j, the element of the list D with the smallest index which has not been attained yet. In this way, finally, we will assign each element of R to some q_n

exactly once, and each distance listed in D will be attained for some $|q_{n+1} - q_n|$ exactly once.

Let $M = \max\{q_1, q_2, \ldots, q_n, r_k, d_j\}$ then any rational number $x > M$ is an element of R which has not been assigned yet, and it is also a distance in D which has not been attained yet. We choose $d, d' \in \mathbb{Q}^+$ such that $M < d < d'$ and we define $d'' = q_n + d + d_j + d' - r_k \in \mathbb{Q}^+$.

Then we set

$$q_{n+1} = q_n + d,$$
$$q_{n+2} = q_{n+1} + d_j,$$
$$q_{n+3} = q_{n+2} + d',$$
$$q_{n+4} = q_{n+3} - d'' = r_k,$$

where $q_{n+1}, q_{n+2}, q_{n+3}$ are unassigned elements of R different from $q_{n+4} = r_k$ because

$$r_k \leq M < d < q_{n+1} < q_{n+2} < q_{n+3}.$$

Moreover, as regards distances, we have

$$|q_{n+1} - q_n| = d,$$
$$|q_{n+2} - q_{n+1}| = d_j,$$
$$|q_{n+3} - q_{n+2}| = d',$$
$$|q_{n+4} - q_{n+3}| = d'',$$

and d, d', d'' are unattained distances in D different from d_j because

$$d_j \leq M < d < d' < d' + \underbrace{d - r_k}_{>0} < d'',$$

So no differences are duplicated. $\qquad\qquad\square$

Remark. It is well known that the rationals are countable. But the standard proofs of this fact do not give an explicit enumeration. We have already encountered the sequence the enumeration of all positive rationals (6.8) generated by the Stern's diatomic sequence. Another way to construct an explicit enumeration of $\mathbb{Q}^+$ is through the fundamental theorem of arithmetic. In fact given an integer $n > 1$, it can be represented uniquely as a product of primes. Now let us regroup the prime factors by even and odd multiplicity, say

$$n = \left(p_1^{2a_1} \cdots p_r^{2a_r}\right) \left(q_1^{2b_1 - 1} \cdots q_s^{2b_s - 1}\right).$$

Then we define the function f, by letting $f(1) = 1$ and

$$f(n) = \frac{p_1^{a_1} \cdots p_r^{a_r}}{q_1^{b_1} \cdots q_s^{b_s}}.$$

For example, for $f(8) = f(2^3) = 1/2^2 = 1/4$, $f(360) = f(2^3 \cdot 3^2 \cdot 5) = 3/(2^2 \cdot 5) = 3/20$. The uniqueness of the integer factorization ensures that f is indeed a bijection from $\mathbb{N}$ to $\mathbb{Q}^+$.

Additional problems for practice.

1. Does there exist an enumeration of the rationals $\{q_n\}_{n \geq 1}$ such that

$$\bigcup_{n=1}^{\infty} \left(q_n - \frac{1}{n}, q_n + \frac{1}{n} \right) \neq \mathbb{R}?$$

2. **MM Problem 1239** (Proposed by B. Hanson, 59(2), 1986). Let $\{r_n\}_{n \geq 1}$ be any enumeration of the rationals in the interval $(0, 1)$. For each $n \geq 1$, let $r_n = 0.a_{n,1}a_{n,2}a_{n,3}\ldots$ be a decimal representation of r_n. Prove that the *main diagonal* $0.a_{1,1}a_{2,2}a_{3,3}\ldots$ is irrational.

3. **MM Problem 1262** (Proposed by E. Just, 60(2), 1987). Show that it is possible to find an enumeration $\{r_n\}_{n \geq 1}$ of the rationals in the interval $(0, 1)$ so that in their decimal expansions

$$r_1 = 0.a_{1,1}a_{1,2}a_{1,3}a_{1,4}\ldots$$
$$r_2 = 0.a_{2,1}a_{2,2}a_{2,3}a_{2,4}\ldots$$
$$r_3 = 0.a_{3,1}a_{3,2}a_{3,3}a_{3,4}\ldots$$
$$\ldots$$

for any $k \geq 1$, the *column* $c_k = 0.a_{1,k}a_{2,k}a_{3,k}a_{4,k}\ldots$ is rational.

4. **Problem 4851** (Proposed by Z. A. Melzak, 66(6), 1959). Show that there exists an enumeration $\{x_n\}_{n \geq 1}$ of the rational numbers of the open interval $(0, 1)$ such that

$$\sum_{n=1}^{\infty} \prod_{k=1}^{n} x_k$$

diverges.

5. **Problem 11852** (Proposed by S. Northshield, 122(7), 2015). For $n \in \mathbb{Z}^+$, let $\nu_3(n) = k$ if 3^k divides n but 3^{k+1} does not. Let $X_1 = 2$, and for $n \geq 2$ let

$$X_n = 4\nu_3(n) + 2 - \frac{2}{X_{n-1}},$$

so that $(X_n)_{n \geq 1}$ begins with $2, 1, 4, 3/2, 2/3, 3, \ldots$. Show that every positive rational number appears exactly once in the list $X_1, X_2, X_3 \ldots$.

6. **Problem 12282** (Proposed by G. Stoica, 128(9), 2021). Prove that the multiplicative group generated by

$$\left\{ \frac{\lfloor \sqrt{2}\, n \rfloor}{n} : n \in \mathbb{Z}^+ \right\}$$

is the group of positive rational numbers.

7. **Problem 11068** (Proposed by H. Wilf, 111(3), 2004). For a rational number x that equals a/b in lowest terms, let $f(x) = ab$.

(a) Show that

$$\sum_{x \in \mathbb{Q}^+} \frac{1}{f(x)^2} = \frac{5}{2}.$$

(b) More generally, exhibit an infinite sequence of distinct rational exponents s such that

$$\sum_{x \in \mathbb{Q}^+} \frac{1}{f(x)^s}$$

is rational.

List of Problems

In this list, bold face denotes the *Monthly* problem featured solution in each section.

P = Putnam Problem, M = *Mathematics Magazine*,
C = *The College Mathematics Journal*.

(Problems from *The American Mathematical Monthly*, *Mathematics Magazine*, and *The College Mathematics Journal*© Mathematical Association of America, 2025. All rights reserved.)

Bibliography

[1] M. Aigner, Moving into the Desert with Fibonacci, *Math. Mag.*, **70**:11-21, 1997.

[2] M. Aigner and G. Ziegler, *Proofs from THE BOOK* (6th edition), Berlin, Springer-Verlag, 2018.

[3] A. V. Akopyan and A. A. Zaslawsky, *Geometry of conics*, Mathematical World **26**, American Mathematical Society, Providence RI, 2007.

[4] O. J. Alabi and G. Dresden, Fault-free tilings of the $3 \times n$ rectangle with squares and dominos, *J. Integer Seq.*, **24**, Article 21.1.2, 2021.

[5] J.-P. Allouche and J. Shallit, *The ubiquitous Prouhet-Thue-Morse sequence*, in Ding, C. (ed.) et al., *Sequences and their applications. Proceedings of the international conference*, SETA '98, Singapore, December 14-17, 1998. Springer Series in Discrete Mathematics and Theoretical Computer Science. 1-16, 1999.

[6] J.-P. Allouche and J. Shallit, *Automatic Sequences*, Cambridge, 2003.

[7] C. Alsina and R. B. Nelsen, *Charming proofs. A journey into elegant mathematics* The Dolciani Mathematical Expositions **42**, The Mathematical Association of America, Washington D.C., 2010.

[8] T. Amdeberhan and R. Tauraso, Congruences for sums of MacMahon's q-Catalan polynomials, *Bull. Aust. Math. Soc.*, First published online, 2024.

[9] T. Amdeberhan, G. E. Andrews and R. Tauraso, Extensions of MacMahon's sums of divisors, *Res. Math. Sci.*, **11**, Article 8, 2024.

[10] G. E. Andrews, A simple proof of Jacobi's triple product identity, *Proc. Amer. Math. Soc.*, **16**:333-334, 1965.

[11] G. E. Andrews, *Theory of partitions*, Cambridge University Press, 1998.

[12] G. E. Andrews, q-Analogs of the binomial coefficient congruences of Babbage, Wolstenholme and Glaisher, *Discrete Math.*, **204**:15-25, 1999.

[13] G. E. Andrews and K. Eriksson, *Integer Partitions*, Cambridge University Press, 2004.

[14] G. E. Andrews and M. Merca, The truncated pentagonal number theorem, *J. Comb. Theory, Ser. A*, **119**:1639-1643, 2012.

[15] T. M. Apostol, *Introduction to Analytic Number Theory*, Springer, 1976.

[16] M. Beresin, E. Levine and J. Winn, A Chessboard Coloring Problem, *Coll. Math. J.*, **20**:106-114, 1989.

[17] E. R. Berlekamp, J. H. Conway and R. K. Guy, *Winning Ways*, Academic Press, 1982.

[18] B. C. Berndt, *Number Theory in the Spirit of Ramanujan*, American Mathematical Society, Providence RI, 2006.

[19] F. Beukers, F. Luca and F. Oort, Power Values of Divisor Sums, *Amer. Math. Monthly*, **119**:373-380, 2012.

[20] R. Blecksmith, M. McCallum and J. L. Selfridge, 3-smooth representations of integers, *Amer. Math. Monthly*, **105**:529-543, 1998.

[21] B. Bollobás, *The Art of Mathematics. Coffee Time in Memphis*, Cambridge, Cambridge University Press, 2006.

[22] B. Bollobás, *The Art of Mathematics - Take Two. Tea Time in Cambridge*, Cambridge, Cambridge University Press, 2022.

[23] G. M. Boros and V. H. Moll, *Irresistible Integrals*, Cambridge, Cambridge University Press, 2004.

[24] P. B. Borwein, *The Prouhet-Tarry-Escott problem, Computational Excursions in Analysis and Number Theory*, CMS Books in Mathematics, Springer-Verlag, 2002.

[25] K. N. Boyadzhiev, Series with central binomial coefficients, Catalan numbers, and harmonic numbers, *J. Integer Seq.*, **15**, Article 12.1.7, 2012.

[26] D. M. Burton, *Elementary Number Theory*, McGraw Hill, 2010.

[27] N. Calkin and H. S. Wilf, Recounting the rationals, *Amer. Math. Monthly*, **107**:360-363, 2000.

[28] E. Chen, *Euclidean Geometry in Mathematical Olympiads*, The Mathematical Association of America, Washington D.C., 2016.

[29] H. Chen, *Excursions in Classical Analysis: Pathways to Advanced Problem Solving and Undergraduate Research*, The Mathematical Association of America, Washington D.C., 2010.

[30] H. Chen, *Monthly Problem Gems*, CRC Press, 2022.

[31] H. Chen, Interesting Series Associated with Central Binomial Coefficients, Catalan Numbers and Harmonic Numbers, *J. Integer Seq.*, **19**, Article 16.1.5, 2016.

[32] H. Chen, Interesting Ramanujan-Like Series Associated with Powers of Central Binomial Coefficients, *J. Integer Seq.*, **25**, Article 22.1.8, 2022.

[33] H. Chen, Another extension of Lobachevsky's formula, *Elem. der Math*, **78**:93-100, 2023.

[34] H. S. M. Coxeter and S. L. Greitzer, *Geometry Revisited*, **19**, Mathematical Association of America, 1967.

[35] B. Csákány and R. Juhász, The Solitaire Army reinspected, *Math. Mag.*, **73**:354-362, 2000.

[36] R. Donaghey and L. W. Shapiro, Motzkin numbers, *J. Comb. Theory, Ser. A*, **23**:291-301, 1977.

[37] J. Dyson, N. E. Frankel, and M. L. Glasser, Lehmer's Interesting Series, *Amer. Math. Monthly*, **120**:116-130, 2013.

[38] H. M. Edwards, *Riemann's Zeta Function*, Dover Publications, Inc., New York, 1974.

[39] B. Farhi, An identity involving the least common multiple of binomial coefficients and its application, *Amer. Math. Monthly*, **116**:836-839, 2009.

[40] P. Flajolet and R. Sedgewick, *Analytic Combinatorics*, Cambridge University Press, 2009.

[41] M. Gardner, *Hexaflexagons, Probability Paradoxes, and the Tower of Hanoi*, Cambridge University Press and Mathematical Association of America, 2008.

[42] I. M. Gessel, Wolstenholme Revisited, *Amer. Math. Monthly*, **105**:657-658, 1998.

[43] W. Gilbert, Fractal geometry derived from complex bases, *Math. Intell.*, **4**:78-86, 1982.

[44] H. W. Gould, *Combinatorial Identities*, published by the author, revised edition, 1972.

[45] I. Gradshteyn and I. Ryzhik, *Table of Integrals, Series, and Product*, 8th edition, edited by A. Jeffrey and D. Zwillinger, Academic Press, New York, 2007.

[46] R. L. Graham, Fault-free tilings of rectangles, in D. A. Klarner, ed., *The Mathematical Gardner: A Collection in Honor of Martin Gardner*, Wadsworth, 120-126, 1981.

[47] R. L. Graham, D. E. Knuth and O. Patashnik, *Concrete Mathematics*, 2nd edition, Addison-Wesley, 1994.

[48] A. Granville, Zaphod Beeblebrox's brain and the fifty-ninth row of Pascal's triangle, *Amer. Math. Monthly*, **99**:318-331, 1992.

[49] A. Granville, Arithmetic properties of binomial coefficients. I. Binomial coefficients modulo prime powers, in *Organic Mathematics (Burnaby, BC, 1995)*, CMS Conf. Proc., vol. 20, American Mathematical Society, Providence RI, 253-275, 1997.

[50] A. Granville, Smooth numbers: computational number theory and beyond, Buhler, J. P. (ed.) et al., *Algorithmic Number Theory. Lattices, Number Fields, Curves and Cryptography*, Cambridge University Press, Mathematical Sciences Research Institute Publications, **44**:267-323, 2008.

[51] B. Green and T. Tao, On sets defining few ordinary lines, *Discrete Comput. Geom.*, **59**:409-468, 2013.

[52] C. M. Grinstead and J. L. Snell, *Introduction to Probability*, Second Revised Edition, American Mathematical Society, 2012.

[53] L. Gualà, S. Leucci, E. Natale, and R. Tauraso, Large peg-army maneuvers, *Proceedings of the 8th International Conference on Fun with Algorithms (FUN'16)*, vol. 49 of *LIPIcs*, 1-15, 2016

[54] L.-S. Hahn, *Complex Numbers and Geometry*, The Mathematical Association of America, Washington D.C., 1994.

[55] L. Hall and S. Wagon, Roads and wheels, *Math. Mag.*, **65**:283-301, 1992.

[56] G. H. Hardy and E. M. Wright, *An Introduction to the Theory of Numbers*, 6th edn., Oxford University Press, 2008.

[57] H. Iwaniec, *Lectures on the Riemann Zeta Function*, vol. 62 of University Lecture Series, American Mathematical Society, Providence RI, 2014.

[58] A. Jacquemot, T. Randall-Page, A. Slavík and S. Wagon, A rolling square bridge: reimagining the wheel, *Math. Intell.*, **46**:171-182, 2024.

[59] J. H. Jaroma, On expanding $4/n$ into three Egyptian fractions, *Crux Math.*, **30**:36-37, 2014.

[60] C. Jeffrey C. (ed.), *The Ultimate Challenge. The $3x+1$ Problem*, American Mathematical Society, Providence RI, 2010.

[61] V. Klee and S. Wagon, *Old and new unsolved problems in plane geometry and number theory*, The Dolciani Mathematical Expositions, **11**, Mathematical Association of America, 1991.

[62] F. Kuczmarski, Roads and Wheels, Roulettes and Pedals, *Amer. Math. Monthly*, **118**:479-496, 2011.

[63] E. Landau, *Handbuch der Lehre von der Verteilung der Primzahlen*, vol. 2, Taubner, Leipzig, 1909.

[64] D. H. Lehmer, Two nonexistence theorems on partitions, *Bull. Amer. Math. Soc.*, **52**:538-544, 1946.

[65] D. H. Lehmer, Interesting series involving the central binomial coefficient, *Amer. Math. Monthly*, **92**:449-457, 1985.

[66] A. Levitin and M. Levitin, *Algorithmic Puzzles*, Oxford University Press, 2011.

[67] C. Li and W. Chu, Infinite series concerning harmonic numbers and quintic central binomial coefficients, *Bull. Aust. Math. Soc.*, **109**:225-241, 2024.

[68] J.-C. Liu, On a congruence involving q-Catalan numbers, *C. R. Math. Acad. Sci. Paris*, **358**:211-215, 2020.

[69] L. Lovász, *Combinatorial Problems and Exercises*, 2nd edn., AMS Chelsea Publishing, Providence, RI, 2007.

[70] M. Marden, *Geometry of Polynomials*, Mathematical Surveys and Monographs. Vol. 3 (2nd ed.), American Mathematical Society, Providence, RI, 1966.

[71] J. Mason, L. Burton and K. Stacey, *Thinking Mathematically*, Pearson Education, 2010.

[72] S. Mattarei and R. Tauraso, Congruences of multiple sums involving sequences invariant under the binomial transform, *J. Integer Seq.*, **13**, Article 10.5.1, 2010.

[73] S. Mattarei and R. Tauraso, Congruences for central binomial sums and finite polylogarithms, *J. Number Theory*, **133**:131-157, 2013.

[74] M. Mehrabi and K. Andersen, Problem 12318, Problems and Solutions, *Amer. Math. Monthly*, **131**:173-174, 2024.

[75] J. Nagura, On the interval containing at least one prime number, *Proc. Japan Acad.*, **28**:177-181, 1952.

[76] I. Niven, Integers of quadratic fields as sums of squares, *Trans. Amer. Math. Soc.*, **48**:405-417, 1940.

[77] S. Northshield, Stern's diatomic sequence $0, 1, 1, 2, 1, 3, 2, 3, 1, 4, \ldots$ *Amer. Math. Monthly*, **117**:581-598, 2010.

[78] H. Ohtsuka and Li Zhou, Problem 12361, Problems and Solutions, *Amer. Math. Monthly*, **131**:634, 2024.

[79] E. Passow and J. Roulier, Monotone and Convex Spline Interpolation, *SIAM Journal on Numerical Analysis*, **14**:904-909, 1977.

[80] M. Petkovsek, H. Wilf, and D. Zeilberger, How to do monthly problems with your computer, *Amer. Math. Monthly*, **104**:506-519, 1997.

[81] K. H. Pilehrood, T. Pilehrood, R. Tauraso New properties of multiple harmonic sums modulo p and p-analogues of Leshchiner's series, *Trans. Am. Math. Soc.*, **366**:3131-3159, 2014.

[82] G. W. Reitwiesner, *Binary arithmetic*, Advances in Computers, Academic Press, **1**:231-308, 1960.

[83] L. Schumaker, On Shape Preserving Quadratic Spline Interpolation, *SIAM Journal on Numerical Analysis*, **20**:854-864, 1983.

[84] G. I. Senum and S.-J. Bang, Problem E3352, Problems and Solutions, *Amer. Math. Monthly*, **98**:369-370, 1991.

[85] J. Silverman, Taxicabs and Sums of Two Cubes, *Amer. Math. Monthly*, **100**:331-340, 1993.

[86] N. J. A. Sloane, The On-Line Encyclopedia of Integer Sequences© (OEIS©), Available at http://oeis.org

[87] A. Soifer, *The Mathematical Coloring Book. Mathematics of Coloring and the Colorful Life of Its Creators*, Springer, 2009.

[88] R. P. Stanley, *Enumerative Combinatorics*, Vol. 1., 2nd edn., Camb. Stud. Adv. Math., vol. 49, Cambridge University Press, 2012.

[89] R. P. Stanley, *Catalan Numbers*, Cambridge University Press, 2015.

[90] R. P. Stanley, *Conversational Problem Solving*, American Mathematical Society, Providence, RI, 2020.

[91] R. P. Stanley and H. Kwong, Problem 12113, Problems and Solutions, *Amer. Math. Monthly*, **131**:89-90, 2021.

[92] E. M. Stein and R. Shakarchi, *Complex Analysis*, Princeton University Press, 2003.

[93] Z.-W. Sun and R. Tauraso, On some new congruences for binomial coefficients, *Int. J. Number Theory*, **7**:645-662, 2011.

[94] L. Szalay, On the diophantine equation $(2^n - 1)(3^n - 1) = x^2$, *Publ. Math. Debrecen*, **57**:1-9, 2000.

[95] T. Tao, *Solving Mathematical Problems - A Personal Perspective*, Oxford University Press, 2006.

[96] R. Tauraso, A new domino tiling sequence, *J. Integer Seq.*, **7**, Article 4.2.3, 2004.

[97] R. Tauraso, q-Analogs of some congruences involving Catalan numbers, *Adv. Appl. Math.*, **48**:603-614, 2012.

[98] I. Tomescu, *Problems in Combinatorics and Graph Theory*, Wiley-Interscience, New York, 1985.

[99] L. Tóth, Linear Combinations of Dirichlet Series Associated with the Thue-Morse Sequence, *Integers*, **22**, A98, 2022.

[100] C. I. Vălean, *(Almost) Impossible Integrals, Sums, and Series*, Problem Books in Mathematics, Springer, 2019.

[101] J. G. Wendel, Note on the Gamma Function, *Amer. Math. Monthly*, **55**:563-564, 1948.

[102] H. Wilf, *Generatingfunctionology*, A. K. Peters, Ltd., 1994. Available at http://www.math.upenn.edu/~wilf/DownldGF.html

[103] B. Williamson *An Elementary Treatise on The Integral Calculus*, Longmans, Green (London), 1888.

[104] D. R. Woods and D. Robbins, Problem E2692, Problems and Solutions, *Amer. Math. Monthly*, **85**:394-395, 1979.

[105] I. M. Yaglom and V. G. Boltyanskii, *Convex Figures*, Rinehart and Winston, New York, 1961.

[106] P. Yiu, Heronian triangles are lattice triangles, *Amer. Math. Monthly*, **108**:261-263, 2001.

[107] J. Zhao, Bernoulli numbers, Wolstenholme's theorem, and p^5 variations of Lucas' theorem, *J. Number Theory*, **123**:18-26, 2007.

Index

For Product Safety Concerns and Information please contact our EU
representative GPSR@taylorandfrancis.com
Taylor & Francis Verlag GmbH, Kaufingerstraße 24, 80331 München, Germany

www.ingramcontent.com/pod-product-compliance
Lightning Source LLC
Chambersburg PA
CBHW060329220326
41598CB00023B/2656

9 781041 003212